T0182033

The Innovators Behind Leonardo

Plinio Innocenzi

The Innovators Behind Leonardo

The True Story of the Scientific and Technological Renaissance

Foreword by Edward Burman

Plinio Innocenzi
University of Sassari
Alghero, Italy

ISBN 978-3-030-08018-1 ISBN 978-3-319-90449-8 (eBook)
https://doi.org/10.1007/978-3-319-90449-8

Softcover re-print of the Hardcover 1st edition 2019

Cover caption: Spring system for a clock. Leonardo da Vinci. Codex of Madrid I, folio 4r, (detail). BNE. (ca. 1493-1497).

Printed on acid-free paper

This Springer imprint is published by the registered company Springer International Publishing AG part of Springer Nature.
The registered company address is: Gewerbestrasse 11, 6330 Cham, Switzerland

To Ilio and Daria

Foreword

Leonardo da Vinci is like one of those authors—Dante, Proust, Joyce—whom everybody "knows," but few have actually read in their entirety. How good it is therefore to have a book about his drawings by an author who is not only a connoisseur of art in general, but also a professor of physics capable of explaining what the elaborate constructions Leonardo depicted are supposed to do. Most of us know about the parachute, the attempts at flight, the putative helicopter, and of course the Vitruvian Man, but few of us can truly understand the myriad machines which throng the six thousand pages of his surviving notebooks. That feat requires someone in possession of specialist scientific knowledge, familiarity with the arcane vocabulary of modern engineering, and a thorough understanding of the technical terms used in Renaissance Italy, both in Latin and Italian.

But perhaps the most fascinating aspect of Plinio Innocenzi's study is his convincing investigations into the origins of those magnificent machines. For Leonardo was very much a man of his time. He was not averse to recognizing a good idea in someone else's work and making it his own with superior draftsmanship and the targeted use of techniques such as unusual perspective. He was part of a community of thinkers exploring the world in new ways, pushing the boundaries of knowledge together. He could perhaps be considered *primus inter pares* as a result of his immense talent, but not a towering solitary genius who sprung from nowhere, as he is often made out to be. In fact, Plinio goes well beyond a straightforward list of Leonardo's artistic and technical achievements, instead informing the reader that he has used Leonardo "as a lens through which to understand a number of stories of technological inquiry and inspiration that originated during the Renaissance."

This leads to a book that is far broader in scope and relevance than a list of works in a museum catalog.

This method is exemplified by the opening chapter, which introduces a number of Leonardo's brilliant predecessors and contemporaries, who worked on similar problems and themes. A reader who has worked through a good biography of Leonardo knows about the key influence of one of the greatest and most original of the Florentine engineers and architects, Filippo Brunelleschi, and how a young Leonardo learned from designs for the Cupola of the cathedral of Florence. But who, to take just a single example, knows Mariano Daniello di Jacopo, known by his nickname Taccola ("Jackdaw")? I checked one of the best-known biographies of Leonardo in English, six-hundred pages long, and found no mention. Yet Taccola is one of the most fascinating minor characters in Plinio's story. This Sienese hydraulic engineer, painter, sculptor, and author of a ten-volume treatise on machines was the source of many of Leonardo's ideas. He deserves to be better known, and this book provides a good start, including drawings that were a genuine revelation for me. In fact, the choice of illustrations by those other than the main subject is exemplary throughout the book.

Taccola also appears in the chapter on the Vitruvian Man, along with his fellow Sienese engineer Francesco di Giorgio, whose translation of part of Vitruvius's *De Architectura* was used by Leonardo himself. It was in fact Taccola who attempted the first Renaissance rendering of the Roman architect's instructions on how to create a human figure simultaneously within a circle and a square. His friend Francesco di Giorgio also made an attempt. The results are both less attractive than Leonardo's famous drawing, but work well enough on a technical level. Still other versions are discussed and illustrated in this fascinating chapter, which manages to throw fresh light (for all but specialist scholars) on one of history's most famous drawings.

Another unexpected and unusual delight for me was the chapter on "Flotation, Walking on Water, and Diving Under the Sea." Leonardo's sketches of vortices of water and mill-rushes are celebrated. They were also the drawings that most impressed me in a volume that arrived as a childhood Christmas present—and when I saw some of the originals in Milan and London. But once again, Plinio takes us far beyond those famous sketches into a little-known world of life jackets with inflatable cavities, like the ones flight attendants teach us to use before take-off. He also covers snorkels for surface diving, with masks and goggles; and scuba-like systems, with tubes and other breathing apparatus for going deeper under water for longer periods. Once again there were precedents, but Leonardo's designs were often more sophisticated from a technical point of view—and certainly better drawn, an important consideration in view of their military application, for then as now technological innovation was often driven by the needs of warfare. But Leonardo's liberal adaptation of previous ideas does not detract from his genius. As Plinio writes, "He may not always have used original ideas, but he had the extraordinary ability to go beyond the general representation of a concept to design detailed working projects that we are still able to reproduce and test today."

This book is full of such unexpected pleasures and insights, and I could go on with fascinating examples from each chapter, but it would be better for readers to plunge into this astonishing world and let themselves be guided by the author through its maze of invention. In his conclusion, Plinio comments on Leonardo's delight in the world of machines and dreams, arguing that he "was happy only when he could spend his time with mathematics, geometry, and mechanics, exploring the natural world..."

That, it seems to me, might equally apply to the author of this volume.

Beijing, China
November 2017

Edward Burman

Edward Burman is currently Executive Director for Mediapolis Engineering in China. He also serves as a Visiting Professor at Xian Jiaotong—Liverpool University at Suzhou and as a Trustee of the Xian City Wall Protection Foundation. In addition, Burman is the author of a number of popular books about history for the general public, including *Emperor to Emperor: Italy Before the Renaissance* and *Italian Dynasties: The Great Families of Italy from the Renaissance to the Present Day*, among numerous other volumes. His most recent work, *Terracotta Warriors: History, Mystery and the Latest Discoveries*, has been published in February 2018.

Preface

Another book about Leonardo? I can image your first reaction looking at the title of this volume. The scholarly and popular literature about Leonardo is seemingly endless, and every tiny detail of his extraordinary life has been analyzed and described, from his artistic work to the famous notebooks. So it might seem that there is not much remaining to discover and tell.

However, frequently, the book you end up writing is the one you would like to read, but cannot find... This has been the case for me in deciding to write *The Innovators Behind Leonardo*.

The greatest portion of publications about Leonardo is academic biographies, catalogs, and specialized articles written by art historians or experts who have dedicated significant time and effort to investigating in detail particular aspects of Leonardo's work in his various and varied fields of interest.

However, the focus has gradually extended to encompass science as well as art, and this has allowed us to gain deeper insight into a very complex personality with many interests in different fields—an approach which nowadays we like to define as multidisciplinary. This has manifested in books and studies by nonspecialists, scientists, and engineers who have brought some fresh air to the field. Particularly notable because of their original critical approach focused on Leonardo's technological legacy are Fritjof Capra's books[1,2] and a handful of articles written by mathematicians and engineers, which we will encounter in more detail in the chapters to come. A very nice illustration of Leonardo's scientific and engineering work can be found in *Leonardo Da*

[1] Capra F (2013) Learning from Leonardo. Berret-Koehler Publishers, San Francisco.

[2] Capra F (2007) The science of Leonardo. Anchor Books, New York.

Vinci: Experience, Experiment and Design by Martin Kemp[3]. Likewise, readers can find a vibrant description of Leonardo's studies in mechanics and mathematics, with a deep analysis of his work in anatomy, in Kenneth Keele's *Da Vinci's Elements of the Science of Man*[4]; this book, published in 1983, is still considered a fundamental work on the subject.

Leonardo's machines, such as his robot, flying machines, and musical instruments, have also been the subject of several detailed reconstructions, sometimes with a certain dose of fantasy—as in the case of Leonardo's purported bicycle, which most scholars believe to have been the work of a fanciful prankster. In recent years, with the aid of computer graphics, an extraordinary level of accuracy in these reproductions has been achieved. The machines' operation and operating principles can now be understood with an unprecedented level of precision, even when only fragments appear in Leonardo's notebooks[5].

Leonardo's extensive notes on mechanics, anatomy, geometry, physics, and optics (to name just a few) bear witness to his efforts to achieve a deep understanding of nature in its entirety. Examining the results of his incredible efforts without an awareness of his scientific predecessors and contemporaries, however, only provides a small part of the "true" story. Leonardo's achievements are prefaced by the appearance of a new generation of innovative thinkers, engineers, mathematicians, and architects who all contributed to creating the fertile environment that nurtured Leonardo's studies and ideas. To omit this backdrop gives only a fragmentary knowledge of a complex process. Leonardo appears as a kind of isolated genius—an attractive image, perhaps, but one which does not correspond to the historical reality. On the contrary, Leonardo was part of a much broader historical trend—the Renaissance evolution of science and technology in the Western world.

During Leonardo's lifetime, from the middle of the fifteenth century to the beginning of the sixteenth, Italy was experiencing one of the most turbulent periods of its long history. The region was divided into a mosaic of small Signorias and city-states, which engaged in a continuous struggle for power while simultaneously facing the threat of invasion by other European countries. The people of this time lived hard lives, which necessitated daily efforts for survival. Despite these difficulties, however, some individuals managed to develop significant technical and artistic skills.

---

[3] Kemp M (2011) Leonardo Da Vinci. Experience, experiment and design. V&A Publications. London.

[4] Keele KD (1983) Elements of the science of man. Academic Press.

[5] Taddei M, Zanon E, Domenico Laurenza D (2005) Le macchine di Leonardo. Giunti Editore. Firenze.

Though it may seem unlikely, the Renaissance was born from precisely those turbulent times and in that small geographical region of the world. Political instability did not represent an insurmountable obstacle for the brilliant minds that flourished in a unique and vibrant combination in the same place within a limited range of time. In addition to the arts, banking and accounting systems, architecture, medicine, engineering, and technology found a fertile ground for rapid development. This avalanche of innovation was triggered by the rediscovery of the Latin and Greek classics. An enormous past knowledge, which had remained buried for centuries, thanks to the translations of dedicated scholars, suddenly became available as an invaluable source of inspiration for new discoveries in a wide range of fields.

Leonardo was born in the midst of this wonderful moment, and as a child was a direct witness to those extraordinary times. Very close to the workshop of Verrocchio in Florence, where Leonardo was apprenticed, the Brunelleschi Cupola (dome) of the cathedral was already part of the city panorama, making an enormous impression on residents. The realization of the dome, with its challenges of design and implementation, can be considered the beginning of the technological Renaissance. It was an unprecedented architectural and engineering masterpiece that continues to impress, even in the modern age. It was also a sign of new times to come, when new challenges would be tackled by men rich in talent and imagination and hungry for knowledge.

Leonardo, better than anybody else, represents the true essence of the Renaissance man. As a curious child, he observed the complexity of the world from a new point of view. At the same time, he tried to employ this knowledge to adapt nature to his needs. These attempts remained in part only fragments of a world of illusions and dreams—a world of flying machines and robots which was, however, destined to become part of our modern reality. Despite, or perhaps because of, their extraordinary farsightedness, the Renaissance men were frequently simply too advanced for the times in which they lived.

The vision of a new technologically enhanced world, where machines would be able to do most of the work, thus releasing man from daily drudgery, may have appeared at that time as a childish game for naïve dreamers. As we will see, however, Leonardo and his fellow visionaries went far beyond mere dreams, developing concrete machines and innovative technologies.

These were a special type of men, not yet the super-specialized experts that we are familiar with from the present day, but rather artist-engineers with multifaceted interests. They were able to pass from painting to mathematics to architecture, all the while assuming that this multidisciplinary capability was absolutely normal. This was a distinctive sign of the times. They had a true thirst for knowledge which acted as the driver for interest in just about everything, combined with a confidence that, given sufficient scholarship and effort, they could conquer nature.

During Leonardo's lifetime, and even much before it, there developed an extraordinary cultural environment for innovators. All of them shared with Leonardo this unique capability of being at the same time artists, engineers, and inventors. As we shall see over the course of the pages to come, Leonardo is enormously indebted to this intellectual community. They not only showed him how to design machines and devices but, even more importantly, they also introduced him to the possibility of imagining a new world.

It is a history not well known outside of specialist circles and not fully explored until now. Without being aware of this unique story, however, it is almost impossible to understand the genesis of Leonardo's ideas, particularly his work on mechanics and machines. From a scholarly perspective, this context has been well described by Bertrand Gille in the *Engineers of the Renaissance*[6] and in Paolo Galluzzi's *Renaissance Engineers: From Brunelleschi to Leonardo da Vinci*, and dedicated readers are encouraged to see out those texts.[7] We will see that the link between many of Leonardo's most famous ideas and those of these precursors is very direct; indeed, many of them simply look like direct copies or rearrangements of his predecessors' work. Ironically, while today the impulse toward seeing Leonardo as an isolated genius has led to a kind of collective self-censorship where the original sources of his work are concerned, for centuries the scientific and technological work of Leonardo itself went almost unnoticed, before suddenly becoming part of a popular cult which styles him as an iconic and mystical figure in movies and bestsellers.

To put Leonardo in the right perspective does not minimize his achievements. On the contrary, it makes even more evident his difference from his contemporaries, which can essentially be summed up in a single concept: the scientific method. As we will see in more detail throughout this book, Leonardo's studies marked a change, since they were based for the first time on *sperienza* (experience), and this represented a real turning point in the history of science and technology. The experimental method itself would come later, with Galileo Galilei, and with it the foundation of modern science, but it was already clear to Leonardo that only direct knowledge through experiments could catalyze true advancement of human comprehension of nature.

[6] Gille G (1996) Engineers of the Renaissance. MIT press, Cambridge, Massachusetts.

[7] Galluzzi P (1996) Renaissance engineers. From Brunelleschi to Leonardo da Vinci. Giunti Editore. Firenze. It is a well-illustrated catalog of the exhibition *The Art of Invention: Leonardo and Renaissance Engineers* and contains several images and descriptions of Renaissance machines and their computer reconstructions. http://brunelleschi.imss.fi.it/ingrin/index.html.

Many of Leonardo's works are a kind of "thought experiment" or "proof of concept" in our modern terminology. They contain the seed of real scientific and engineering innovation largely in advance of their time. Leonardo's machines appear as true exercises of design. His unique capability of complex spatial representation allowed him to visualize projects with a detailed precision that, in many cases, would only be possible to reproduce with the help of computers.

Leonardo also had another important skill, one which has been significantly underestimated: he drew and elaborated upon previous knowledge. In this, too, Leonardo used the same method that every scientific researcher uses today. He consulted libraries, taking note of everything that he considered important; he made field trips to see in person what could be of interest to know; and he exchanged ideas with some of the most brilliant minds of the time. He had a thirst for knowledge and was lucky to live in one of the most incredible periods of human history during an explosion of creativity, which has left behind some of the most precious human legacies.

Putting Leonardo in the context of his scientific predecessors and contemporaries makes him feel more human and much closer to us; the thin rope that links the world before to Leonardo's own life is also directly connected to our present. The road to reach our technologically advanced age would be, however, an uneven one. History is never an easy, linear development of events. Failures are part of the learning process, creating problems that require long collective effort to solve, sometimes over the course of centuries. Several routes would turn out to be simply dead ends—such as the obsessive search for perpetual motion, or squaring the circle. Other dreams such as flight, which for a long time appeared impossible, would become reality only some centuries later.

In his notebooks (about 6000 pages of which are available today), Leonardo noted and sketched everything that looked interesting to his curious mind with the intention of making these notes the basis for his studies. The contents of these pages are not limited to drawings and concepts, but also include exercises in Latin and arithmetic, shopping lists, and lists of books to collect. They contain observations on mechanics, paintings, flight, physics, anatomy, and mathematics, but also jokes and personal memos. Also to be found are several machines and ideas taken from outside sources—from manuscripts in Leonardo's possession or from notes taken during visits to the rich libraries of Italian cities such as Florence and Pavia, which he visited quite often.

Leonardo's notes are, however, not easy to read. As is well known, he wrote from right to left; in addition, the calligraphy and his Italian are difficult to interpret. The incredible efforts of several generations of scholars have been necessary to decipher his enormous volume of notes. We are sincerely indebted to them for the access we now have to the extensive patrimony of knowledge that is Leonardo's legacy. A beautiful account of this patient and systematic work can be found in the book *Leonardo & io,* written by one of the most famous Leonardo scholars, Carlo Pedretti.[8] Thanks to this sustained collective enterprise, all surviving pages have been digitalized, and a diplomatic (i.e., a transcription of the text that reports the characters as they appear, with minimal or no editorial intervention or interpretation) and critical transcription is now available online.[9]

The digitization of the technical knowledge of the Renaissance, most of it contained in delicate manuscripts with very few copies, is not limited to Leonardo's work. Today, online versions of Leonardo's main sources of inspiration, such as the treatises of the Siena engineers, Francesco di Giorgio[10] and Taccola,[11] are also available. This has allowed for much easier consultation, and critical comparison of the different original sources, which only a few years ago would have required years of patient exploration in European libraries.

Starting from these manuscripts it is possible to take a long journey back into the past, which will carry us into a world full of contradictions that is quite difficult to understand. In this strange world, the same person who sketches odd war machines capable of mutilating people in horrible ways is also able to represent on the same page a picturesque scene with a peaceful fisherman sitting on the shore of a river. Such unexpected juxtapositions are not unique to Leonardo—they were in the nature of what were, from our perspective, difficult and contradictory times.

In this book, I have used Leonardo da Vinci as a lens through which to understand a number of stories of technological inquiry and inspiration that originated during the Renaissance. Following the traces of these stories makes it easier for us to understand Leonardo and his time. Every chapter is dedicated to a different story, though they always relate to scientific and technological subjects. The first chapter introduces some of the "others"—Leonardo's scientific predecessors and contemporaries. Foremost among them, the Siena Engineers will figure prominently in the specific stories that follow. The

[8] Pedretti C (2008) Leonardo & iO. (in Italian) Mondadori, Milano.

[9] The *Biblioteca Comunale Leonardiana di Vinci* has realized a digital archive, which contains the diplomatic transcription of all the written work of Leonardo da Vinci. www.leonardodigitale.com.

[10] Four codices of Francesco di Giorgio (Codex Ashburnham 361, Ms. Regg. A 46/1/9 bis, Vat. Urb. Lat. 1757, Ms. 197.b.21) are available online also at www.leonardodigitale.com.

[11] A copy of Taccola's De Ingeneis can be downloaded at the Cornell Library web site: http://ebooks.library.cornell.edu/k/kmoddl/pdf/037_001.pdf.

remaining chapters chart the origin and genesis of a range of ideas, from flying machines to diving apparatus, that captured the imaginations of Leonardo and the other great minds of his time, and whose pursuit continues into the present day. Because each chapter charts a different area of inquiry, they need not be read strictly in the order in which they appear. As Leonardo himself was used to do, readers should feel free to jump from one topic to another as the interest strikes them. The subjects have been selected to capture the most interesting ideas, and those with the deepest connections between Leonardo's world and our present reality.

Rather than providing simply an illustrated catalog of Leonardo's machines, it is hoped that this topical approach will give readers a taste of the scientific excitement of Leonardo's time, making it possible to experience the dreams of the "hungry and foolish"[12] (to borrow Steve Jobs's turn of phrase) that allowed Leonardo, his predecessors, contemporaries, and successors to imagine and shape the future.

Alghero, Italy Plinio Innocenzi

---

[12] "Stay hungry, stay foolish". Steve Jobs's speech at Stanford, on 12 June 2005.

Note to the Readers

The completion of this manuscript in its current form would not have been possible without the digitalization of the invaluable cultural heritage represented by Renaissance and Medieval manuscripts related to science and technology. Only a few years ago, consulting these materials, dispersed as they are among libraries across Europe and the USA, would have been a task that required enormous amounts of time and financial support. Thanks to digitalization efforts, it has been possible for me to consult the original versions of many of the manuscripts mentioned in this book. In particular, all of Leonardo's written work is now easy to view from the comfort of one's own home. When available, I have noted the sources I consulted in the body of the text. I have likewise cited the sources of all the images I present in the figure captions. I have preferred to use such images in edited form to allow readers not accustomed to the peculiarities of original manuscripts to appreciate the technical aspects of these drawings in detail. I have also included the original (untranslated) version of all quotations in the footnotes. It would be quite difficult for most non-Italian readers (or, for that matter, most Italian readers) to understand the language used at the time. It is, however, fascinating to have a look at the original words used by Leonardo; and any reader can have the pleasure of seeing the explanations in Leonardo's own words while imagining the sound of his voice.

Acknowledgements

I am so much in debt with Sara Kate Heukerott and Sanaa Ali-Virani from Springer for their incredible support. Without them I could not have realized this book. Grazie! I am also in debt with prof. Attilio Mastino for the translations from Latin and with Edward Burman for his support in reading the manuscript. A very special acknowledgment to my family, my wife Tongjit and my sons Jacopo and Lavinia, because they are my main source of inspiration.

Contents

About the Author

Plinio Innocenzi is a full professor of Materials Science and Materials Technology at the University of Sassari and has a special interest in science popularization at different levels. He has a doctorate in physics but attended a Classic Lyceum dedicated to Humanities, Art, Latin, and Greek, which is why he has always maintained a multidisciplinary vision and developed a special passion for the connections between science and art. He has dedicated particular attention in recent years to Leonardo's scientific work and has participated in numerous conferences and festivals on the topic. This book is the result of Prof. Innocenzi's desire to make general readers more aware of the multifaceted origins of scientific and technological knowledge.

Abbreviations

BAM	Biblioteca Ambrosiana, Milan (Italy)
BAV	Biblioteca Apostolica Vaticana, Rome (Italy)
BIF	Bibliothèque de l'Institut de France, Paris (France)
BL	The British Library of London (UK)
BML	British Museum, London, (UK)
BMLF	Biblioteca Medicea Laurenziana, Florence (Italy)
BNCF	Biblioteca Nazionale Centrale di Firenze, Florence (Italy)
BNE	Biblioteca Nacional de España, Madrid (Spain)
BNF	Bibliothèque Nationale de France, Paris (France)
BRT	Biblioteca Reale, Turin (Italy)
BSBM	Bayerische Staatsbibliothek, München (Germany)
GDSU	Gabinetto dei Disegni e delle Stampe degli Uffizi, Florence (Italy)
ULJCS	University Library, Johann Christian Senckenberg, Frankfurt Am Main (Germany)
VAM	Library of Victoria and Albert Museum, London (UK)

1

Leonardo and the "Others": Engineers and Inventors of the Early Italian Renaissance

The general perception of Leonardo da Vinci as a self-educated genius who suddenly appeared in Florence during the Renaissance is highly fascinating and romantic, but unfortunately quite far from the reality. This popular image does not take into account the extraordinarily fertile environment of the time, when versatile artists and architects were able to innovate in many different fields, contributing to that unique season of human history. Leonardo is very much in debt to these "others": a significant portion of his work on mechanics and machines is largely a derivation or development of the studies of his predecessors, who had already planted the seeds for a scientific and technological Renaissance.

From the very beginning of his apprenticeship in Andrea del Verrocchio's workshop, Leonardo came into contact with the wonderful world of machines, which exerted on him a fascination which lasted all his life. Some scholars observe that he loved his studies on mechanics more than his artistic work as painter, which gave him fame and success when he was alive. As we can see in his notebooks, Leonardo carefully drafted and studied the machines created by Filippo Brunelleschi (1377–1446) to build the Cupola of the Duomo of Florence (Cathedral of Saint Mary of the Flowers). At the same time, he used the patrimony of knowledge accumulated by a small group of talented engineers, mostly from the Republic of Siena, to develop visionary projects of unprecedented ambition.

Leonardo's body of work on mechanics is not diminished if we put it in this context. We simply begin to appreciate it, not as the work of an isolated genius but as the synthesis of a lengthy process that started in the mid-fourteenth century in Italy with Brunelleschi and in greater Europe with the

P. Innocenzi, *The Innovators Behind Leonardo*, https://doi.org/10.1007/978-3-319-90449-8_1

German school. It is during this period that we see dramatic strides in scientific and technological knowledge. The contributions to this season of innovation came from a variety of different sources and places. It was also a collective process, furthered in part by anonymous artisans, such as those who participated in actualizing Brunelleschi's machines, whose work has never been fully credited.

In this chapter, the protagonists of this creative season will be introduced. They are the "others:" figures largely unknown to nonspecialists, who developed the foundation for Leonardo's own work and ideas.

While this chapter focuses only on the main group of Italian artist-engineers, largely from Tuscany (and Siena in particular), since they stand out as a group for their strong influence on Leonardo, Roman and medieval knowledge also played a significant role, as will be discussed in future chapters where we delve into the development of individual ideas and devices.

The "Sienese Archimedes": Mariano di Jacopo–Called Taccola

As we will see, two individuals play an outsized and fundamental role in the history of the technological Renaissance: Mariano Daniello di Jacopo called Taccola ("jackdaw")[1,2] (1381–c. 1458[3]) and Francesco di Giorgio Martini (1439–1501), both from Siena. Not very well known outside of scholarly circles, their work and its impact on the history of technology has been studied in depth only recently.[4] They are some of the main sources of inspiration for the machines, mechanics, and military technology that have been attributed to Leonardo. Often, in museums and exhibitions, some of their inventions that were only reproduced or slightly modified by Leonardo are wrongly attributed to him. Taccola and Francesco di Giorgio were not simply

[1] The nickname Taccola (Jackdaw) came from his father; his certificate of baptism reads, "Mariano Danniello of Jacomo detto Tàcchola." For some time it was thought that the nickname was due to his very pronounced aquiline nose.

[2] Doti G (2008) Entry on Mariano di Iacopo, In: Treccani. Dizionario Biografico degli Italiani. Volume 70.

[3] The date of death is not known with accuracy but was sometime between 1453 and 1458.

[4] The manuscripts of Taccola were rediscovered in the second half of the eighteenth century, but were then forgotten again in the libraries of München in Germany and Florence in Italy. It was only in the 1960s that a critical edition of his work was published. The rediscovery and publication in facsimile is due to the work of Gustina Scaglia and Frank Prager.

Frank D. Prager, Gustina Scaglia. Mariano Taccola and his book "De ingeneis." Cambridge (Mass.), MIT Press, 1972. Mariano di Iacopo detto il Taccola, Liber primus leonis, liber secondus draconis,... and addenda, edited by Gustina Scaglia, Frank D. Prager, Ulrich Montag, Wiesbaden, L. Reichert, 1984, 2 v. (I text, II facsmile).

precursors, they were much more—innovators, engineers, and artists with multifaceted personalities—true Renaissance men who represent a fundamental stepping stone between the knowledge of the Middle Ages, the German school of machines, and the new age of the Renaissance.

Taccola and Francesco di Giorgio were singular individuals, not only because they played a fundamental role in the history of engineering but also for their innovative way of diffusing knowledge beyond the closed environment of the workshops of the Middle Ages and early Renaissance. Their treatises represent the first systematic attempt to elaborate and reproduce technical knowledge through explanatory illustrations. On the other hand, identifying Taccola and Francesco di Giorgio only as engineers (*ingeniarii*) is actually quite reductive: they had interests which spanned a variety of fields, from arts such as painting and sculpture to engineering and architecture. They represent the cultural element in the progression between Brunelleschi and Leonardo in a kind of ideal handover between different generations of artist-engineers.

Taccola survived Brunelleschi by 12 years. He completed his treatise *De Ingeneis* just as the Florentine architect finished his Cupola and died (Leonardo was also born around this time). Francesco di Giorgio had the opportunity to read and study Taccola's books, and Leonardo, in turn, would read and annotate Francesco di Giorgio's own manuscripts. This direct line of intellectual descent among the different protagonists of Renaissance is notable; it represents an ideal passage of knowledge that marks the transition from the Middle Ages and the new age to come. If we look at the years in which each lived—Brunelleschi (1377–1446), Taccola (1381–1458), Francesco di Giorgio (1439–1501), and Leonardo da Vinci (1452–1519)—we immediately notice that their lives overlap such a way as to allow a direct connection between each successive generation. Records show that there was a direct transfer of knowledge from Brunelleschi to Taccola: the Sienese engineer described in his manuscripts his encounter with the Florentine architect and the discussions they shared on different subjects.[5] They shared a common passion for engineering and machines, which was likely the basis of their friendship. A detailed account of this event can be found in *De Ingeneis*, where Taccola writes about Brunelleschi with great respect. The discussion between the two engineers included the problem of plagiarism and intellectual property. Ironically, many of the machines invented by Brunelleschi for the construction of the Cupola, such as a special boat for shipping marble, are reported in Taccola's notebooks,

[5] Prager FD (1968) A Manuscript of Taccola, Quoting Brunelleschi, on Problems of Inventors and Builders. Proceedings of the American Philosophical Society, 112:131–149.

only to be later wrongly attributed to Leonardo. Leonardo would have been well aware of these innovations, thanks to the efforts of Taccola and Francesco di Giorgio; their "catalogues" of technologies remained a benchmark for several generations.

The Sienese engineers shared with Leonardo not only a common basis of technological knowledge but also a working method: all three used as an essential instrument of work a notebook, in which they annotated and traced ideas and projects. These notes made it possible to extract and re-elaborate a second time what they needed in order to fully realize a given project (see, e.g., Taccola's machines and techniques for sieges in Fig. 1.1). Another

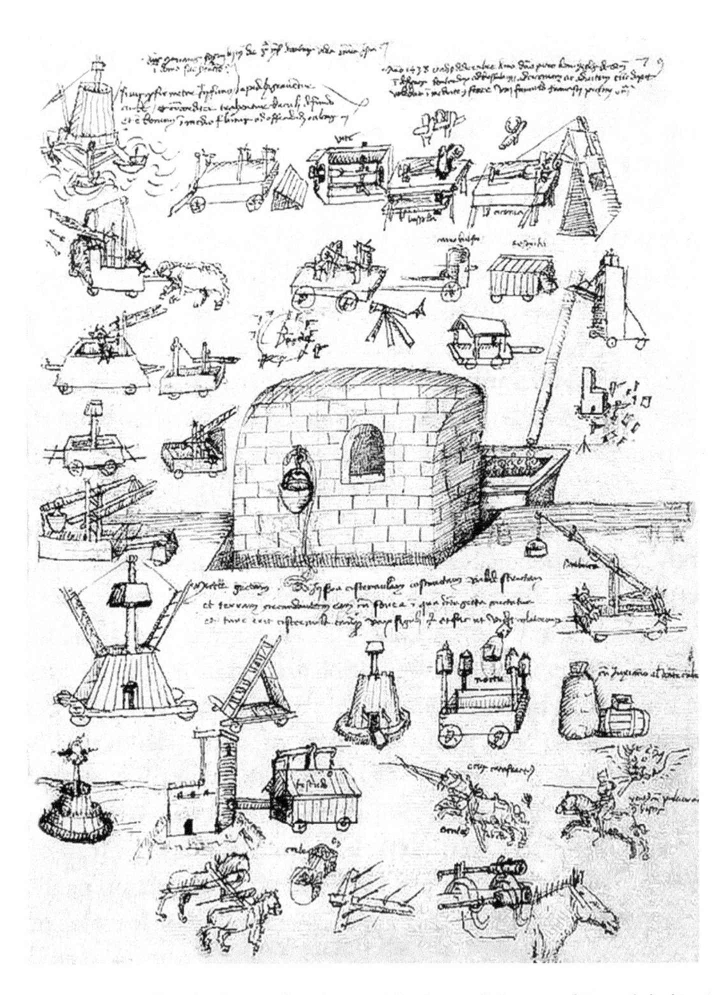

Fig. 1.1 Machines and techniques for sieges. Mariano di Jacopo (Taccola). (De Ingeneis, Cod. Lat. Monacensis 197 II, folio 82r. (BSBM))

common element is the wide use of graphical representations and notes to illustrate ideas and practical projects. Their drawings are, in most cases, clear and easy to understand, even if they sometimes exhibit a naive simplicity. They also share with Leonardo's notebooks the will to elucidate rather than hide knowledge. It is an unexpected "open access" approach, which was quite surprising for a time in which technical know-how generally remained carefully confined within the boundaries of the workshop.

Another distinctive element is the illustrative and didactic use of drawings that reached, especially in the case of Francesco di Giorgio, an unprecedented level of technical quality. The sketches were used to explain projects in detail, allowing them to be reproduced. The drawings in their manuscripts are clear and elegant, a significant advancement with respect to the simple and sometimes naive sketches that appeared during the Middle Ages or in the texts of the German school. These drawings mark a definitively new approach to engineering based on significant attention to graphical representation of projects and the requisite technical explanation to realize them. There is also another remarkable difference, which is the explosion of creativity and the remarkable ability to innovate that can be found in the manuscripts of the Italian engineers with respect to the past. These figures exhibit an awareness that they were not alone. They were living and working with some of the most creative artists in history, competing and cross-fertilizing to develop new ideas and new visions of the future.

Taccola wrote two treatises. The first, *De Ingeneis*[6] ("Concerning engineering"), was a four-volume work written over a long period, between 1419 and 1450, and was dedicated to projects of civil and military engineering, with special attention to hydraulics.

His second, more ambitious collection is *De Machinis* ("Concerning machines"), a ten-volume work written between 1430 and 1449. This treatise is mainly composed of descriptions of military technology—machines for sieges and fortifications. Taccola, the son of a winemaker who nonetheless received a good literary education, wrote in Latin. In contrast, Leonardo, an illegitimate son of a wealthy Florentine notary, had only a basic education and

[6] The autograph copies of books I and II of *De Ingeneis* are conserved in the National Library of München, Codex Latinus 197; book I is composed of folii 1–21, 30–75, while folii 22–29 are bound in a separate file. Book II includes folii 76–96 which continue in the books I–IV, folii 97–137. Books III and IV are conserved in the National Library of Florence, Codex Palatinus 766, book III folii 27–57, and IV folii 58–76.

A copy of *De Machinis* is conserved in the Bayerische Staatsbibliothek as Codex Latinus 28,800.

At least other three copies of *De Machinis* survive: one at the New York Public Library as Codex Spencer 136, another at the Biblioteca Marciana of Venice as Codex Latinus 2941, and the final copy, illustrated by Paolo Santini, is conserved at the Bibliothèque Nationale de France as Codex Latinus 7239.

developed his knowledge only thanks to his autodidactic efforts. Leonardo did not know Latin, and, because of this, he modestly termed himself *omo sanza lettere*: a man without education. In time, Leonardo was able to achieve a reasonable mastery of Latin which, during the Renaissance, was still considered the language of the educated. He studied Latin with dedication, and several pages of his manuscripts are filled with sentences and lists of words with their translation in Italian, which he compiled for practice.

Taccola, on the other hand, was much more self-confident about his erudition and his capability as an inventor and man of genius (*ingeniarius*). In *De Machinis*, he dared to term himself the Archimedes of the magnificent and powerful city of Siena: *Ser Mariano Taccole alias archimedes vocatus de magnifica ac potente civitate Senarum.*

Taccola was a tranquil person with no special ambition for advancement (though, as we have just seen, he was quite proud of his skills as an inventor and engineer), who spent most of his life in Siena in a relatively peaceful setting, at least considering the turbulent times. He had a successful career in the administrative ranks of the small republic, became administrator of the Domus Sapienta (one of the main cultural institutions of the city), and was later employed as a hydraulic and military engineer by the Republic of Siena. His true interests lay outside of administration and law, and he developed a particular passion for art and engineering. We have evidence of some artistic works in wood that were commissioned from him. One of these works was requested in 1437 by the famous *Maestro* of Siena, Jacopo della Quercia, indicating that in Tuscany Taccola was appreciated for his artistic as well as engineering talents.

Taccola lived in very dangerous times, and the small Republic of Siena had to work hard to survive, bouncing between military confrontation and strategic political alliances with other states in order to maintain its independence and democratic government. From the year 1386, just 5 years after Taccola's birth, until 1487, Siena was ruled by the elected council of Priori, with the exception of a brief period between 1399 and 1404 when Gian Galeazzo Visconti,[7] the Duke of Milan, was nominated to be Signore of the city.[8,9]

[7] Gian Galeazzo Visconti (1351–1402), was the son of Galeazzo Visconti II and Bianca of Savoy and became in 1395 the first Duke of Milan.

[8] During the Middle Ages, the Italian cities gradually transformed into city states, mostly oligarchic republics. Over time, some of the oligarchs tried to seize power as individuals, and transformed the cities into "Signorie." This is, for instance, what happened to Florence when the Medici family obtained the control of the city. In the case of the Republic of Siena, the people were afraid that the city was too weak to face an attack from Florence, so they spontaneously nominated Gian Galeazzo Visconti as Signore of the city state.

[9] The Priori were generally selected or elected from within the noble families and the city corporations. Between 1386 and 1399, several different "priorati" ruled Siena, 10 "priori" (1386–1387), 11 "priori" (1388–1398), and 12 "priori" (1398–1399). Between the years 1404 and1487, the governing body of Siena had 10 priori.

From 1487 to 1525, Siena was ruled by the Petrucci family, who took power from the Priori (Signoria dei Petrucci). Throughout both periods, Siena had a very powerful neighbor, the city of Florence, and conflicts for supremacy were almost unavoidable.

These confrontations were ongoing, culminating in Siena's capitulation in 1555 following a long siege. This loss marked not only the end of the small Republic of Siena but also the start of a trend whereby foreign powers would, little by little, take control of the mosaic of small Italian states, ending the region's period of democracy and government of free people.

Sigismund of Luxemburg (1368–1437), Emperor of the Holy Roman Empire, supported Siena against Florence, so it is easy to understand Taccola's devotion to him. In 1432, Taccola put himself under Sigismund's protection, offering his services as a hydraulic engineer. In exchange for his devotion, Sigismund personally granted Taccola the important title of Count Palatine. In spite of the high rewards and important positions he reached within the city administration, at the death of his wife, Taccola became a friar and refused any other appointment until his death, sometime between 1453 and 1458.

The search for a powerful patron was a common and fundamental need not only for artists but also for architects and engineers throughout the Renaissance. Many of Leonardo's choices were dictated by the necessity of seeking financial support and protection. This explains his frequent moves from one Italian city to another as the political situation changed. Notably, Leonardo offered his services to the Duke of Milan on the basis of his ability as military engineer rather than for his artistic skill, though it was for the latter that he was already famous. In a time of continuous war, this type of knowledge was highly prized by the Italian city states, which valued not only military engineering but also the ability to design fortresses and castles, siege machines, and new weapons.

As mentioned above, Taccola became famous in Siena as a hydraulic engineer. Traces of this interest survive in his treatises, where several projects for pumps, water mills, and dams are described. Taccola's engineering work was well known in the sixteenth century, and numerous copies of his manuscripts were made. Notwithstanding the popularity of Taccola's treatises during the Renaissance, at least within the circle of people interested in civil and military engineering, his work quickly felt into oblivion. The copies of *De Ingeneis* and *De Machinis* remained dispersed and forgotten in European libraries, a fate similar to that experienced by Leonardo's notes. Their recent rediscovery has finally allowed for a more complete understanding of the influence of Taccola's work on figures like Leonardo and, more broadly, on the history of technology.[10]

---

[10] Fane L (2003) The Invented World of Mariano Taccola: Revisiting a Once-Famous Artist-Engineer of fifteenth Century Italy. Leonardo 36:135–143.

Taccola used notebooks to record and sketch his projects, exactly as Leonardo would do many years later. Several copies of his original manuscripts were made, but by hand (printing was not yet the norm), which limited the diffusion of his work. Each copy differed slightly from the original, especially in the quality of drawings. A nicely colored version of *De Machinis*[11] from Paolo Santini was donated to the powerful Sultan Mahomet II (1432–1481). It was likely brought to Istanbul by a friend of Taccola, Ciriaco d'Ancona,[12] a merchant and traveler of liberal education who came into possession of the manuscript during one of his stays in Siena. This copy would later be bought by the Ambassador of the King of France and sent to Paris, where it is still conserved at the National Library.

The Notebooks of Leonardo and the Siena Engineers

As we have noted, Taccola's notebooks contain drawings and explanatory comments in Latin. Some pages look extraordinarily similar to those from Leonardo's own notebooks, at least in their general appearance. Both contain a chaotic overlap of illustrations of different subjects accompanied by explanatory notes and comments. The main differences between the two are the language, which for Taccola is still Latin, and the fact that Taccola wrote left to right as opposed to Leonardo's mirror writing.

In Taccola's manuscripts, the schematics of the machines and their details are still relatively simple, without use of perspective and *chiaroscuro*[13] to represent details, as was later done by Leonardo. Sometimes the illustrations feature cute or even amusing details (see, e.g., Fig. 1.2). As we have said, the quality of the figures differs from copy to copy; much would have depended on the skill of the individual illustrator or copyist. Some versions have good quality drawings and transmit a vivid representation of the world imagined by Taccola. In general, however, the drawings are far from achieving the beauty and technical quality of Leonardo's work. They are, nonetheless, inspiring and

[11] Manuscript Lat. 7239, Bibliothèque Nationale de France. Copy of Paolo Santini, second middle of XV century. It is composed of 166 pages with dimension 35 × 24.5 cm.

[12] Cirìaco d'Ancona was a traveler, antique dealer, and humanist scholar of the de' Pizzicolli family (Ancona 1391–Cremona 1452). He traveled extensively through Italy, Greece, Egypt, and Constantinople putting together a rich collection of antiques. His collection of epigraph transcriptions (Commentarii) was destroyed during a fire, and only a few notebooks remain of it.

[13] Chiaroscuro is a drawing technique that is used to enhance the contrast between light and dark. It was used for the first time during the Renaissance, and Leonardo was one of the main artists to contribute to its development.

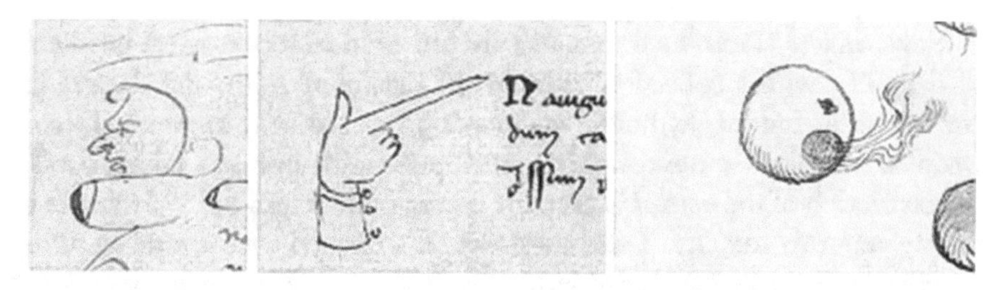

Fig. 1.2 Details from De Ingeneis by Mariano di Jacopo (Taccola): (from left to right) a funny face, the pointing hand, and the smiling bomb

charming for their ability to simply and empathetically represent a bizarre world made up of strange machines and original inventions.

A creative, if relatively simple, personality emerges from the manuscripts. Taccola did not possess Leonardo's sophisticated design capability, but he was undoubtedly a prolific source of inspiration for subsequent generations. On the other hand, we do not know with precision how much Taccola took inspiration from preceding sources, although certain similarities to and overlap in content with manuscripts from the German school suggests that he may have been aware of previous works in the field.

Notwithstanding their creativity, Taccola's drawings possessed limited capability to depict the technical details of mechanical devices. This represented an almost insurmountable obstacle to transforming his ideas into reality. It is only with Leonardo that the graphical representation of mechanics enters a new era; Leonardo introduced three-dimensional perspective and axonometric representations with an extremely sophisticated level of detail. Today, we are able to reproduce most of Leonardo's projects despite their age, even when they are fragmented in several parts. The possibility of reproducing a project, a mainstay of the scientific method, represents a notable advancement in the history of science and technology.

The notebooks of the Siena engineers would become a method of work, used by Leonardo and several others, with notes and illustrations to record ideas and projects. However, if we compare Taccola's notebooks to those of Leonardo, it is clear that Taccola's ability to think and represent the details of mechanical projects is limited. Leonardo's capability of visualizing a mechanical device puts his work on another level, but, as we will see in more detail below, already in Francesco di Giorgio's work, for instance, in the projects for a self-propelled cart, it is possible to find the signs of a big change in respect to the past. Francesco di Giorgio, in fact, demonstrated a technical knowledge and a natural talent to design machines and mechanical devices which could even rival that of Leonardo.

Nevertheless, Taccola's notebooks are full of ideas to solve real problems of daily life, with a collection of hydraulic machines, mills, and devices for construction. Indeed, hydraulic engineering was the field to which Taccola was able to make his most original contributions, for example, taking part in the realization of the *bottini* in Siena, a system of underground repositories to supply water to the city. Their construction stretched over several hundred years and was finally concluded in the fifteenth century. It was an exceptional engineering chef d'oeuvre that is still functioning today.

War Machines and Techniques

Another frequently recurring subject in Taccola's notebooks is war. The notebooks do not limit themselves to machines but also include techniques and strategies for siege. The solutions envisioned are often imaginative, and some odd ideas are represented without much grounding in reality, such as an armored dog carrying fire to attack a knight. The theme of war later also captivated Leonardo and reflects a time when war was common. The creativity of the engineer-architects was put to the service of the art of war, and innovative solutions were proposed for the design of war machines, armored carts, multiple bombards, and more. This work was also dictated, as we have noticed, by the need for powerful protectors. The war machines and other military technologies developed by Taccola, Francesco di Giorgio, and Leonardo represent a moment of transition from a medieval world to a more modern time. Their machines were in great part inspired by the Roman military tradition known through classic treatises such as Vegetius's *Epitoma Rei Militari*. Taccola's works alone include hundreds of war machines and vivid illustrations depicting strategies and techniques for warfare, especially for the siege of cities and fortresses. In addition to the "classic" inventory of traditional Roman weapons, different types of firearms such as bombards, culverins, and more updated cannons are also described in his manuscripts.

This mix of medieval weapons and new technologies can be also found in the work of Leonardo, who, retracing the footsteps of the Siena school, would also dedicate himself to the design of numerous machines and weapons for war that were never realized.

In fact, both the Siena engineers and Leonardo were not particularly innovative in this field. For this reason, this area is by far the least interesting, at least from our modern perspective, of the various subjects to which they devoted their efforts, and in later chapters, we will largely focus our attention elsewhere. The majority of their ideas for warfare were still tightly bound to tradition, with few exceptions such as Taccola's mines and Leonardo's tank.

Working Methods of Leonardo and the "Others"

Another interesting aspect that emerges from the work of Leonardo and the Siena engineers is the eclecticism of the ideas and themes that stretch across so many different fields. On the same page, for example, it is possible to find plans for a war machine and a water well (Fig. 1.3). Peace and war seem inextricably bound up with one another. Every page of their notes can be a surprise, but the attempt to make a systematic study of a specific subject is rare.

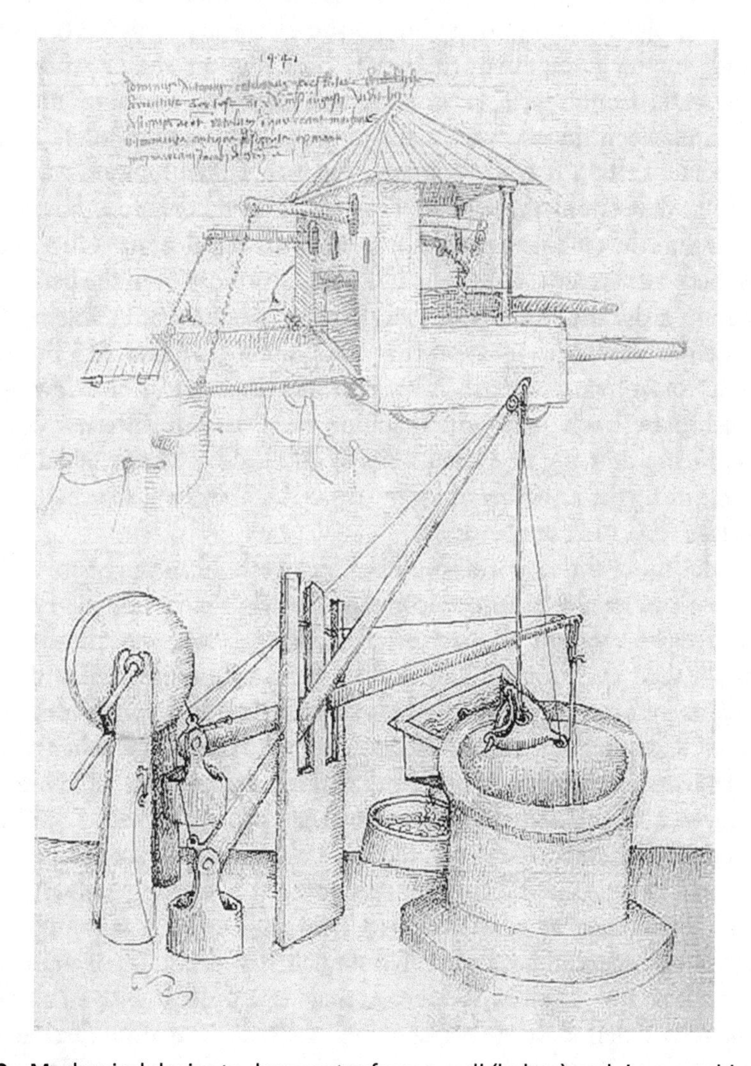

Fig. 1.3 Mechanical device to draw water from a well (below) and siege machine with a drawbridge to cross trenches (above). (De ingeneis, Books I-II, Cod. Lat. Monacensis 197 II, folio 96r. (BSBM))

It seems that they were animated by an inexhaustible curiosity for the novel world that they were anticipating and did not bother to waste time systematizing knowledge. The fever of discovery was given priority—organization could be done later. This is part of the system of thinking which is a peculiar characteristic of the Renaissance *ingeniarii*. They were basically attracted to and interested in everything and had the ambition to master all knowledge without boundaries.

There is an interesting difference, however, between the Siena engineers Francesco di Giorgio and Taccola on the one hand and Leonardo on the other. The illustrations in Francesco di Giorgio and Taccola's notebooks often depict men and women going about their daily life. They are shown while fishing, drawing water from a well, or working at a mill. The machines are used to pose technical solutions whose main purpose is alleviating man's struggles; this reflects a certain friendly empathy for their fellow human, still poor in technology that could make their lives easier. For Leonardo, however, the machines and the technical problems assume a more abstract value, and man himself becomes an object of study. His humanity remains in the background, confined in a distant world. Only in painting was Leonardo able to convey his particular sensibility by capturing the human character and the endless ambiguity of human behavior in his portraits. In some of his drawings, the relationship between man and machine even assume infernal contours, such as in the drawing of *Codex Windsor RL 12647*, where naked men are depicted during the assembly of a gun (Fig. 1.4). They look so exhausted as to almost lose their human appearance.

In underlining the importance of the Siena school in the history of technology, we cannot forget to acknowledge that they were not the first to use graphical representation of mechanical projects. There was already a fairly well-established tradition in Italy and in Europe more generally, thanks to several pioneers, such as Villard de Honnecourt, Guido da Vigevano, Roberto Valturio, Giacomo Fontana, and Konrad Kyeser, to name a few. We will dedicate a detailed description of their work and influence in later chapters. There is, however, a remarkable difference between their manuscripts, still deeply anchored to medieval knowledge, and the work of the Siena engineers. As we have noted above, the Siena engineers marked a turning point, breaking boundaries and creating a fantastic new world where projects were presented with an unprecedented level of skill and quality of detail. Their drawings are far from the simple sketches of the past; they were realized with an ability and level of detail that only an artist could achieve. The multifaceted artist-engineers of the Renaissance were moved by the rediscovery of Roman and

Fig. 1.4 Study with a hoist for a cannon in a foundry. c. 1487. (Pen and ink. Leonardo da Vinci. Windsor castle, Royal Library (R.L. 12,647))

Greek technologies and scientific knowledge. Using this as their starting point, they ignited a small technological revolution whose beginning is marked by the realization of Filippo Brunelleschi's Cupola in Florence,[14] which we will discuss in greater detail in the next chapter.

[14] Galluzzi P (2001). Entry on: Gli ingegneri del Rinascimento: dalla tecnica alla tecnologia. In: Storia della Scienza. Enciclopedia Treccani.

Francesco di Giorgio Martini

After Taccola's death, his work was continued by another talented young Sienese artist, who was destined to have a strong impact in several fields: Francesco di Giorgio Martini (1439–1501). Francesco di Giorgio was around 18 when Taccola died and may have been his pupil. In any case, he was certainly in close proximity to him and his work. He was deeply influenced by it and can be regarded as Taccola's true intellectual heir. Compared to his *Maestro*, Francesco was gifted with uncommon artistic talent in various fields, eventually becoming an excellent architect, engineer, painter, and sculptor.

The autographed copy of Taccola's *De Ingeneis I* and *II*, now conserved in Florence, contains some notes and sketches that were drawn on them by Francesco di Giorgio, providing direct evidence of the continuity between their work. As we have noted, Leonardo in turn would annotate Francesco di Giorgio's own manuscript to complete the chain of this ideal transfer of knowledge between generations.

Francesco di Giorgio's most famous book is the *Trattato di Architettura Civile e Militare*[15] (Treatise of civilian and military architecture), which had a very strong influence on generations of architects who used it as a reference textbook. It was, however, in the *Codicetto vaticano*[16] that Francesco di Giorgio reproduced several of Taccola's concepts, weaving together original and re-elaborated devices and machines. Be that as it may, Francesco's fame as an engineer is due to yet another manuscript, the *Opusculum de architectura*,[17] a collection of technical drawings of very high quality that reproduce various machines and devices without annotations (see, e.g., Fig. 1.5). The details of the projects are shown with a technical facility clearly derived from his skill as architect and artist. The graphical method used to represent the projects is unique, and even Leonardo could not match some examples of his technical ability. Francesco di Giorgio drew his machines introducing a three-dimensional representation and a detailed graphical reproduction of the mechanical parts (Fig. 1.6).

Without a doubt, these drawings represent a breakthrough in terms of technical capability and originality of content. In some ways, Francesco di

[15] Merrill EM (2013). The Trattato as Textbook: Francesco di Giorgio's Vision for the Renaissance Architect. Architectural Histories, 1(1):Art. 20. doi: doi.org/10.5334/ah.at.

[16] Francesco di Giorgio Martini. *Codicetto Vaticano*. Città del Vaticano, Biblioteca Apostolica Vaticana, manuscript Urb. Lat. 1757, (ca. 1465–1470?). It is composed of 190 pages of very small dimensions, 8 × 6 cm.

[17] Francesco di Giorgio Martini. *Opusculum de architectura*, autograph copy composed of 80 folii, written between 1475 and 1478. Londra, British Museum, MS 197.b.21.

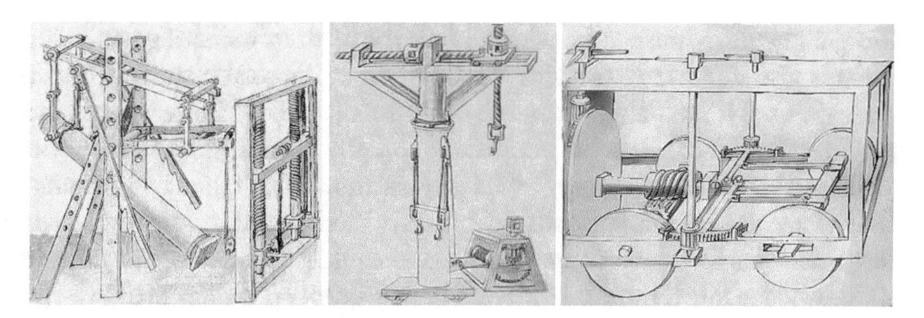

Fig. 1.5 Drawings of machines: machine to lift columns, folio 7v (left), crane, folio 12r (center), automobile, folio 31r (right). (Francesco di Giorgio. Opusculum de Architectura (Ms. 197.b.21). BML)

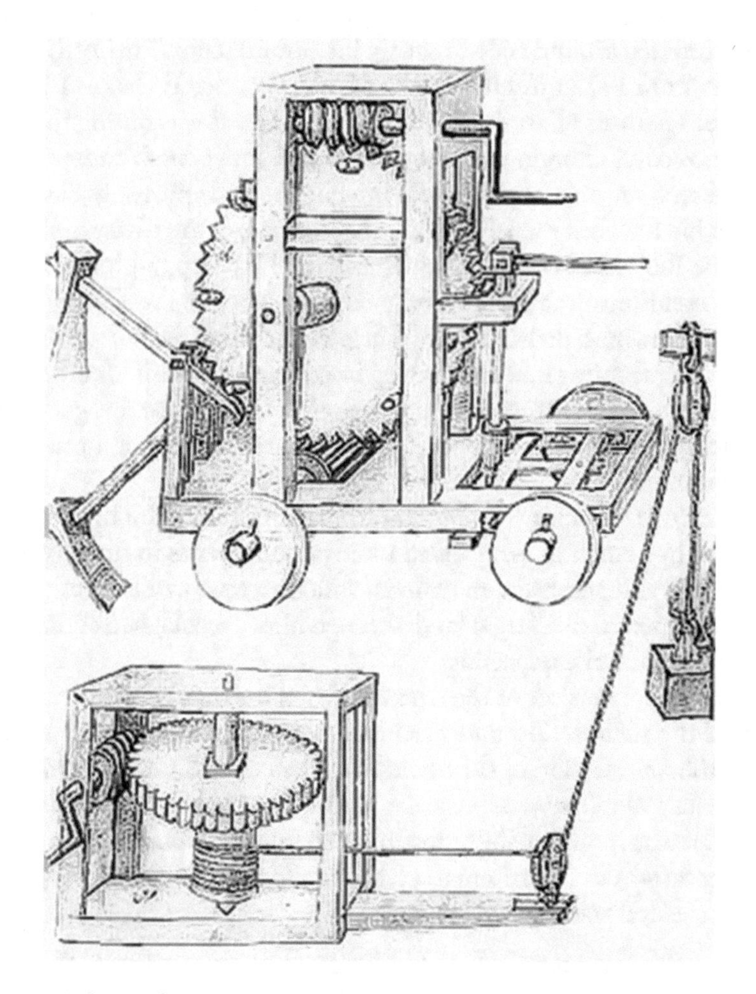

Fig. 1.6 Machines from Francesco di Giorgio, a mechanical hoe and a winch. (Opusculum de Architectura (Ms. 197.b.21), folio 5r. BML)

Giorgio's manuscripts can be said to mark the birth of modern engineering. There is also another remarkable departure from Taccola's manuscripts: the language used is no longer Latin, but instead Italian. This choice was not dictated by Francesco's lack of education—on the contrary, he was so well acquainted with Latin that he edited a translation into Italian of Vitruvius's *De Architectura*. The use of Italian was a calculated choice to employ the language that was used by common people and could therefore help to circulate his knowledge much more easily in the Italian-speaking regions than by using Latin. However, the choice also had some negative consequences: while Latin was read and spoken across Europe by well-educated people, Italian was only used in a limited area.

Francesco di Giorgio was in high demand for his skills as a military architect and traveled around Italy to satisfy his commissions. He finally stopped in Urbino from 1477 until the death of Duke Federico II (1422–1482), one of the main patrons of art during the Renaissance. It was during this period that Francesco di Giorgio wrote the *Trattato di architettura civile e militare*. Two different versions of this work remain: the first, probably dating back to the period between the 1478 and 1481, survives as the *Codex Ashburnham 361* in the Biblioteca Medicea Laurenziana of Florence (54 folios). Though the title mentions architecture, only the first part of the Codex provides texts and drawings dedicated to that subject. The second section deals instead with machines and mechanics. In comparison with his other manuscripts, the topics are arranged systematically. The manuscript goes beyond the structure of a simple notebook to assume the dignity of a pre-encyclopedic handbook.

This *Codex* is particularly important from a historical point of view because it was read by Leonardo, who added some personal notes to the original text. He wrote 12 autographed annotations, which serve as evidence of the influence that Francesco di Giorgio (and, through him, Taccola) had on Leonardo's study of mechanical engineering.

In the second version of the *Trattato*,[18] architecture becomes the central theme of the manuscript, and mechanics play only a secondary role. It is here, in the final section of the book, that Francesco di Giorgio's translation of Vitruvius's *De Architectura* can be found. This translation is particularly notable because it presumably afforded Leonardo an opportunity to read this fundamental work of the Roman architect and to be more closely in touch with the classical tradition.

[18] Francesco di Giorgio Martini. *Trattato di Architettura Civile e Militare* (around 1490), Manuscript II.I.141, Biblioteca Centrale of Florence (dimensions 43.5 × 29 cm).

Another element underscores the importance of this translation. Both Taccola and Francesco reproduced an illustrated version of the *Vitruvian Man* (see Chap. 9). Leonardo's rendering of this subject would become one of his most famous. As such, the direct influence they exerted on him is undeniable. As we will see later in more detail, the Vitruvian ideal of the harmony of dimensions in architecture based on the proportions of the human body would be deeply studied by Leonardo; he devoted himself to the subject with all his intellectual curiosity. As we will also see, Leonardo's considerable analytical capability allowed him to go far beyond the simple attempts and schematics of the two Sienese engineers.

Francesco di Giorgio was never moved by the same broad interest that motivated Leonardo—to understand nature and its basic laws. His work clearly confines itself to the boundaries of the architect-engineer perspective, without the ambition to also be a scientist. Leonardo was different—his curiosity moved him far beyond the frontier of established knowledge. For him, machines and mechanics were just part of a comprehensive exploration of the basic principles of nature.

Leonardo and Francesco di Giorgio had the opportunity to work together when they were both called for a survey of the construction site of the Duomo of Pavia. The two "ingeniarii" arrived in Pavia on 8 June 1490 and lodged in the same place, a small inn called the Saracino (Saracen), as shown by a receipt of payment dated 21 June.[19]

Francesco di Giorgio's legacy is massive, not only in terms of civilian and military architecture, with his castles, palaces, fortresses, and citadels, but also as an artist, painter, and sculptor. It is still possible to admire some of his architectural works, such as the Palazzo della Signoria in Gubbio, and his paintings, which may be found in major museums across the world.

The destiny of his treatises on machines is quite strange. The *Opusculum*, in fact, experienced far-reaching success from the very beginning. It was widely copied fully or in part and was used as a source of inspiration and as a model by generations of engineers. His work, together with that of Taccola, was largely an object of copying. This activity was undertaken without recognizing the original authors, who were forgotten over time, though their machines continued to appear in many books published long after their deaths. This is a strange destiny. The authors were forgotten but not their work, which experienced wide distribution beyond Italy and enjoyed, through citations and copies,

[19] Item die XXI Junii Johanni Augustino de Barberiis hospiti ad signum Saracini Papiae pro expensis sibifactis per Dominos Franciscus senensem [Francesco di Giorgio Martini] et Leonardum Florentinum [Leonardo da Vinci] ingeniarios cum sociiset famulis suis et cum equis, qui ambospecialiter vocati fuerunt pro consultatione suprascriptae fabbricae in summa lib. XX.

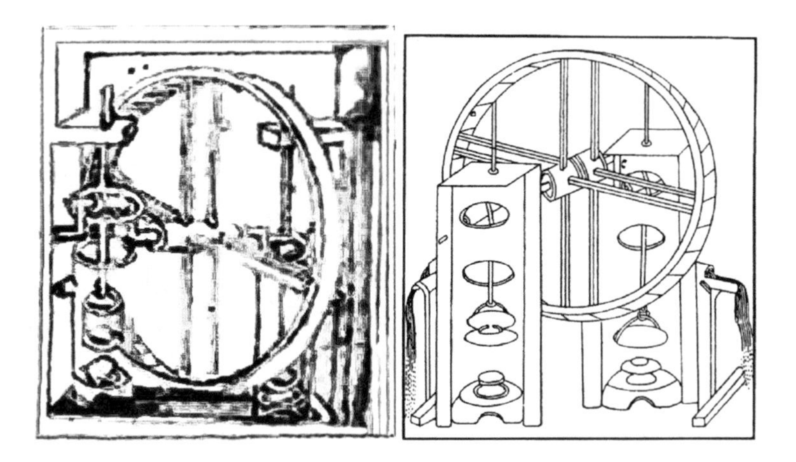

Fig. 1.7 Water mill with a system of double pumps. (Original version from Francesco di Giorgio (left), Codex Ashburnham 361, folio 43r. BMLF (1480 ca). Copy of the mill from Francesco di Giorgio in the Chinese Encyclopedia (right), (1726))

long-lasting popularity. The activity of these plagiarists has been partially reconstructed, showing that the influence of the Siena school of mechanics extended surprising distances across the globe.[20] Taccola and Francesco di Giorgio's technical drawings and machines were known as far as China, where they were taken by the assistant of Matteo Ricci, Sabbatinus de Ursis (1575–1620). The illustrations of machines by the Italians were added without any credit to the Chinese Encyclopedia,[21] compiled in 1725 during the Qing dynasty. When Western scholars started to be interested in Chinese technologies, they were surprised to find these advanced-looking machines. Only much later did it become clear that this was another effect of the sustained copying of the Siena engineers' work (Fig. 1.7).

The circulation of knowledge is something that follows strange paths. It is important to note that, despite the higher (if not necessarily as original as once thought) quality of Leonardo's work on mechanics, it remained completely unnoticed for centuries. The real impact of his work on the history of technology is close to zero. It was only with the systematic study and reorganization of his notebooks and the rediscovery of the manuscripts of Madrid that the importance of his studies in the field was recognized. Taccola and Francesco di Giorgio were also forgotten with time, but their work, through

[20] Reti L, di Giorgio Martini F (1963) Francesco di Giorgio Martini's Treatise on Engineering and Its Plagiarists. Technology and Culture, 4: 287–298.

[21] *Tu Shu Ji Cheng* (The Complete Collection of Illustrations and Writings from the Earliest to Current Times). A vast encyclopedia compiled in China during the Qing dynasty Kangxi and Yongzheng. The compilation began in 1700 and was terminated in 1725 with the first printed edition in 1726.

copies of their manuscripts, continued to be known and exerted a relatively strong influence for a long time. The rediscovery of Leonardo's engineering work in the last century has focused all our attention on his machines and knowledge of mechanics. The Siena engineers, as we have underlined several times, are known only within a restricted circle of specialists. We should clearly recognize that our huge debt to them for the advancement of technology is much more than that owed to Leonardo. If Leonardo's manuscripts had been published, the history of technology might have been completely different, but, instead, it is through the work of Taccola and Francesco di Giorgio that innovations in mechanics managed to spread beyond the closed circle of the Sienese engineers.

The relationship of the Sienese engineers with China does not stop there. In fact, in 2008, an English writer, Gavin Menzies, published a book that immediately became highly controversial—*1434: The Year a Magnificent Chinese Fleet Sailed to Italy and Ignited the Renaissance*. According to Menzies, in 1434, when Taccola was already 52 years old, a large Chinese fleet led by Admiral Zheng arrived in Italy, where they brought Chinese technological knowledge. Menzies argues that this event ignited, through Taccola and Leonardo da Vinci, the technological Renaissance in Italy.[22] To the scholarly eye, the theory appears quite unfounded, as there is no documented trace of the arrival of the Chinese fleet in Tuscany (it seems never to have reached beyond Madagascar). Moreover, Taccola completed much of his work long before the purported arrival of the Chinese expedition. However, the big splash that Menzies's publication made requires us to mention it here, if only to dismiss its unfounded claims.

The Sangallo Family

Francesco di Giorgio was not alone in combining his activity as an architect and engineer with his interest in mechanics and construction machines. Another giant of the Renaissance, Giuliano da Sangallo (1443–1516), shared a very similar career. Giuliano da Sangallo came from a large and famous family of architects and engineers whose work had a strong impact in their time. His brother, Antonio da Sangallo the Elder (c. 1453–1534), and his nephew, Antonio da Sangallo the Younger (1484–1546), were also successful architects and military engineers.

[22] Menzies, G. (2008) 1434. Harper Perennial.

Giuliano da Sangallo was much appreciated as an architect at his time and became the favorite of Lorenzo the Magnificent, ruler of Florence, and a major patron of arts during the Renaissance. Aside from his engineering and architectural work, Giuliano da Sangallo's main legacy is represented by two notebooks, the *Taccuino Sienese* (*Sienese Sketchbook*, 1490–1516) and the *Codex Barberini*.[23] They contain an impressive collection of architectural drawings, projects, and designs of machines. The similarity of his sketches to those of his contemporary, Francesco di Giorgio, is very clear. Several of Brunelleschi's machines are also reproduced and display an incredible similarity to Francesco di Giorgio and Leonardo's sketches of the same subject. For Giuliano da Sangallo as well, the notebook was a working tool, used to make copies of monuments and buildings and to fix ideas and projects. It is thanks to such notebooks that we can still follow the ideas and ways of thinking of these Renaissance men.

Giuliano da Sangallo's nephew, Antonio da Sangallo the Younger, mostly dedicated his time to architecture and hydraulic and military engineering.[24] He was, however, still very much attracted to the well-established tradition of mechanics, likely inspired by his uncle Giovanni's *Taccuino Senese*. Antonio da Sangallo and his workshop left an impressive collection of around 1200 drawings, now in the Gabinetto dei disegni e stampe of the Uffizi Gallery Museum in Florence. In these drawings, he continued to explore and reproduce the repertory of machines inherited from Brunelleschi, Taccola, and Francesco di Giorgio. At the same time, some of his drawings display originality and inventiveness and represent very well his wide range of interests beyond his main work as an architect. Construction technologies, mechanics, and hydraulic and naval engineering are also a part of the eclectic professional knowledge of this family of artists.

Bonaccorso Ghiberti

The figure of Brunelleschi represented a catalyst of the technological innovation process in which, as we have just learned, the Siena school played a fundamental role. The design and construction of machines necessary to solve the construction problems connected with the realization of the Cupola left a deep impression in Brunelleschi's contemporaries. As we will analyze in detail

[23] *Codex Barberini* lat. 4424 (post 1464 – ante 1516). Biblioteca Apostolica Vaticana, Rome.

[24] Frommel CL (1994) The architectural drawing of Antonio da Sangallo the Younger and his circle: fortifications, machines and festival architecture: 001. Adams ND (ed.) MIT Press.

in the next chapter, traces of the models of these machines can be found in manuscripts by Taccola and Francesco di Giorgio and in the notebooks of Leonardo.

Bonaccorso Ghiberti (1451–1516) is another artist-engineer worth mentioning in connection with Brunelleschi. He wrote a miscellaneous collection, *Zibaldone* (scrap-book) (usually dated to c. 1500) a manuscript copy of which is conserved in the National Library of Florence.[25] The *Zibaldone* contains elegant drawings of architecture, designs for cannons, and studies of machines and metallurgy.

Ghiberti[26] had an important ancestor in his grandfather Lorenzo Ghiberti (1378–1455), a major Renaissance artist, who initially worked with Filippo Brunelleschi on the Cupola. Lorenzo Ghiberti had a very antagonistic relationship with Brunelleschi, who was not at all an easy person to work with and was eventually removed from the construction site of the Cupola.

The Ghiberti family, of Florentine origin, was for several generations dedicated with much success to several forms of art, including painting, sculpture, jewelry, and architecture.

There is little biographical information about Bonaccorso Ghiberti. He worked for a time in the workshop of his father Vittorio, who succeeded Lorenzo, during which time he dedicated himself to jewelry, studies of mechanics, and metallurgy.[27] In particular, he gained expert knowledge of the techniques for producing bells and firearms and later dedicated several pages of the *Zibaldone* to describing this art of fabrication. He must have become famous for his ability in metallurgy because he is referred to variously as *scultore et fabro* (sculptor and smith), *maestro ingegniere* (master engineer), and *maestro di gietto* (master of casting). In Fig. 1.8 a system for lifting a cannon barrel from the *Zibaldone* is shown. This drawing is very similar to Leonardo's hoist for a cannon shown in Fig. 1.4, so much so that it is hard to believe that Leonardo was not inspired by Ghiberti's work.

When he was only 4 years old, Ghiberti inherited all his grandfather's possessions, which included the books and drawing of his *scriptorium* (atelier). This material is likely the original source for the illustrations of the construction hoists and cranes which are shown in the *Zibaldone*. This is the most important part of the manuscript because it facilitated the transmission of the

[25] Codice Banco Rari 228. Biblioteca Nazionale Centrale di Firenze. The manuscript is composed of 239 folios of 14.5 × 19.5 cm dimension.

[26] La Bella C (2000). Entry on Ghiberti. In: Dizionario Biografico degli Italiani, Treccani, Volume 53.

[27] Scaglia G (1976). A Miscellany of Bronze Works and Texts in the "Zibaldone" of Buonaccorso Ghiberti. Proceedings of the American Philosophical Society, 120: 485–513.

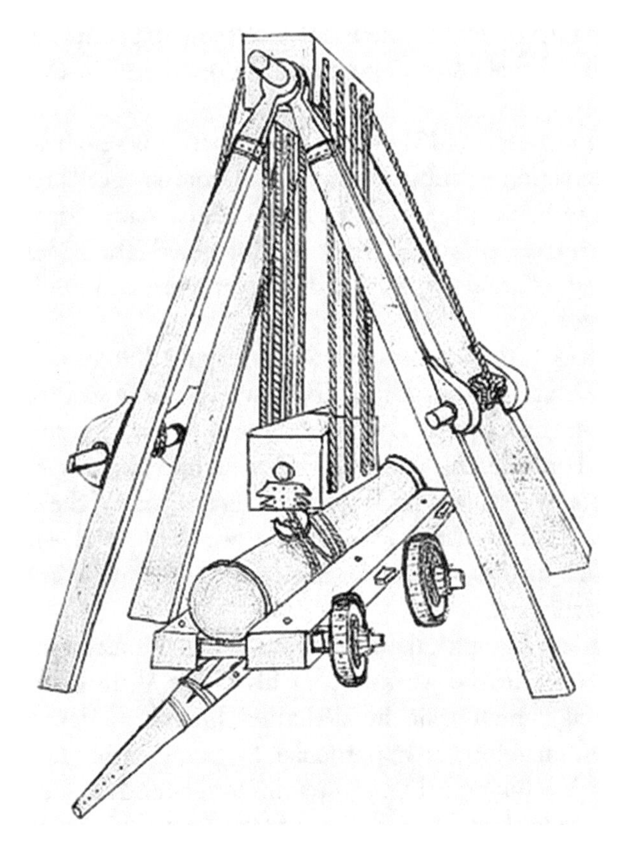

Fig. 1.8 System for lifting a cannon barrel. (Bonaccorso Ghiberti. Zibaldone, Codice Banco Rari 228, BNCF)

patrimony of knowledge represented by the innovative machines invented by Brunelleschi.[28]

Unlike the copies of the Florentine architect's devices, the projects for cannons in the *Zibaldone* are original, the direct result of his experience as a smelter. Other illustrations of a military nature are directly copied from Valturio's *De Re Militari* (see Chap. 6). In spite of the great interest shown by Italian engineers, their knowledge of artillery and firearms was not particularly advanced, and their projects remained confined within the boundaries of then-current Renaissance knowledge. The Italian school of firearms was quickly surpassed by the French school, which introduced significant advances. The studies in this field by Taccola, Francesco di Giorgio, Bonaccorso Ghiberti, and Leonardo were not particularly effective or innovative.

[28] Prager FD, Scaglia G (1970) Brunelleschi. Studies of his technology and inventions. Cambridge, MA.

The *Zibaldone* was written within the same context of eclectic artists and engineers that made up the cultural milieu of the Renaissance. As we see again and again, the interests of these open-minded men spanned a plethora of fields, from architecture and military warfare to metallurgy and mechanics. The *Zibaldone* represents another example of their creative and profound way of thinking, and we retain the same fascination with the *Zibaldone* as we do for Taccola, Francesco di Giorgio, and Leonardo's more famous manuscripts.

Vannoccio Biringuccio

There is one last figure to add to the group of extraordinary Renaissance architects and engineers who served as the primary references for Leonardo's fantastic world of machines. Vannoccio Biringuccio[29] (c. 1480–c. 1539), yet another native of Siena, was a chemist and a master of metallurgy. We have seen that the art of metallurgy was fundamental for the fabrication of firearms. As such, skilled craftspeople in this field were highly sought after. All the artist-engineers introduced above were involved in the production of cannons and bombards or designed new solutions for employing firearms. Long descriptions of the fabrication of firearms can also be found in Leonardo's notebooks, along with the drawings of his multiple guns and applications of bombards for besieging cities. Biringuccio, like Francesco di Giorgio and Buonaccorso Ghiberti, was directly involved in the manufacture of artillery, while Leonardo and Taccola never acquired direct experience in this field.[30] This common interest in metallurgy, weapon design, and manufacturing links these different generations of engineers.

Between the years 1534 and 1535, Vannoccio Biringuccio wrote a manuscript in Italian, which was destined to have a deep impact on the history of science. The *Pirotechnia*[31] was dedicated to metallurgy, as stated in the frontispiece of the first edition published posthumously in Venice in 1540. "About Pirotechnia: ten books that broadly treat the subject of mines, but are also

[29]Tucci U (1968) Entry on: Biringucci (Bernigucio) Vannoccio. In Dizionario Biografico degli Italiani. Enciclopedia Treccani, Volume 10.

[30]Bernardoni A (2014) Artisanal Processes and Epistemological Debate in the Works of Leonardo Da Vinci and Vannoccio Biringuccio, in Laboratories of Art. Art Alchemy and Art Technology from Antiquity to the eighteenth Century. Edited by S. Dupre, Volume 37 of the series Archimedes. 53–78 Springer.

[31]The Pirotechnia of Vannoccio Biringuccio. The classic Sixteenth Century Treatise on Metals and Metallurgy. Translated and edited by Cyril Stanley Smith and Martha-Teach Gnudi. 1990 Dover Publications.

about the research and practice of the art of fusing and casting metals (and related arts)."[32]

Biringuccio had a rather eventful life: he was involved in the struggle for power in Siena and was banished from the city as a result. He took the opportunity to travel around Italy and visited numerous mines, accumulating important knowledge to refine his technical capabilities. Finally, thanks to the eventual stabilization of the political situation in Siena, he returned home to become senator of the city. His fame reached as far as Rome, and he was appointed director of foundry and munitions by the Pope. The precise date and place of his death is not known, but was likely no later than 1539.

Biringuccio is considered one of the founders of modern chemistry and metallurgy because of his contributions to the field. He used a practical and experimental approach that was foreign to medieval alchemy; he reported only "what I have seen and also what… I have done by myself or I ordered to do". This new method can be found not only in the work of Biringuccio but also in that of Leonardo. Leonardo also widely described metallurgical processes and technologies, though admittedly in a much less organized way than Biringuccio.

As an aside, it is worth noting that Leonardo had no direct experience with metallurgy, but he demonstrated a great interest in firearms, particularly mortars and bombards. He dedicated several extensive studies to ballistic methods and the manufacture of arms. Leonardo also meticulously described procedures such as the metal fusion process for preparing a bombard, from the choice of wood for the fire to the selection of the alloy components: "The metal can be made in the usual way--up to 6 or 8 percent: i.e. using six parts of tin with one hundred of copper. The less you use, the safer the bombard will be." (*Codex Trivulziano*, folio 15v).

It is difficult to assess how much the work of Biringuccio had a direct influence on Leonardo. It is very likely that they met in Milan during one of Biringuccio's trips and discussed Leonardo's big project of the bronze horse commissioned by Duke Ludovico il Moro. It was an ambitious and long-term project that kept Leonardo very busy for a long time, from the design to solving the technical problems connected with the production of such a large bronze statue. His final plan called for a statue 7 meters high and requiring about 100 tons of bronze to cast. Sadly, Leonardo never saw the final result of his efforts because, after all the preparations were complete, the bronze for casting was instead used to produce cannons necessary to defend Milan.

[32] Original text: "De la Pirotechnia, libri dieci dove ampiamente si tratta non solo di ogni sorte e diversità di miniere, ma ancora quanto si ricerca intorno alla pratica di quelle cose di quel che si appartiene a l'arte de la fusione over gitto de metalli come d'ogni altra cosa simile a questa."

Leonardo and Biringuccio share in any case a common approach which is based on direct experience and empirical evidence, which is the beginning of the foundation of modern Galilean science.

As we have begun to see here, Leonardo's interest in mechanics and engineering was clearly triggered by the many engineers and inventors who contributed to create the very fertile and creative environment for the Renaissance of technology and scientific knowledge. We shall delve deeper into these subjects and the work of these individuals and others in the pages to come.

2

Cupolas and Machines

The Cupola

The construction of the Cupola of the cathedral in Florence (Cattedrale di Santa Maria del Fiore) by the Florentine architect Filippo Brunelleschi[1] (1377–1446) represents one of those influencing events that are capable of affecting the course of history—in this case the history of art and technology.[2] The Cupola is not only one of the greatest masterworks of Renaissance architecture but also marks the beginning of a new era beyond the known frontiers of traditional knowledge (Fig. 2.1). The technical challenges involved in building the Cupola were of extraordinary difficulty and needed creative and innovative solutions. The difficulties faced were not only connected to the design of the project but also to its practical realization; no such construction had ever been built before.[3] The sheer dimensions of the Cupola, which remains the biggest masonry Cupola in the world, raised completely new structural and construction problems.[4]

Construction on the cathedral started in 1296, and it took more than a century before the main body was completed, including the drum for the Cupola. The architect Arnolfo di Cambio (c. 1240–c. 1310) planned the project and was in charge of the construction. For several years no attempts were made to face the challenge of completing the church with a cupola

[1] Battisti E (2012) Filippo Brunelleschi. Phaidon Press.

[2] King R (2000) Brunelleschi's Cupola. Bloomsbury, USA.

[3] Fanelli G, Fanelli M (2006) Brunelleschi's Cupola: Past and Present of an Architectural Masterpiece. Mandragora.

[4] Prager FD, Scaglia G (2004) Brunelleschi. Studies of his technology and inventions. Dover publications, New York, USA.

P. Innocenzi, *The Innovators Behind Leonardo*, https://doi.org/10.1007/978-3-319-90449-8_2

Fig. 2.1 Details of Brunelleschi's Cupola in Florence

because it was clear that placing such a heavy weight on the main body might compromise the stability of the whole structure. Finally, on 19 August 1418, an open competition was launched, and 200 gold florins (a considerable amount of money for the time) were offered as a prize for the best proposal.

In 1420, after a careful evaluation of all of the submissions, the committee chosen by the *Opera del Duomo* for the selection, the "citizens for the construction of the Cupola" (*cives pro constructione cupola*), entrusted the work to Filippo Brunelleschi and Lorenzo Ghiberti (1378–1455, grandfather of Bonaccorso, the author of *Zibaldone).*

Brunelleschi, who had an unusual personality and liked mocking and ridiculing people, did not think much of Ghiberti as an architect and believed he was incompetent. In 1425, Brunelleschi finally succeeded in kicking out his Florentine colleague and became the sole person in charge of construction. In late 1429, to convince the committee and the citizens of Florence, Brunelleschi performed a practical demonstration in the Piazza Duomo, building a small model with a very innovative method, which did not utilize any centering structure. The work on the Cupola was completed in 17 years, from 1420 to 1436, with the exception of the lantern at the top of the Cupola, which was completed in 1461. The completion of the Cupola ended the skepticism that had surrounded Brunelleschi's venture from the very beginning and proved that he had indeed been the right person for the challenge.

The main construction problem was presented by the centering, the temporary structure of support commonly used as a template for the construction of arches and vaults. Centerings are generally made of wood and make it possible to place the bricks or stones of an arch which has no mechanical strength until the keystone is added. The centering is normally reinforced by a supporting structure and is maintained until all elements have been positioned and the mortar has hardened. It is removed immediately afterward.

The first problem Brunelleschi had to face was the dimension of the drum to be used as support for the centering; it was around 43 m in width and was positioned 54 m above the floor.

The second problem posed an even bigger challenge—the drum had an irregular structure, octagonal in shape. To the general surprise and incredulity of the Florentines, Brunelleschi decided not to use any centering structure at all and to instead proceed with the construction directly from the drum. It appeared an almost impossible task, but the Florentine architect accomplished his work without any fault. To this day, the question of how he was able to achieve this is still debated.

Brunelleschi's Cupola is composed of eight sides divided by decorative angle ribs. A double brick hull forms the structure: an inner shell of even thickness and an outer shell that gradually narrows from the bottom to the top.

Brunelleschi combined a herringbone brick pattern (*spina di pesce*) and careful calculation of the inclination of the structure to build the Cupola without a centering. The solution he adopted was creative and innovative; it allowed him to solve the problems of the weight on the structure and the irregularity of the drum. Another advantage of Brunelleschi's idea was also clear: It saved both time and money. The Florentines, mainly merchants and bankers, were not indifferent to this matter and were pleased with the architect's work.

After making the decision to work without any centering structure, Brunelleschi still had to face another significant problem: how to carry the materials up to the construction site and position them with the required precision. This necessitated the design of special devices for these operations featuring an unprecedented level of complexity. These are exactly the machines whose memory has been transmitted to us through the illustrations of Taccola, Francesco di Giorgio, Bonaccorso Ghiberti, Giuliano da Sangallo, and Leonardo da Vinci to fascinate generations of engineers and architects up to our own times.

When the construction of the Cupola was completed, the wonder and astonishment were great; the impossible mission had been accomplished, and Brunelleschi's masterpiece was in place to witness the arrival of a new era.

The words of Leon Battista Alberti in his *De Pictura* ably capture the admiration of Brunelleschi's contemporaries: "Who can be so envious that will not praise Filippo looking to such a big structure, over the skies, so wide as to cover all the people of Tuscany. [It was] done without any help of structures or wood templates, such kind of artifice, that if I judge well, is incredible to realize in our time and also unknown to the ancients?"[5]

The Machines of Brunelleschi

The many illustrations of Brunelleschi's machines and devices for the construction of the Cupola give us a fairly robust picture of the way they operated. Beginning with Taccola, the first person to sketch Brunelleschi's machines, these devices held an irresistible attraction for Renaissance architects. The technical quality of Taccola's illustrations is not perfect, and sometimes they even contain errors in reproducing the working system. Nevertheless, because they are the earliest sources, they remain of great interest.

One of the most famous of these sketches is the hoist driven by a horse (Fig. 2.2), designed for lifting weights up to the worksite of the Cupola. The original machine was likely moved by a pair of oxen; Taccola's reproduction is a schematic simplification.

The hoist does not permit any variation of speed but makes it possible to switch from raising to lowering the weight without the need to unhitch and reposition the horse: The helicoidal screw at the base of the rotating central shaft can be lifted up and down, allowing the operator to select which one of the two horizontal wheels will be locked to the vertical wheel with the drum moving the rope. By a simple change of the interlocked wheel, it is possible to raise or lower the material without the need to unhitch the animals, saving time and making the device simpler to operate. This is considered to be the first reverse gear in history and is the result of Brunelleschi's creativity.

Once the problem of how to raise loads up to the worksite had been solved with this simple and time-saving device, there were other questions to face. Changing the direction of the hoist was a nice invention, but what about also changing the lifting speed? Controlling the lifting speed has the advantage of making it possible to adjust the time to raise the loads as a function of their weight. This is what Brunelleschi achieved with another of his hoists, which was equipped not only with a reverse gear but also with a three-speed gear.

[5] Original text: "Chi mai sì duro o sì invido non lodasse Pippo architetto vedendo qui struttura sì grande, erta sopra e' cieli, ampla da coprire con sua ombra tutti e' popoli toscani, fatta sanza alcuno aiuto di travamenti o di copia di legname, quale artificio certo, se io ben iudico, come a questi tempi era incredibile potersi, così forse appresso gli antichi fu non saputo né conosciuto?"

Fig. 2.2 Hoist driven by a horse. Mariano di Jacopo (Taccola). De Ingeneis III. Manuscript Palatino 766, folio 10r. BNCF

This system is captured better in the more sophisticated drawings of Bonaccorso Ghiberti, Giuliano da Sangallo, and Leonardo da Vinci (Fig. 2.3) than in Taccola's simple sketch.

Brunelleschi's three-speed hoist was a giant, heavy machine powered by oxen to raise loads of different weights. A sense of its dimensions can be obtained by looking at the system for attaching the animals under the vertical shaft. In Giuliano da Sangallo's sketch, up to four animals can be used for driving the hoist.

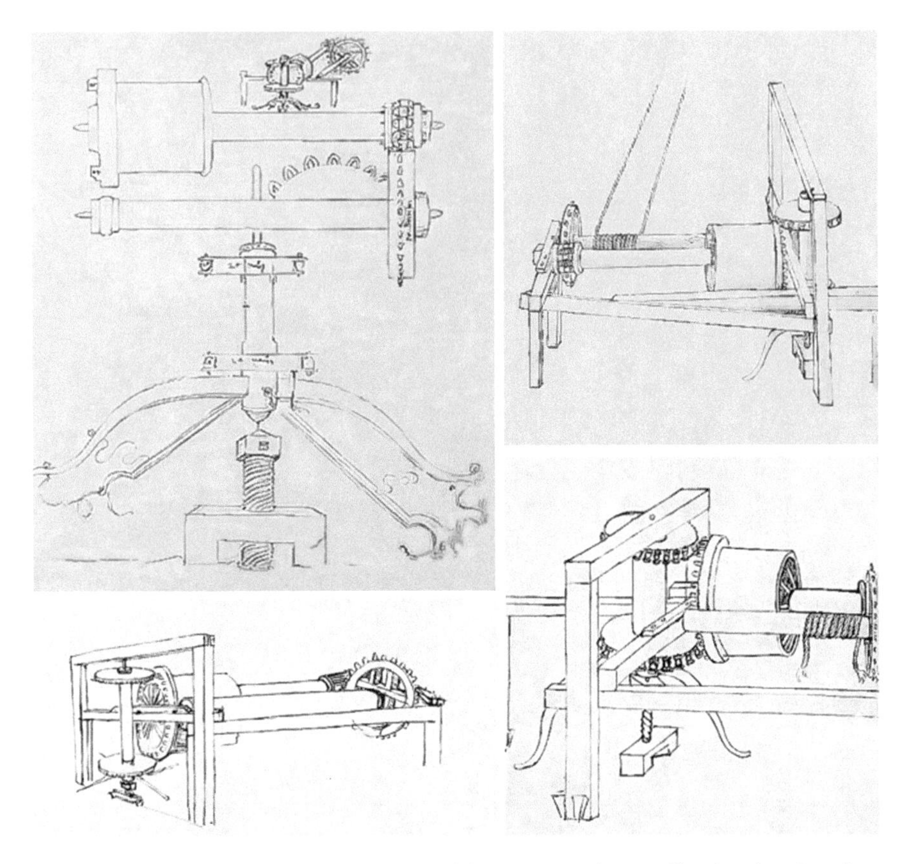

Fig. 2.3 Reproduction of the three-speed hoist gear of Brunelleschi. Sketches from Giuliano da Sangallo (*left top*), Taccuino Senese, Ms. S.IV.8, folios. 47v, 48f, Biblioteca degli Intronati di Siena; Leonardo da Vinci (*bottom left*), Codex Atlanticus, folio 1083v, BAM; Bonaccorso Ghiberti (*right top* and *bottom*), Zibaldone, B. R. 228, folios. 102r, 103v, BNCF

The available illustrations provide a fairly comprehensive description of the device. It was designed as a rectangular wooden frame with four legs firmly fixed to the ground to assure the necessary stability to the system. The entire apparatus appears as a giant clocklike system with controlled and coordinated movement of the parts.

The vertical driving shaft moved by the animals has two horizontal wheels to transmit the movement to the drum. A worm screw at the base of the shaft, as we have seen in Taccola's simpler sketch, makes it possible to change the horizontal wheel to match the drum teeth. This is the gear system that changes the lifting direction by reversing the rotation of the horizontal cylindrical shafts. The two joined, horizontal wheels operate on an alternating basis while the horizontal shafts rotate at three different speeds, which can be selected as a function of the load. The smallest cylinder was used for the heavier loads and the largest one for the lighter loads. This giant hoist was extremely efficient, allowing several dozen loads per day to be raised but also easy to operate and requiring only minor maintenance for the entirety of its operation.

Figure 2.3 also offers a good opportunity to compare the artists' different methods of technical reproduction. Bonaccorso Ghiberti's design is extremely detailed and features 3D perspective, while Leonard's sketch appears to be more of a draft done without much attention. Sangallo, on the other hand, seems to have instead preferred to concentrate on the details of the machine.

After having raised the material to the worksite, there remained the problem of positioning it with the proper degree of precision and efficiency. For this purpose Brunelleschi created another custom-made device called the *castello* (castle), which is similar to contemporary cranes. Bonaccorso Ghiberti and Leonardo both reproduced the machine in their sketches (Fig. 2.4). The

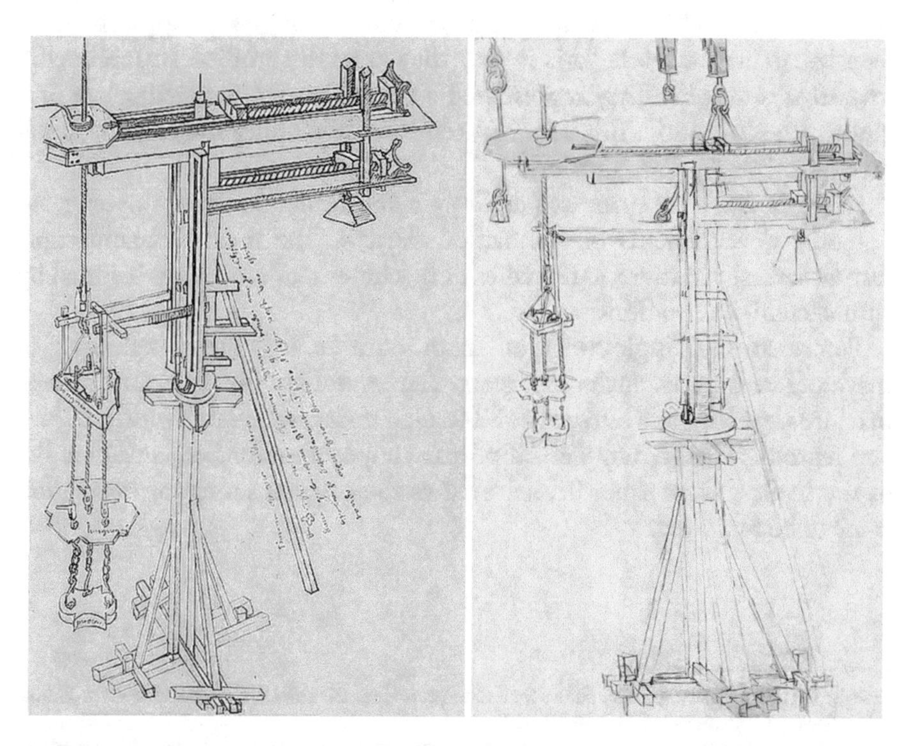

Fig. 2.4 Rotating crane from Brunelleschi. Bonaccorso Ghiberti., Zibaldone, folio 106r. Biblioteca Rari 228, BNCF (*left*). Leonardo's version of the rotating crane. Leonardo da Vinci. Codex Atlanticus, folio 965r. BAM (*right*)

crane could rotate 360 degrees and allowed the load to be moved horizontally using a system of counterweights to ensure its stability.

The machines for supporting the construction of the Cupola did not remain at the worksite for very long. They were probably changed quickly as soon as a new construction phase began. Different systems were therefore designed and updated in response to need.

The two machines that we have just described, the hoist and the crane, are some of the most famous machines that Brunelleschi created for the construction of the Cupola. They were reproduced first by Taccola, and then by Francesco di Giorgio[6] who, starting from these models, developed several improved versions. These machines also held great fascination for Francesco Bonaccorso, Giuliano da Sangallo, and Leonardo da Vinci.[7] It is not by chance that all of these figures harbored such great interest in Brunelleschi's machines. If we recall their biographies and skills, we realize that these men all shared a similar background. They were multifaceted artists with broad interests in several fields, and the innovation introduced by Brunelleschi represented a unique opportunity for them to improve their skills as architects and to learn new technologies to apply in their own construction work. This is why they carefully studied Brunelleschi's inventive solutions: They represented a patrimony of knowledge too precious to be lost and a fundamental tool for developing their professional activities.

In particular, both Francesco di Giorgio and Leonardo deeply investigated potential developments of mechanical devices. The former's manuscripts feature several variations and studies of machines that are clearly inspired by Brunelleschi's innovations.

Francesco di Giorgio went on to produce an impressive collection of machines and parts, including gears, cranes, and hoists (Fig. 2.5). These machines, which were also part of Taccola's legacy, were later widely copied and reproduced, and they were of primary importance for Leonardo's studies of mechanics. Over time, Brunelleschi as the original source of inspiration came to be forgotten.

[6] Scaglia G (1996) Drawings of Machines for Architecture from the Early Quattrocento in Italy. Journal of the Society of Architectural Historians. 25:90–114.

[7] Reti L (1964) Tracce dei progetti perduti di Filippo Brunelleschi nel Codice Atlantico di Leonardo da Vinci. IV Lettura Vinciana, Vinci, Bibl. Leonardiana.

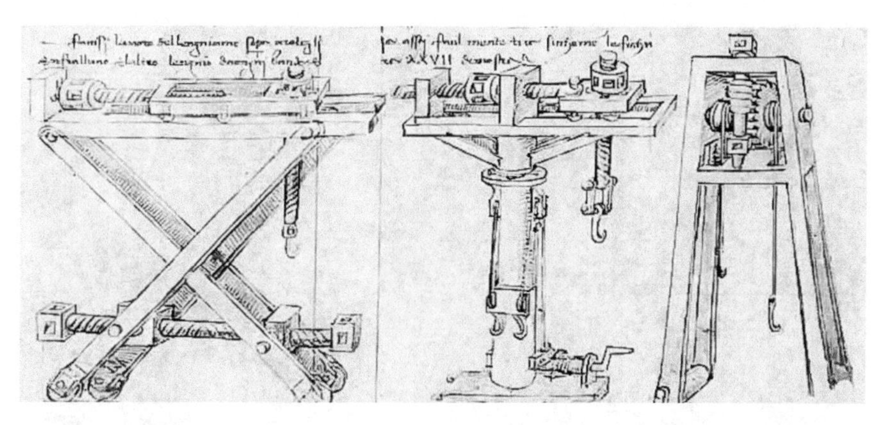

Fig. 2.5 Studies of construction machines. Francesco di Giorgio. Codex Ashburnham 361, folio 46r (detail). BMLF

The Lantern

On 30 August 1436 Brunelleschi's Cupola was finally completed. However, one final task remained before Brunelleschi could put the word "end" to the big endeavor: The domed structure was meant to be finished with a lantern-shaped *oculum* (or skylight), like the Pantheon in Rome. To choose the designer, another public competition was held. It may seem strange that the work was not given directly to Brunelleschi. In any case, not without some discussion, Brunelleschi's project was again the one selected for the lantern. However, the actual construction did not begin until several years later in 1446, just a few months before his death. The construction was continued by his student Michelozzo di Bartolomeo and then by Antonio Manetti, finally ending in April 1461. The lantern's completion was extremely important because it not only played an aesthetic function, but, with its 750-ton weight, also guaranteed the static stability of the full structure.

Several illustrations of the machines and techniques used for the construction of the lantern still remain to this day. The most detailed sketches were made by Bonaccorso Ghiberti, Giuliano da Sangallo, and Leonardo. If we compare their depictions, their similarity is impressive. If we exclude the possibility that they copied one another's work, it is very likely that these machines, like the others used to make the Cupola, remained near the construction site several years later.

The machines were likely still in good condition to allow their direct reproduction through sketches; this could explain the similarity of the drawings from different sources. In particular, if we compare the sketches of

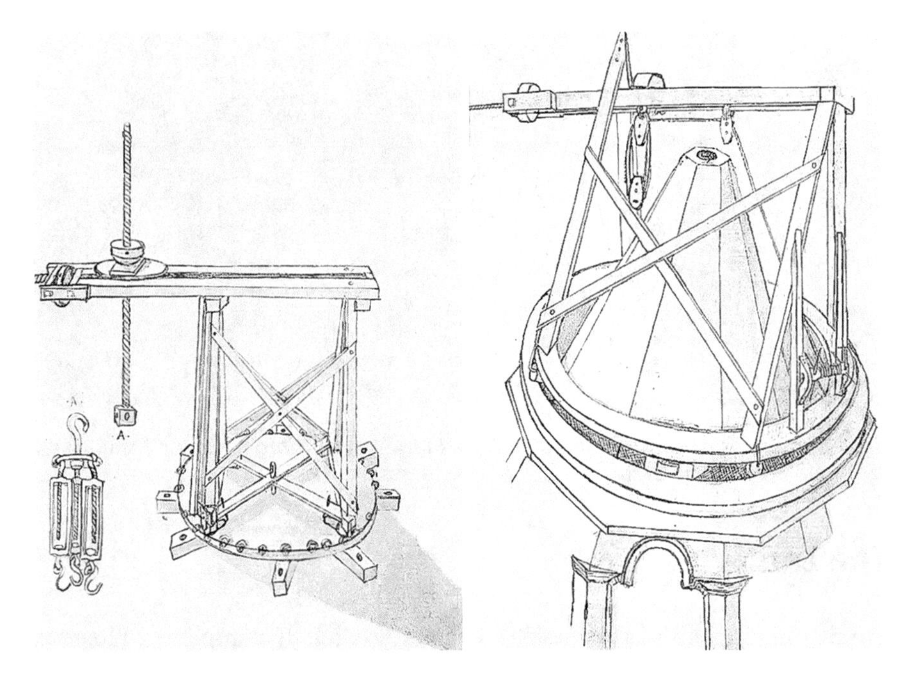

Fig. 2.6 Rotating crane from Brunelleschi. Bonaccorso Ghiberti. Zibaldone, folio 104 (*left*). Rotating crane to use on the top of the lantern construction site. Zibaldone Folio 105 (*right*). Biblioteca Rari 228, BNCF

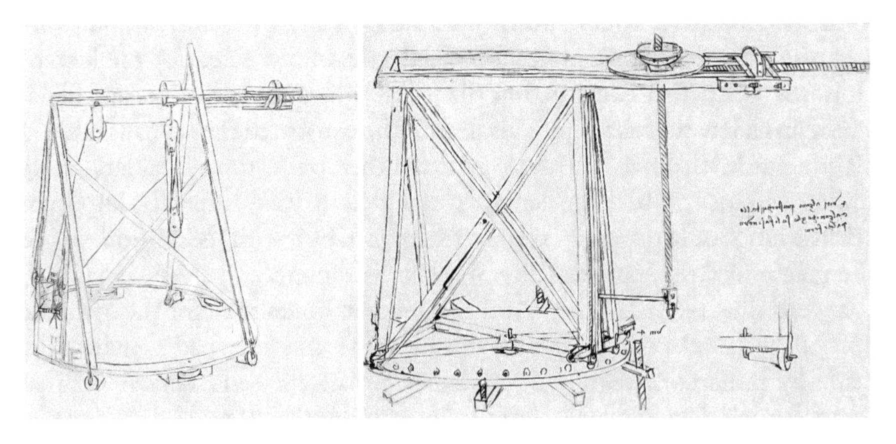

Fig. 2.7 Rotating cranes from Brunelleschi. Leonardo da Vinci. Codex Atlanticus, folio 808v (*left*) e 808r (*right*). BAM

Bonaccorso Ghiberti (Fig. 2.6) and Leonardo (Fig. 2.7), they are so similar that they overlap almost perfectly. The same could be said of Giuliano da Sangallo's drawing (Fig. 2.8).

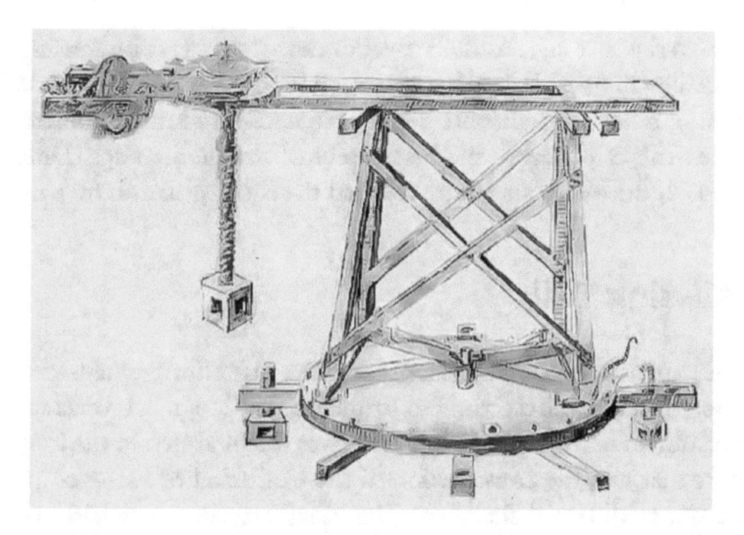

Fig. 2.8 Rotating crane. Giuliano da Sangallo. Taccuini senesi, folio 12 f. Biblioteca degli Intronati, Siena, Italy

The machines reproduced in Fig. 2.7, which are in folio 808v and 808r of *Codex Atlanticus*, have been widely advertised as an expression of the inventiveness of Leonardo; they represent a couple of rotating cranes, which were built and installed at the top of the Cupola to build the lantern. This allows an exact positioning of the materials, as clearly shown in Ghiberti's sketch (Fig. 2.8).

The first type of crane is based on a circular platform which permits 360 degree rotation while a system of screws makes it possible to move the load horizontally with great precision (left image Fig. 2.7). The second crane was designed to place the material using a block and tackle system. The small wheel at the base allows for the rotation of the crane around the lantern to place the material in the right place, as shown in Ghiberti's drawing (right image in Fig. 2.6).

The rotating crane shown in Fig. 2.7 was commonly credited to Leonardo, but as we can see, his role was simply to copy the original model. Leonardo reproduced several of Brunelleschi's construction machines and took inspiration from them for the design of several others, such as hoists and cranes. Unfortunately, he did not cite the original source of his ideas, and for a long time it was difficult to identify the true origin of these machines, particularly because Brunelleschi, who, as we know from his reported discussions with Taccola, was very protective of his work, which he tried to keep secret as much as he could, did not leave behind any written records. In this age of inventors, the problem of patents and intellectual property started to become a rather serious worry, a sure sign of the arrival of new times. It is interesting to note that Brunelleschi in 1421 had been granted for 3 years the exclusive use of his invention of a boat to transport the marble necessary for the construction of the dome. This is probably the first patent of the modern age. Later, on 19 March 1474, the Venetian Senate released the first legislation on patents.

The Missing Ball

After the lantern was put in place, only one last effort remained to finally complete the Cupola. According to Brunelleschi's design, a bronze sphere had to be positioned on top of the Cupola. It was the final step in the long, collective effort that was the construction of the Cathedral of Florence and represented a not at all trivial challenge: The heavy bronze sphere had to be raised to the top of the Cupola, carefully welded, and affixed to the structure. This task fell to one of the most famous artists of Florence, Andrea Verrocchio (1435–1488), who also happened to be the *maestro* ("master") of Leonardo da Vinci. In 1471, 25 years after the death of Brunelleschi, the bronze sphere was put in place on the Cupola, finally marking the very end of its construction. It is not clear if Leonardo played a role in the realization of the sphere-topped dome. At the time, he was already 19 years old and an experienced pupil of Verrocchio. Regardless, he carefully noted all of the phases of the construction and left a clear reference to this experience in one of his notebooks: "Remember the weldings used for the ball of Santa Maria del Fiore."[8]

The *Maestro*

This very common and very important term, translated inadequately into English as "master" and still used today in modern Italian, designates an individual, most frequently an artist or musician, who has achieved the highest level in his work. It is a title of extreme respect.

The story of the sphere does not stop there, however. On 27 January 1600 at five in the morning, the sphere was hit by lightning during a heavy storm. It was detached from its position and fell, bouncing off and damaging the Cathedral in several places. A new bronze sphere was quickly put in place, and in 1602 it was possible to admire the Cupola in all its magnificence once

[8] Original text: "Ricordati delle saldature con che si saldò la palla di Santa Maria del Fiore." Manuscript G, folio 84v, BNF.

more. Walking along the east side of the *Piazza Duomo* in Florence, it is still possible to observe the exact point where the ball fell, which is indicated by a circular epigraph in white marble.

A Secret Machine

Looking carefully at the illustrations of machines in the *Zibaldone* of Bonaccorso Ghiberti, a couple of drawings attract special attention. They contain a peculiarity that creates an air of mystery—a cipher, or code, used to describe the device. Two non-consecutive sketches, 95r and 98r, depict without any doubt the same "secret" machine, which appears to be a hoist with toothed wheels. The drawing referred to as 98r is a top view of the device reproduced in the previous folio, 95r (Fig. 2.9).

The cipher is fairly simple and is based on the substitution of every letter with the preceding one, for example, c instead of d. Even after decoding the text, however, the true operation and use of the machine remain a mystery. The purpose of the device appears to be to move light weights such as a horizontal drum with a pulley rope system. However, the description recorded in

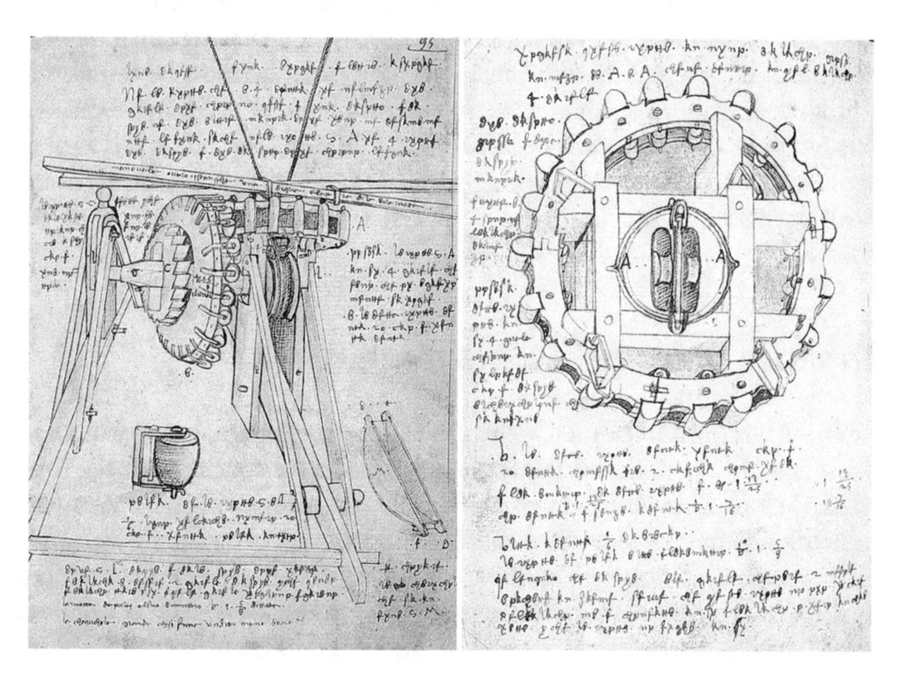

Fig. 2.9 The secret machine in the Zibaldone of Bonaccorso Ghiberti, side view (folio 95r) and top view (folio 98r). Banco rari 228, BNCF

the text is confusing and without any technical value; many details remain obscure, such as the true dimensions of the machine. Several parts of the device even seem to have been deliberately omitted or hidden.

There is no trace of the machine in the documents of the *Opera del Duomo*, and perhaps it was not one of the machines designed by Brunelleschi.

The machine is relatively simple, consisting of a man- or animal-powered horizontal wheel, which is moved by two rods. The motion is transmitted to a toothed vertical wheel, which moves a drum through a shaft where a lifting rope is wound. The top view of the machine shows that a fixed pulley is positioned in the middle of the horizontal wheel to pass the rope for lifting loads.

The mysteries around these sketches do not stop here; in fact an almost exact copy of Ghiberti's device appears in the bottom part of folio 105v of Leonardo's *Codex Atlanticus* (Fig. 2.10). Leonardo's drawing is a much less detailed sketched draft, perhaps a simple memo, as he was used to make in his notebooks. What is puzzling is how Leonardo knew about this machine, which was not perhaps as secret as Bonaccorso Ghiberti had liked to think. At any rate, unlike in many other cases, Leonardo did not add any note or comment to this particular sketch.

In the same folio 105v of the *Codex Atlanticus*, just above the secret machine, Leonardo sketched a different device, this one from Brunelleschi (Fig. 2.10, *top*). The system is a crane for placing construction material. In this case, Leonardo did not sketch with precision; the system can also be found well reproduced, with some differences, in the *Zibaldone*, where a more detailed version of the crane is shown by Bonaccorso Ghiberti (Fig. 2.11). It is a rotating crane to be used at the working site at the top of the Cupola for raising small weights and moving them sideways, with a function similar to the *castello*.

The Fate of Brunelleschi's Machines

After the completion of the Cupola, including the lantern, the machines of Brunelleschi were removed and likely left close to the construction site for a relatively long time. This, as we have already commented, could explain why so many different sources were able to report the same machines with such high precision. Some of the machines, however, were likely never reproduced in sketches, and we have lost all trace of them forever, because, as we mentioned before, Brunelleschi never left any direct written evidence of his projects.

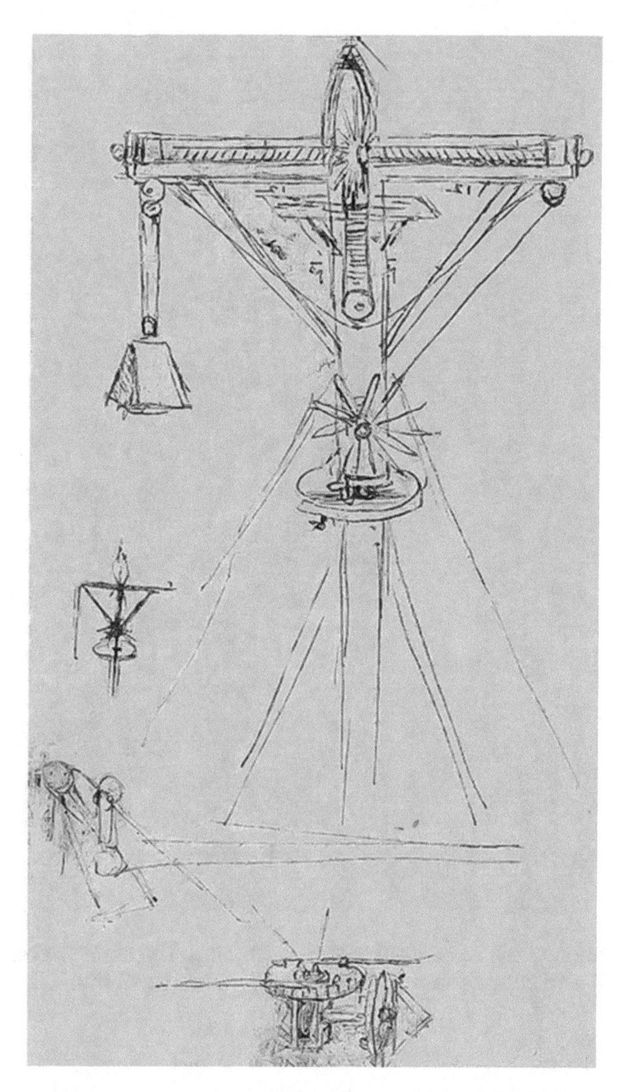

Fig. 2.10 Leonardo da Vinci, Codex Atlanticus, BAM, folio 105v. In the lower part, it is possible to see the "secret machine" reproduced in the Zibaldone of Bonaccorso

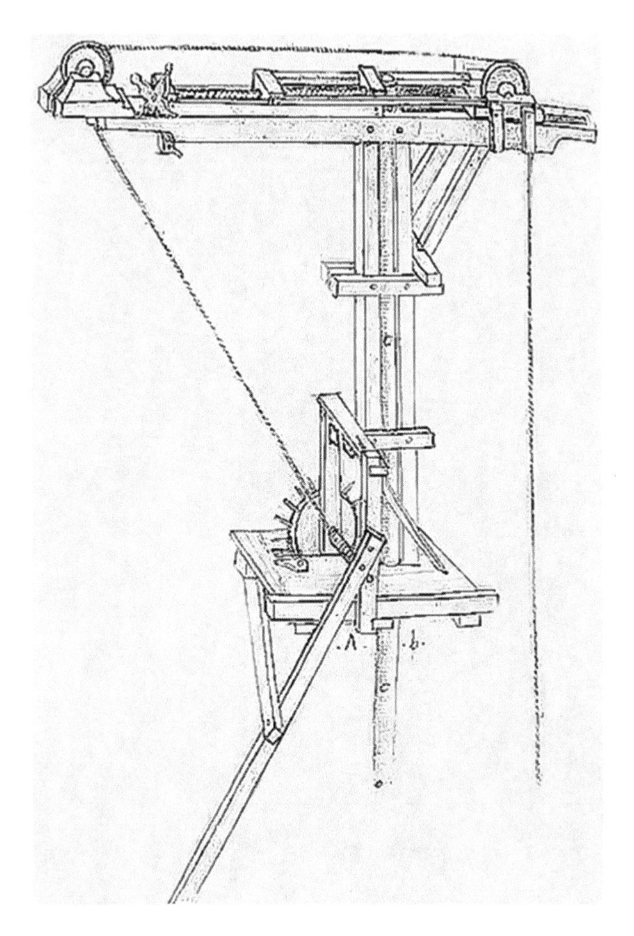

Fig. 2.11 Rotating crane to be used in the scaffolding. The crane makes it possible to lift small loads and move them horizontally. Bonaccorso Ghiberti, Zibaldone, folio 107v. BNCF

The construction machines for the Cupola were truly revolutionary for the time and introduced several innovative ideas, such as the lifting gears and speed gears that made the operation easier and faster. The sad truth, however, is that, following their innovative use in the construction of Brunelleschi's Cupola, they were quickly forgotten, at least from a practical point of view, and had no further impact in advancing construction technology. Brunelleschi's use of such machines to aid construction remained a unique case, and, with the exception of the architects and engineers, discussed above who studied and transmitted the memory of his innovative machines, we do not have evidence that any others tried to take advantage of his work in a practical sense.

It would be necessary to wait for the beginning of the industrial revolution in the late eighteenth century to see the development of such advanced construction devices again. This was a lost opportunity. But it was not the only one. As we will see, the greatest part of Leonardo's work also went essentially unnoticed for centuries. Were these two innovators too advanced their time? This is the belief of Prager and Scaglia, who, in their monograph[4] on Brunelleschi and his technology write that, "It is conceivable that Filippo was too far ahead of his time and therefore unable to influence his time with regard to machines…."

A Man of Great Genius

On 15 April 1446, Filippo Brunelleschi suddenly died in Florence at the age of 69, a fairly advanced age for the time. Only 1 month earlier, the Archbishop of Florence had consecrated the beginning of the work for the completion of the last part of the Cupola, the *lantern*. The *maestro* at least had the satisfaction of seeing the completion of his Cupola, the lifetime achievement for which he worked without rest for more than 25 years of his life. Francesco di Giorgio was 6 years old when Brunelleschi died, while Leonardo was born only 6 years later. As we have seen, these two figures would take on the responsibility of building upon and transmitting Brunelleschi's lessons in engineering and architecture. Brunelleschi's curiosity, innovative vision of the world, and love of mechanics functioned as catalytic elements for the growth of modern engineering. Brunelleschi was, however, more than just an architect and engineer. He was one of the best examples of a *Renaissance man*, with wide interests in many fields, similar to Leonardo and Francesco di Giorgio. In addition to architecture, he dedicated his time to a variety of activities such as studies of perspective, machines (not only for construction but also for theatrical representations), mechanics, and more fundamental studies in disciplines such as mathematics, astronomy, and hydraulics. As we have discussed, he did not like to put his ideas to paper, probably because he was afraid that these could be stolen 1 day, but he left behind numerous inventions and innovations which became some of the most important triggers for the technological Renaissance.

At his death, *Maestro* Brunelleschi received an enormous tribute from his fellow citizens (who were not always so keen to recognize his genius when he was alive). After deep deliberation, the Florentines decided to grant him a special honor, to be buried in the very cathedral that he had helped to make a masterpiece of the Renaissance. By a strange twist of fate, the tomb was forgotten for centuries, but Brunelleschi's fame as an architect and engineer has nonetheless endured.

In 1972, during restoration work in the cathedral, Brunelleschi's tomb was rediscovered by chance. On the gravestone the Florentines had written some simple words that Brunelleschi would certainly have appreciated: *Corpus Magni Ingenii Viri Philippo Brunelleschi Fiorentini* ("The body of the great genius Filippo Brunelleschi of Florence").

3

A *Codex Atlanticus*: A Book Thief, Jokes, and Fake Bicycles

One of the best preserved treasures of the *Veneranda Biblioteca Ambrosiana*[1] (The Ambrosian Library) in Milan, one of the oldest libraries in the world, is an incredible collection of 1110 folios written by Leonardo da Vinci. They form the so-called *Codex Atlanticus*. It is the largest surviving collection of notes from Leonardo and has a complicated story connected to the strange destiny of Leonardo's notebooks.

The name *Atlanticus*, though seemingly evocative, has a rather mundane origin: it was derived from the dimensions of the pages which were used to mount and collect Leonardo's notes and sketches some years after his death. The dimensions of these pages, by chance, were the same as those used for making the geographical atlases of the time, hence the name *Atlanticus*.

The author of the *Codex Atlanticus* in its present form was not in fact Leonardo da Vinci, but Pompeo Leoni (1530–1608), an artist in his own right and an important art collector. In order to create the *Codex*, he used pages from Leonardo's notes, which he obtained from Orazio Melzi, son of Francesco Melzi, Leonardo's pupil and the inheritor of the notebooks.

The *Codex Atlanticus* appears as a miscellany of notes, drawings, and projects on different subjects, which were written by Leonardo between 1478 and 1519, a period almost of 40 years.[2] The adventure-filled story of the *Codex*, like that of other Leonardo manuscripts, has consumed the attention of generations of scholars who have tried to retrace its path.[3] This story is, indeed, so

[1] The Ambrosian Library was founded by Cardinal Federico Borromeo in 1607 and opened to the public in 1609. The library took its name Ambrosiana from the patron saint and protector of Milan, Ambrogio.

[2] Galluzzi P (1974). The strange vicissitudes of Leonardo's manuscripts. UNESCO Courier.

[3] Pedretti C, Cianchi M (1995) Leonardo. I codici. Art Dossier, Giunti.

P. Innocenzi, *The Innovators Behind Leonardo*, https://doi.org/10.1007/978-3-319-90449-8_3

intricate and fascinating that it has often been used as a plot for movies, novels, and TV series.

Collectors and Scissors

The disaster of the dispersion of Leonardo's notebooks begins with the death of Milanese count Francesco Melzi (c. 1491–1568/1570), Leonardo's loyal and beloved pupil, who inherited the *Maestro*'s written legacy. Melzi followed Leonardo on his last trip to France and remained with him until his last moment along with another favorite pupil, Gian Giacomo Caprotti da Oreno (1480–1524), known as Salaì, who instead inherited some of his paintings.

Francesco Melzi was very close to Leonardo and harbored for him a great affection and admiration. When the *Maestro* died, he wrote a very touching letter to inform his brothers. He felt responsible for preserving and reorganizing Leonardo's written work and then scattered across thousands of pages without clear organization of subjects or chronology. After an initial reorganization of the work, Melzi remained in France for several months after Leonardo's death just to prepare the transportation of the notebooks to his villa in Vaprio d'Adda, close to Milan. Melzi, who was also a talented painter, dedicated the rest of his life to the attempt to systematize Leonardo's notes, trying to group them topically.

It was a very complex operation, which the nobleman likely underestimated. Leonardo's notebooks, besides being annotated using mirror writing, were an incredible medley of just about everything: illustrations of machines, shopping lists, Latin homework, mathematics and geometry calculations, studies of physics, and architecture projects. It was a unique collection of knowledge that covered the entire natural world explored throughout his life by Leonardo. It was very easy to get lost inside these wide-ranging memories, and it would take the efforts of generations of dedicated and talented scholars to describe, organize, translate, and make them widely available.

The disordered and chaotic notebooks have been used by some authors as a demonstration of Leonardo's inability to think in a systematic and organized way or as a sign of changeable character. This is only partially true, since within the apparent disorder of the notebooks, we can find a common thread represented by certain ideas and projects that were always at the center of Leonardo's attention such as the studies on flight or mechanics. We should also not forget that what has come down to us is perhaps only one third of all Leonardo's work, and this has no doubt contributed the apparent fragmentation of the notes. The chronological reconstruction of the notebooks has shown that the different subjects studied by Leonardo were part of a systematic and ambitious project to discover the main secrets of nature. The notebooks were working tools to record impressions, ideas, projects, and interests in a continuous learning process. In this sense, their difference from Taccola and Francesco di Giorgio's notebooks is clear. The latter planned to use their notes as treatises, even if (especially in Taccola's case) the written material was not really systematically organized.

Within Leonardo's notes, it is possible to find all of the many subjects that caught his fancy. The *Maestro* likely considered them a kind of memory chest to be used for drawing what was necessary for a later reorganization of his work. Leonardo himself clearly explained the purpose of his notebooks and his way of using and reading them. "Begun in Florence, in the house of Piero di Braccio Martelli, this 22*nd* day of March, 1508. This has been done without any order, taken from the many pages that I have copied here, with the hope to put them in order according to the subjects they treat. I believe that by the time I finish this, I may replicate the same thing several times. Reader, do not blame me for this, because the things are so many, and the memory cannot keep all and say, 'this I do not want in writing because I already wrote it before.' Even if I were to attempt to avoid such errors, it would be necessary that for every case, I should always read everything preceding, creating a long interval in writing from one time to the other" (*Codex Arundel*, folio 1r).[4]

When Leonardo wrote this note, he was already 56 years old and had started thinking about reorganizing his work that was "*senza ordine* (*without order*)" by subject and using all the notes that he had collected. From his note, it appears he was worried about the efficient management of the enormous mountain of memos he had accumulated over the years. He knew that, even for him, it would be a difficult task. Leonardo was well aware of the fragmentation of his writings and the unavoidable repetition of ideas fixed on the page at different moments.

At the same time, in the *Codex Arundel* (folios 35r and 35v), he wrote extremely detailed lists[5] of the treatises he had in mind to write using his stud-

[4] Original text: "*Cominciato in Firenze, in casa di Piero di Braccio Martelli, addì 22 marzo 1508. e questo fia racolto sanza ordine, tratto di molte carte le quale io ho qui copiate, sperando poi di metterle per ordine alli lochi loro, secondo le materie che esse tratteranno; e credo avanti ch'io sia al fine di questo, io ci arò a riplicare una medesima cosa più volte; sì che, lettore, non mi biasimare, perché le cose son molte, e la memoria non le po' riservare e dire; questa non voglio scrivere, perchè dinanzi la scrissi. E s'io non volessi cadere in tale errore, sarebbe necessario che ogni per caso ch'io ci volessi copiare su, che per non replicarlo, io avessi senpre a rilegere tutto il passato, e massime stanto con lunghi intervalli di tenpo allo scrivere, da una volta a un'altra* (*Codex Arundel*, folio 1r)."

[5] The two pages contain a list of 15 "libri (books)."

ies, for example, "Book about making or repairing bridges on rivers. ... Book about boats which move upstream in the rivers...."

Unfortunately Leonardo would be distracted from such endeavors by the events of his life. In the final part of his life in the castle of Amboise in France, his physical condition would not allow him to organize the enormous amount of notes and drawings accumulated in his lifetime, and nothing of his work would be published for a long time. If at least the studies on mechanics and anatomy had been published, they might have changed the course of the history of science.

At the end of his enormous efforts, Francesco Melzi was able to compile only the *Treatise on Painting* (*Trattato della Pittura*).[6] Melzi had good intentions but probably lacked the knowledge to delve deeply into the more scientific and technical side of Leonardo's fragmented work. Melzi's attempt, though not lacking in dedication, yielded only partial results. At his death, most of the pages were yet to be classified and reorganized.

At this point, Leonardo's written legacy, so much at the center of Francesco Melzi's life, was inherited by the latter's son Orazio. Orazio, however, was not at all interested in Leonardo's notebooks, which on the contrary he considered useless and too voluminous, and they were left to languish in the attic of the Villa Melzi. Indeed, Orazio seemed to harbor a kind of aversion to the very legacy his father had tried so hard to cultivate. Consequently, he tried to dismantle it.

This was just the beginning of the very intricate vicissitudes of Leonardo's codices, which for several centuries would figure in the plots of a succession of ruthless people. It is a story of angels and demons, featuring both people who, with love and determination, tried to preserve the work of the *Maestro* and those who only tried to extract the maximum personal gain. Another category of people, perhaps the worst, comprised those, such as Orazio Melzi, who were completely unable to understand the importance of Leonardo's work and displayed a total disregard for it. At the end of this dispersion of the enormous amount of papers left behind by Leonardo, only one document remained at the Villa Melzi: the *Patente Ducale*, a certificate released by the Duke about his status, presented on 18 August 1502 by Cesare Borgia to Leonardo da Vinci, appointing him "prestantissimo et dilectissimo familiare Architetto et ingegnere generale"[7] (architect and general engineer).

In this game between angels and demons, Pompeo Leoni played one of the most important and most ambiguous roles. Pompeo was the son of an impor-

[6] The first printed edition was realized by Giacomo Langlois in Paris versions in both Italian and French.

[7] Alvisi E (2010) Cesare Borgia: Duca di Romagna. Nabu Press.

Fig. 3.1 The Leoni palace. Serviliano Lattuada—"Descrizione di Milano ornata con molti disegni...(Description of Milan with several drawings...)," Milan 1738, p. 444, V book. In the text, the palace is indicated as the house of the Calchi family. (This work is in the public domain in its country of origin and other countries and areas where the copyright term is the author's life plus 70 years or less. Wikimedia commons, CC-PD. https://upload.wikimedia.org/wikipedia/commons/1/1e/Lattuada_Serviliano_-_Descrizione_di_Milano_ornata_con_molti_disegni...%2C_Milano_1738%2C_p._444%2C_tomo_quinto.jpg. The image has been modified with respect to the original to reduce the noise)

tant artist, Leone Leoni (1509–1590),[8] called Lione Aretino, who was famous as a sculptor and especially as a medal maker. He was not only a famous artist but also a refined art collector. In 1565, after receiving several well-paid commissions, Leone Leoni built a palace in Milan's city center, now known as the *Casa degli Omenoni* (House of the Big Men) due to the sculptures of giants on its façade. Though its original structure has changed a little over time, the building (Fig. 3.1) is still well preserved and can be viewed by visitors to Milan. The Leoni palace became the home of an impressive art collection including sculptures, paintings, and manuscripts. The collection was famous for its originality and eclecticism, as described by Vasari in the *Vite*[9]: "To display the magnificence of his spirit, the beautiful intelligence he had received from nature, and his good luck; Lione, with much expense, transformed a

[8] Cupperi W (2005) Entry on Leone Leoni. In: Dizionario Biografico degli Italiani. Treccani. Volume 64.

[9] Vasari G. Vite dei più eccellenti architetti pittori et scultori italiani da Cimabue insino a' tempi nostri. Vita di Lione Lioni, Aretino.

small house in contrada de' Moroni into a beautiful work of architecture, full of capricious inventions, of which there is no match in Milan."[10]

In 1530, Lione's only male child, Pompeo Leoni, was born; he would share a passion and love for art with his father. He was also a talented artist and, even more than his father, a fanatical collector of art. His own work was so much appreciated that he was appointed the royal sculptor of the King of Spain Felipe II. Pompeo died in Madrid on 9 October 1608, leaving some unfinished works and an impressive collection of art. At the time of his death, part of that collection was in his house in Madrid, and the rest at the *Casa degli Omenoni* in Milan.

Meanwhile, at the in Villa Melzi, a strange movement of Leonardo's notebooks and other written works began, spurred by the few people who knew about their presence there. One of these people was Orazio Melzi's tutor, a certain Lelio Gavardi who, taking advantage of Orazio's lack of interest in his father's inheritance, stole 13 of Leonardo's manuscripts. He intended to sell the notebooks to the Duke of Florence, Francesco de Medici. Unfortunately for him, the duke was advised to ignore things of "such small value," and Gavardi remained without a proper buyer for the books. At this point, Gavardi decided to return the volumes to their rightful owner and asked a student at the University of Pisa, Giovanni Ambrogio Mazzenta, to deliver them to Orazio. Again, Orazio displayed his disregard for Leonardo's notes by simply giving them to the student. However, the manuscripts would not stay with Giovanni for long. In 1590, he joined the religious order of friars known as Bernabiti and, having taken a vow of poverty, was required to give up all his worldly possessions, including the 13 books.

The manuscripts were divided between his two brothers Guido and Alessand, with six going to the form and seven to the latter. Here, as we will see time and again, Leonardo's precious books seemed to have been afflicted by a strange fate that prevented them from remaining in the same hands for long.

It is at this point that Pompeo Leoni enters the story. He also became aware of the presence of Leonardo's treasure in the Villa Melzi and concocted a clever plan to convince Orazio that he could profit from these notebooks. Pompeo proposed that he might prepare a collection of Leonardo's writings as a gift for the King of Spain. In exchange, Orazio could expect to receive honors and benefits from the king. Pompeo's arguments were quite convincing,

[10] Original text: "*Lione, per mostrare la grandezza del suo animo, il bello ingegno che ha avuto dalla natura, e il favore della fortuna, ha con molta spesa condotto di bellissima architettura un casotto nella contrada de' Moroni, pieno in modo di capricciose invenzioni, che non n'è forse un altro simile in tutto Milano.*"

and, this time moved by the possibility of personal gain, Orazio asked the Mazzenta brothers to return the manuscripts. The attempt was partially successful as Pompeo, on Orazio's behalf, immediately received seven volumes, with the Mazzentas promising three more at a later date. Three volumes, however, had already been dispersed: Guido had gifted one volume to Cardinal Federico Borromeo, one to Duke Carlo Emanuele of Savoia, and a third to the painter Ambrogio Figino.

The latter two were lost, while the manuscript that was given to Cardinal Borromeo is now kept at the Biblioteca Ambrosiana in Milan.

Despite these challenges, as we have seen, Pompeo Leoni succeeded in collecting a good number of Leonardo's autographed volumes, along with several scattered drawings and notes. Unfortunately, after having collected such a large number of these notebooks, he had a rather destructive idea: He decided to group the different pages following his own personal, arbitrary vision of their content. Leoni used scissors to cut and paste pages from the different parts of the volumes, losing the original order and chronological development of Leonardo's thinking, which became extremely difficult to reconstruct. At the end of this vandalistic operation, Pompeo Leoni was quite satisfied. The result of the cut-and-paste effort was a voluminous but elegant codex, the *Codex Atlanticus*, to which he gave his imprimatur by engraving in gold letters[11]: "Drawing of machines and secret arts and other things of Leonardo da Vinci, collected by Pompeo Leoni."

Pompeo Leoni's role had mixed results. On the one hand, he acquired possession of Leonardo's work because, as an artist and art collector, he recognized the value of this legacy. It was also clear to him that he could profit in several ways such as selling the notebooks to some rich art collectors. On the other hand, we do not know what would have become of the manuscripts without Pompeo's interference. It is very likely that his intervention helped to save a good part of Leonardo's written work from dispersal and destruction. After gathering the greatest part of the manuscripts included in scattered notes and drawings, Pompeo took them with him to Spain, where the *Codex Atlanticus* (at least the first version) was likely composed.

As is usual in these cases, the real problems began with Pompeo's death. As we have also seen in the case of the Melzi family, sons do not always share the same interests as their father.

Pompeo Leoni died in Madrid in 1608, leaving behind one of the most spectacular art collections of the time, divided between his homes in Madrid

[11] Original: "*Disegni di machine et delle arti secreti et altre cose di Leonardo da Vinci racolti da Pompeo Leoni.*"

and Milan. This rich patrimony went to his son-in-law, Polidoro Calchi, who also received the Casa degli Omenoni, the entire art collection accumulated by the Leonis, and the *Codex Atlanticus*. Pompeo's daughter and son-in-law were much more interested in liquidizing the art collection than preserving it. Unfortunately, almost immediately after Pompeo's death, they began selling off the masterpieces.

In the art collection in Milan alone, Pompeo and his father Leone had been able to collect paintings by the most famous Italian artists of the time, such as Titian, Tintoretto, and Parmigianino. The Madrid collection was so huge as to represent a fundamental contribution to the knowledge and diffusion of Italian art in Spain and the wider world. It included paintings by Parmigianino, Titian, Correggio, Raffaello, Taddeo Zuccari, Federico Barocci, and El Greco. Pompeo's heirs started to sell this collection little by little. Finally in 1628, the *Codex Atlanticus*, which had in the meantime been brought back to Milan, was also sold. The buyer was Count Galeazzo Arconati, a gentleman collector very much attracted to Leonardo's work. He bought not only the *Codex Atlanticus* but also several uncollected pages, including sketches and working drawings. Among these were 11 charcoal drawings of the Last Supper, which are now scattered across different museums. For a while, the *Codex* remained safe in the nobleman's beautiful Villa Castellazzo in Bollate, close to Milan, kept in the good company of other precious artistic masterpieces, such as the original statue of Gnaeus Pompeius Magnus, who by tradition witnessed the homicide of Gaius Julius Caesar from 23 stabbings. In addition to the *Codex Atlanticus*, the nobleman also collected 11 additional Leonardo manuscripts from the group given to Mazzenta. Count Arconati was able to reunite the 1119-page *Codex Atlanticus* and other manuscripts totaling 1008 pages. In 1637, he decided to donate his collection of Leonardo's manuscripts to the Veneranda Biblioteca Ambrosiana in Milan.

For more than a century, this important legacy sat quietly in Milan, until an uninvited visitor came knocking at the door of the city in the person of Napoleon Bonaparte. After conquering most of the Italian peninsula, Napoleon began systematically plundering Italy's artistic treasures. Leonardo's codices were also subjected to the same treatment and were packed and sent to France. The Congress of Vienna and the end of Napoleon's venture in 1815 marked the restoration of the old system of power in Europe, and on the basis of a specific international treaty, the legitimate owners requested the return of the looted treasures. The person in charge of the negotiation on behalf of Milan, then part of the Austro-Hungarian Empire, was an Austrian officer named Baron Ottensfels. He was not a cultivated person and even mistook Leonardo's mirror writing for Chinese characters. In the end, he happened to bring back the *Codex Atlanticus*, but the other manuscripts, mostly small notebooks which were conserved at the *Institut de France*, remained in Paris. These were classified with letters from A to M and are still referred to in this way. These manuscripts were never returned to Milan.

The Book Thief Guglielmo Libri

Despite being well-preserved in Paris, the notebooks did not remain safe. They caught the attention of an Italian scholar with the strangely fitting name Guglielmo Libri ("books" in Italian, Fig. 3.2). As an established professor, Libri was able to obtain official authorization to work on Leonardo's notebooks in the 1840s. With extreme patience and caution, he deliberately tore off dozens of pages from Codices A and B and secreted them out of the library without being discovered. He then sold the stolen pages to an English nobleman, Lord Ashburnham, who had them bound to form two distinct volumes. These two manuscripts, kept in the Lord's residence, fortuitously survived a fire in the house and were later returned to France in 1888.

Fig. 3.2 Portrait of Guglielmo Libri. A. N. Noel, Paris, BN84 C 123273. Phot. Bibl. Nat. Paris. (This work is in the public domain in its country of origin and other countries and areas where the copyright term is the author's life plus 70 years or less. Wikimedia commons, CC-PD. https://upload.wikimedia.org/wikipedia/commons/e/e8/Guglielmo_Libri_Carucci_dalla_Sommaja.jpg)

In this strange story of theft perpetrated by states and individuals, Libri[12,13,14], with his incredible name (in *nomen omen*,[15] as they say), holds a special place worth recounting in greater detail.

Guglielmo Libri

Libri was born in Florence in 1802 to a noble family. At the age of 14, he was already attending the University of Pisa, from which he graduated in 1820. He was a gifted mathematician and published important works that allowed him to get a professorship at the University of Pisa at the young age of 21. He traveled to Paris where he came in contact with the extraordinary community of French mathematicians, including Cauchy, Gay-Lussac, Poisson, and Ampère e Laplace.

After returning to Florence, Libri was appointed director of that city's Accademia dei Georgofili library. It was at this point that he began his career as a book thief. Libri was an expert and passionate book lover, but he was also moved by greed and money. When he removed the pages of Codices A and B at the Institute of France, he made an accurate selection of the material with the highest market value, such as the pages containing good-quality drawings by Leonardo. Illustrations were much more attractive for potential buyers who were probably not as interested in text because of Leonardo's complicated writing system.

During Libri's tenure as director of the Georgofili library in Florence, strange losses of several volumes began to be observed, and he was forced to resign quietly without too much public scandal. In 1830, he returned to Paris just in time to take part in the revolution. Immediately afterward, he returned to Florence with the hope of forcing Grand Duke Leopoldo of Tuscany to proclaim a liberal constitution. The attempt failed, and as a result Libri was expelled from Florence.

Paris was once again his destination, and in the French capital, he found an environment very willing to honor his ability as a mathematician and historian of science. He obtained important honors, such as the *Légion d'honneur*, and admission to the *Collège de France* and *Académie des Sciences*.

The turning point, however, of Libri's career as a thief was his appointment as *inspecteur des bibliotheques publiques*, which was like putting an alcoholic in charge of a liquor shop. Libri immediately used his position to start a systematic looting of the French libraries. His "private collection" of books grew very quickly and came to contain more than 40,000 volumes and 1800 manuscripts by the middle of the 1840s. These came in part from acquisitions at public auction and from private collections but also from his raids of French libraries.

[12] Giacardi L (2005) Entry on Guglielmo Libri. In: Dizionario Biografico degli Italiani. Treccani.Volume 65.

[13] Del Centina A, Fiocca A (2010) Guglielmo Libri, matematico e storico della matematica. L'irresistibile ascesa dall'Ateneo pisano all'Institut de France. Olschki.

[14] Maccioni Ruju PA, Mostert M (1995) The Life and Times of Guglielmo Libri (1802–1869). Scientist, patriot, scholar, journalist and thief. A nineteenth-century story. Verloren Publishers.

[15] *Nomen (est) omen*. The name is a sign. This famous Latin sentence is still commonly used in several European languages to suggest that the name is linked to the fate of a person.

Libri's activities did not go completely unnoticed, however. Several denunciations made their way to the authorities, but his network of friends and protectors was extensive and able to conceal his activities for a time. With the change of the political situation in France and the advent of the Second Republic, Libri was forced to escape quickly to London. Meanwhile, in Paris, he was sentenced to 10 years in prison and lost all the honors he had accumulated, despite his continued denials of any guilt. His request to the French Senate for rehabilitation was also rejected. Nonetheless, he succeeded in getting back his personal library and becoming an English citizen by marrying his rich friend Mélanie Double Collins.

In London, Libri sold his book collection little by little. After his death, an accurate investigation, sponsored by the general administration of the Bibliothèque Nationale, finally reconstructed Libri's thefts and falsifications. The volumes stolen by Libri were returned to France along with the codices in the hands of Lord Ashburnham. Italy bought the rest of the Ashburnham collection, composed of volumes and manuscripts stolen from the Georgofili library, and finally returned them to their original location.

Curiously, traces of Libri's activities have been found even as far away as the United States, for example, in the form of a 27 May 1641 letter by Descartes discovered in 2010 in a library at the University of Pennsylvania and which was on the list of manuscripts stolen by Libri. The letter was subsequently, therefore, returned to the Institute of France.[16]

Bad Jokes and Bicycles

The *Codex Atlanticus* survived the raids of Guglielmo Libri and the trips around Europe and finally returned to Milan in 1815, while the collection of Leonardo's notebooks (*Codices A–M*) remained in France. The *Codex Atlanticus* was carefully kept in the Biblioteca Ambrosiana, but clear signs of degradation started to appear over time. Four centuries after its creation at the hands of Pompeo Leoni, molds and moths attracted by the glue began to degrade the paper and put the *Codex Atlanticus* in danger. It was decided therefore in the 1960s to submit the codex to a careful restoration, which was entrusted to the Monastero Esarchico of Grottaferrata near Rome, which boasted a laboratory for the restoration of ancient books. This highly specialized laboratory was managed by monks who possessed unique experience in the restoration of manuscripts, making it one of the best places in the world for this extremely

[16] Maccioni PA (1991) Guglielmo Libri and The British Museum: A Case of Scandal Averted. The British Library Journal. 17:36–60.

Fig. 3.3 Drawing of a bicycle that appears rotated with respect to the rest of the page. Leonardo da Vinci. Codex Atlanticus, folio 133v. BAM

delicate task. Another advantage was the possibility for the work to be conducted in secrecy and away from prying eyes, as requested by the Prefect of the Veneranda Biblioteca Ambrosiana of Milan.

It was a difficult and very delicate task that took the laboratory more than 10 years, from 1962 to 1972. During the restoration, the single pages written by Leonardo were detached one by one from the supporting paper and carefully restored and repaired. During this process, the monks made a very surprising discovery: a page that was folded and glued shut (*Codex Atlanticus*, folio 133v). After a careful operation to open it, the page revealed strange and unexpected sketches (Fig. 3.3): a caricature of a man, a drawing of a phallus, and… a bicycle.

The discovery astonished the monks, who invited one of the leading contemporary Leonardo scholars, Augusto Marinoni (1911–1997), to try and make sense of what they had found. After viewing the page, Marinoni attributed it to the hand of one of Leonardo's pupils, inspired by a lost drawing of the master. Following that logic, the invention of the bicycle could then be attributed to Leonardo in or after 1493, based on the dating of a chain drawn in the *Codex of Madrid I* (folio 10r, Fig. 3.4).

The evidence of the toothed chain was taken to show that, if Leonardo had invented the chain for circular motion transmission, he could also have imagined a bicycle. As we can see from Fig. 3.4, Leonardo did indeed design several types of chains, some of which could be employed for power transmission in

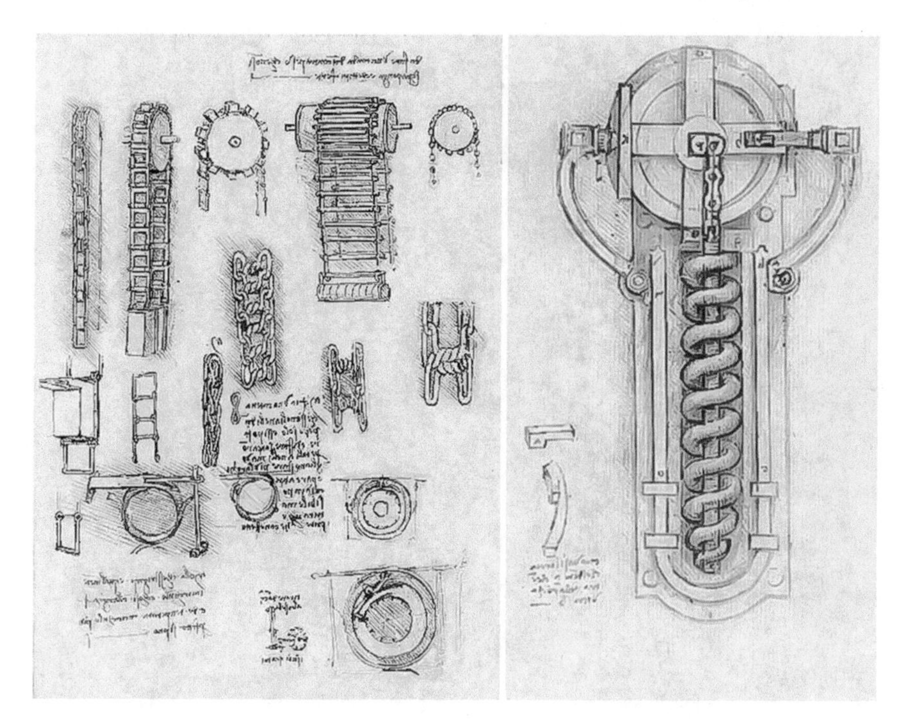

Fig. 3.4 Models of chains (detail). Leonardo da Vinci. Codex of Madrid I, folio 10r (*left image*). BNE. Codex Atlanticus, folio 158r, (*right image*). BAM

a manner similar to that used in modern bicycles and motorcycles. Studies of chains can also be found in the *Codex Atlanticus*, folios 987r and 987v (Fig. 3.5). In another drawing found in the *Codex Atlanticus*, folio 158r, Leonardo depicted a flint for firearms. The spring necessary to operate the flint is coiled by a chain whose tangles appear very much like those of modern bicycle chains. Unfortunately, as with many other of Leonardo's inventions, years would pass before the rediscovery of Leonardo's chains. He was, as usual, ahead of his own time.

The invention of the chain as we know it today is attributed to André Galle, a French medal maker, watchmaker, and inventor (1761–1844). On 29 July 1829 in France, he filed the patent n. 4072 to protect the invention of the metal chain that was named after him ("chain Galle"). Thus, it was over 400 years later that a chain similar to the once Leonardo envisioned was "officially" invented.

The discovery of "Leonardo's bicycle" had a big media impact and ignited the imagination of the public. However, at the same time, it sparked a heated dispute among Leonardo scholars. Other experts challenged Marinoni's explanation and even accused him of authoring the fake in collaboration with the monks of Grottaferrata.

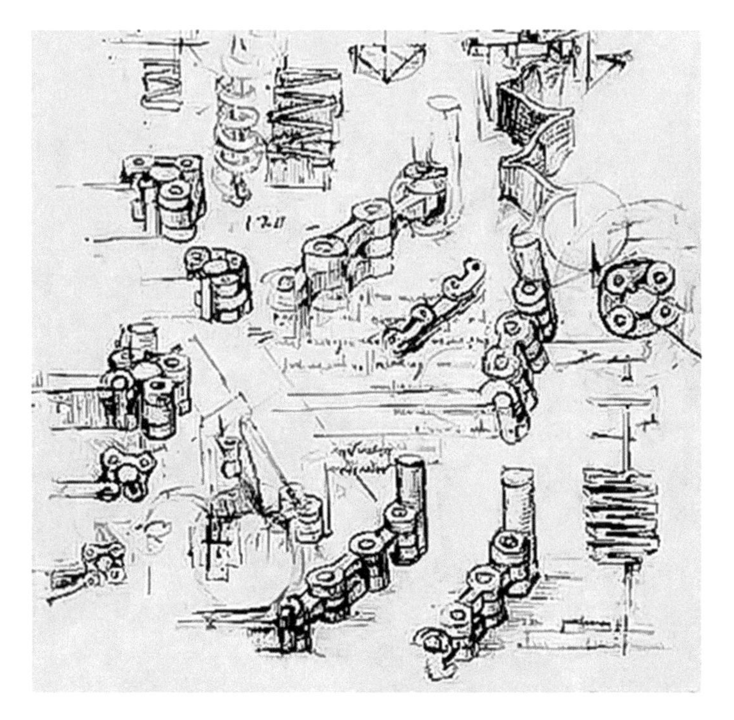

Fig. 3.5 Studies of chains and springs. Leonardo da Vinci. Codex Atlanticus, folio 987r (detail). BAM

The doubts (if not the accusations) were legitimate. As we wrote above, everything seems to suggest that some of Leonardo's students were behind at least some of the sketches. In addition, the artist of the bicycle is definitely not Leonardo, as one can easily verify by comparing it with his confirmed drawings, which display distinctive traits even to the untrained eye.

Regarding the other two drawings on the page, the phallus enters into a hole on which is written the word Salaì, the nickname of Giacomo Caprotti, one of Leonardo's favorite pupils and possible lover. The caricature of a male figure may also refer to Salaì. In addition, these drawings are all done in charcoal, and the name Salaì is written normally, from left to right, rather than in Leonardo's mirror writing.

In addition, Marinoni's conclusions seem to ignore many of the details of the drawing. His mistake was probably to observe the bicycle carefully but outside of the context of the page. The bicycle is oriented differently to the other two sketches and is a two-tone black and brown drawing. Even at first glance, the hand that sketched the "bicycle" appears different from the one that sketched the two jokes.

Fig. 3.6 A printed drawing of a draisine from Wilhelm Siegrist (1817). (This work is in the public domain in its country of origin and other countries and areas where the copyright term is the author's life plus 70 years or less. Wikimedia commons, CC-PD. https://upload.wikimedia.org/wikipedia/commons/4/42/Draisine1817.jpg. The image has been modified with respect to the original to fit the printed page)

As we have seen, all of the evidence tends to indicate a fake, but we may never know when and where it was drawn. Indeed, the model in the drawing is already quite advanced, so the fake may even have been created as late as the beginning of the twentieth century. The design is reminiscent of early models such as the draisine of Karl von Drais, built in 1816 in Germany without any chain for transmitting motion to the wheels (Fig. 3.6). Regarding the dating of the *Codex Atlanticus* bicycle sketch, it is even possible that the forger may have given the bicycle a primitive look to make the joke more credible.

There is one last chapter to this history of hoaxes, forgers, and quarrels among scholars. The world's leading expert on Leonardo da Vinci, Carlo Pedretti, recalled noticing folios 132 and 133 of the *Codex Atlanticus* in 1961. They were glued to each other, so he carefully observed them in transparency. He was able to distinguish only two circles and some lines in the area where the bicycle was later found. Therefore, it seems likely that the forgery occurred after 1961. In this scenario, the restorers would be the only possible suspects but for the following letter written by Nando di Toni, resigned member of the Commissione Vinciana, to the Italian weekly *L'Espresso* on 19 May 1974, shortly after the announcement of the bicycle sketch:

> "As for the bike, I note that on several occasions some folios of the Codex Atlanticus left the Ambrosiana [library], more or less officially, before the restoration requested by André Corbeau and myself. Those who have had the opportunity to take away--to take to Florence and ship from Lugano by post--stolen sheets of Leonardo from the Ambrosiana could have had the opportunity, at

> different times, to make fun of posterity by drawing this rudimentary bicycle. To be possible at the time of Leonardo, it certainly could not have the fenders, chain-guard, brakes, tail light, bell or reflector. It was enough to leave the idea of pedals, saddle, and steering chain drive, though the latter is patently not working."

The letter describes a disturbing scenario, one in which sheets of Leonardo's work disappear and reappear in unexpected places according to a script that does not seem to change over time.

So we will probably never know who the author of the false bicycle is. This "extraordinary discovery" has, however, generated a series of wooden replicas, which have been exhibited in museums and exhibitions, though not without some retroactive embarrassment.

Bicycles Through History

To conclude this story of thieves and false bicycles, let us go back to another of Leonardo's inventions, the pedal drive for a "paddle boat" (*Codex Atlanticus*, folio 945r), in some ways connected with the bicycle (for more on this and other boats, see Chap. 6). This boat can be moved via a pedal drive with reciprocating *linear* motion instead of circular motion (the latter being what we see in a bicycle). Circular drive brings certain advantages, as it generates more power for the same work and, when used in a bicycle, would represent an undeniable improvement.

Inspired by this the pedal motion idea, an Italian engineer, Marco Antonelli, has created an innovative new bicycle (Fig. 3.7) with the help of Charles Rottenbacher, an assistant professor of engineering at the University of Pavia. This model, actualized by the company Dobertec and named the *Twist Bike "Atlantic*," was exhibited at the Milan Triennale in 2012. It is propelled by pedals with linear movement,[17] the alternative way of transmitting motion suggested by Leonardo.

However, this story of bicycles isn't yet over. In fact, on folio 48v of book II of Taccola's *De Ingeneis*, there is a small drawing (Fig. 3.8) that is difficult to interpret in detail, but which definitely suggests the representation of a sort of bicycle.

[17] www.lescienze.it/news/2012/05/15/news/bici_leonardesca_twist_bike_atlantic_triennale_milano-1023293/.

Fig. 3.7 The Dobertec Twist Bike "Atlantic," exposed at the Triennale of Milan. (Courtesy of Dobertec)

Fig. 3.8 The "first" bicycle. Mariano di Jacopo (Taccola), De Ingeneis II folio 48v (detail). Cod. Lat. Monacensis 197 II, BSBM

This sketch has gone unnoticed thus far, but it is likely the first schematic of a bicycle precursor ever documented. It can be considered a kind of precursor to the draisine, with a handlebar steering system and the possibility for the rider to sit in a covered seat such as in a rickshaw. In this simple sketch, there is no mechanism for the transmission of motion with chains, though the concept was well known to Leonardo, as we have seen previously.

Taccola's "bicycle" could in fact have been noticed by Leonardo, but we will never know if he was actually inspired by this particular idea.

The idea of using two wheels as a means of transport should be attributed, then, to Taccola. The *De Ingeneis*'s fall into oblivion in the sixteenth century probably did not favor the spread of the idea, but we have to give Taccola the credit of having been the first to envisage human locomotion based on a two-wheel system.

4

Those Magnificent Men in Their Flying Machines

On folio 1058v of the *Codex Atlanticus*, Leonardo traced a small drawing which has garnered much attention despite its size. It is not very flashy in the context of the page but strikes a contemporary observer at first sight. It is a rough sketch of a man hanging from a pyramidal object seemingly made of cloth (Fig. 4.1).

In just a few lines, Leonardo described the use and purpose of this object: "If a man has a pavilion made out of linen which is 12 braccia[1] [around 23 feet] wide and 12 tall, he will be able to throw himself down from any great height without hurting [himself]."[2,3]

This is clearly the description of a parachute, at least in an embryonic stage. The drawing, however, is not incidental because, on the same page, Leonardo also made studies of parts of winged flying machines operated by people. Thus we see that the study of the parachute was a part of Leonardo's general exploration of the potential for human flight. As we will see in the course of this chapter, Leonardo analyzed the problem of flight from several different perspectives, and some of his ideas anticipated what would be achieved only centuries later.

---

[1] In Italy the units of measurements changed from one city to another. A Florentine *braccio (arm)* (*braccio fiorentino*) corresponds to 0.5836 m; therefore 12 *braccia* is around 7 m. The braccio of Milan was 0.595 m.

[2] Original text: "se uno uomo ha un padiglione di pannolino intasato che sia 12 braccia per faccia e alto 12, potrà gettarsi d'ogni grande altezza sanza danno di sè."

[3] Translation from Encyclopædia Britannica online (www.britannica.com/topic/Leonardo-da-Vincis-parachute-1704849).

P. Innocenzi, *The Innovators Behind Leonardo*, https://doi.org/10.1007/978-3-319-90449-8_4

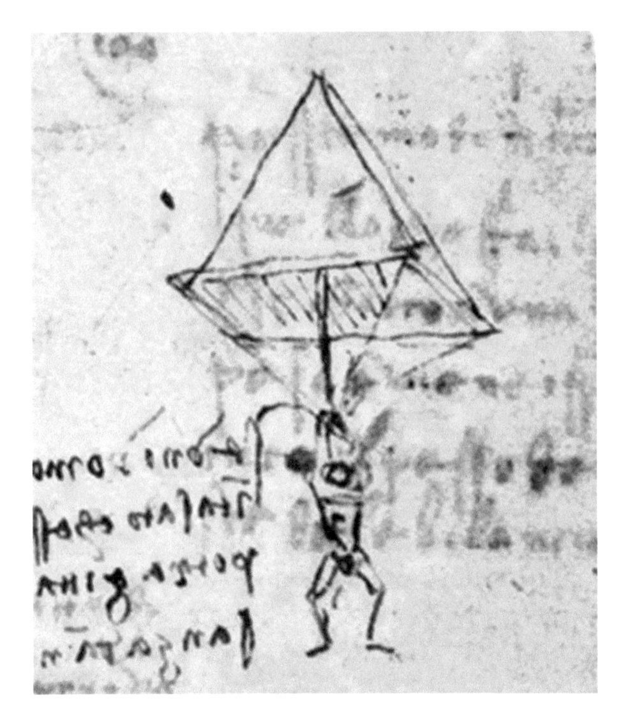

Fig. 4.1 Sketch of a man with a "parachute." Leonardo da Vinci. Codex Atlanticus, folio 1058v, (detail). BAM

Leonardo made his parachute drawing in 1485. This date is important to keep in mind because something similar had been designed just a few years prior. While at first glance it may seem surprising, Leonardo's idea was probably borrowed from a sketch in an anonymous manuscript[4] dated 1470. It was in this treatise that the possibility of creating a system to descend safely after leaping from a height[5] was described for the first time (Fig. 4.2). The treatise's elegant drawings of machines are mostly reproduced from manuscripts written by Taccola and Francesco di Giorgio, who were most likely the real inventors of the parachute, although we cannot absolutely rule out the possibility that the parachute idea was original to the anonymous artist of the manuscript or another source.

The close similarities between this drawing and Leonardo's sketch make it reasonable to deduce that Leonardo was probably inspired by the drawing in

[4] The anonymous manuscript of 261 pages is conserved at the British Library and is largely based on the work of Taccola and Francesco di Giorgio; only some of the machines contained therein cannot be attributed to models of the two Sienese engineers.

[5] White LJ (1968) The Invention of the Parachute. Technology and Culture. 9:462–467. The Johns Hopkins University Press and the Society for the History of Technology.

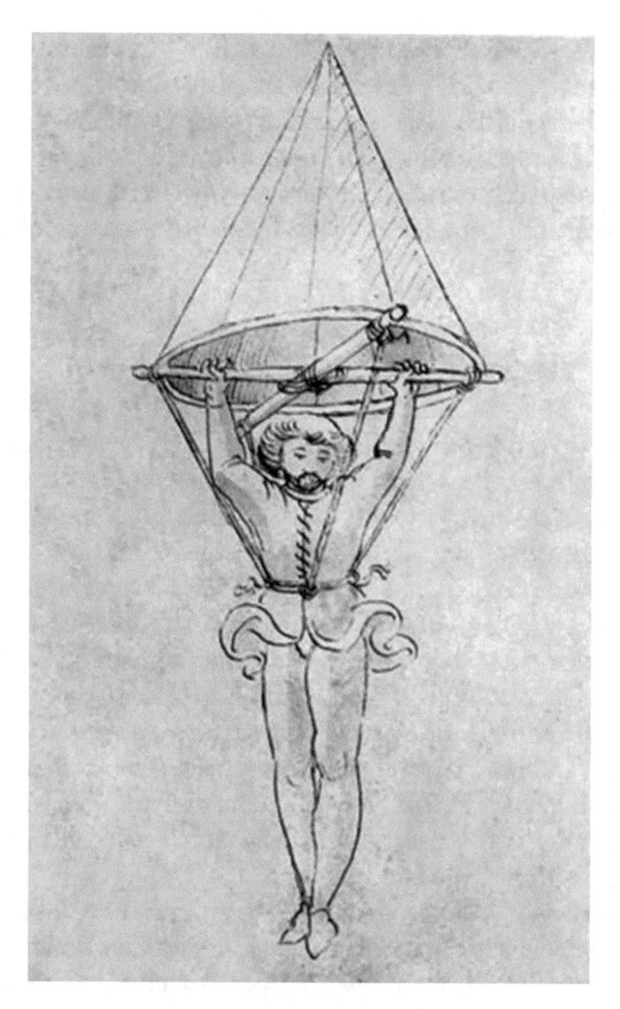

Fig. 4.2 Gentleman with a parachute. Anonymous engineer from Siena. Libro di machine. Additional Manuscript 34, 113, folio 200v. BL (1470)

the anonymous manuscript. The similarity between the conical shape of the unattributed parachute and Leonardo's flying pyramidal device is striking, although the anonymous sketch is more beautiful and better defined. The latter drawing shows an elegantly dressed man clutching a wooden frame that supports the conical canopy cover.

Curiously, the parachutist keeps a sponge in his mouth to prevent damage to his teeth from the kickback of the fall. In addition, four strings attached to the rods are used as a safety belt. The parachute appears, however, woefully inadequate, and the teeth would not be the only body part to be endangered in descent.

Leonardo's model is a lot more realistic, even if it is just a sketch. Another peculiarity is that Leonardo's sketch is accompanied by an instruction booklet of sorts, which details the exact dimensions necessary for the parachute's construction. The changes introduced by Leonardo are actually quite substantial. Instead of the wooden rods that serve as the base for the conical parachute, Leonardo's project's square frame allows for the construction of a pyramidal parachute.

The Later History of the Parachute

For over 100 years the idea of the parachute seems to have fallen into oblivion. However, another inventor and engineer with a passion for machines, the Venetian nobleman Fausto Veranzio, finally extended the life of Leonardo's invention, with an illustration in one of his treatises,[6] *Machinae Novae* (*New Machines*), printed in Venice in 1616,[7] showing an illustration of a flying man (*Homo Volans*) descending from a tower (Fig. 4.3). So we see that, 131 years after Leonardo's parachute, something similar finally appeared in a book. The similarity of this representation to Leonardo's design is surprising. Veranzio's model, however, does not share the pyramidal structure that characterizes Leonardo's project (though it does utilize the same square-shaped base). The body of Veranzio's flying man is safely secured to the parachute with ropes, whereas Leonardo's parachutist simply supports himself with his bare hands.

In a short note, Veranzio described his invention as follows: "Take a square sail, which is stretched between four wooden sticks of the same dimension and affixed [to them] with ropes. Attach a man to the four corners of the sail. With no danger, [the man] can throw [himself] from a tower or any other high place. It is better to go down in the hours when there are no wind puffs. However, the force of the falling man will create wind, which will slow the sail so he does not fall, but will come down

[6] Beside his book dedicated to machines, *Machinae Novae*, he wrote also other different treatises dedicated to lexicography and history such as a dictionary, *Dictionarium quinque nobilissimarum Europæ linguarum, Latinæ, Italicæ, Germanicæ, Dalmatiæ, & Vngaricæ (Dictionary of the Five Most Noble Languages of Europe, Latin, Italian, German, Dalmatian and Hungarian)*, published in Venice in 1595 and a history of Hungarian literature, *Scriptores rerum hungaricum*, published in 1798.

[7] The book *Machinae Novae* of Veranzio has been digitalized by *Google Books* and is available online.

Fig. 4.3 The flying man "Homo Volans." Illustration 38 of Fausto Veranzio's *Machinae Novae*

gradually; be careful to calibrate the dimension of the sail to the weight of the man."[8]

So, in Veranzio's opinion, if there is no wind and the weight of the man is well balanced with the size of the sail, no problems would be observed using his system to jump from a tower. The parachute, as shown in the drawing, was designed for a very specific purpose, namely, to throw oneself from a tower.

[8] Original text: "Piglisi una vela quadra, quale sia distesa tra quatro pertiche uguali, e con le funi ui s'attacchi l'huomo à i quatro cantoni di quella uela, che senza alcun pericolo potria posia gettarsi d'una Torre, o d'ogni altro logo eminente: venir a basso fe benè in quell'hora non spiri vento alcuno, tuttavia l'impeto istesso de l'Homo cadente, eccitarà e cagionar il Vento, quale ritardarà la vela, si che non precipiti, mà descenda à poco à poco; Fa pero' di mistiero commensurare il peso del'homo con la grandeza della vela."

Veranzio himself specifies this in the list of inventions included at the end of his book: "Jumping down from a tower without getting hurt."[9] Although it is often reported that Veranzio personally tested his parachute from St. Mark's Campanile in Venice, there are no documents to prove it. So it seems that the *Homo Volans* project remained a purely theoretical model.

Fausto Veranzio

Fausto Veranzio was a Venetian nobleman, born in Sebenico in the Republic of Venice (now Šibenik in Croatia) in 1551. He studied law as well as physics and mechanics at the University of Padua and soon after began a career as a senior civil servant at the court of King Rudolf II of Hungary. Veranzio was a polyglot and moved with great ease in the different environments of a precursor "Mitteleuropa."[10] For example, he became chancellor for Hungary and Transylvania and likely met with Johannes Kepler and Tycho Brahe.

Veranzio's political career ended abruptly with the death of his wife, immediately after which he decided to take holy vows and was subsequently offered the role of bishop.[11] However, he did not take up the assignment and instead returned as a barnabite friar to Venice where he indulged his passion for mechanical studies.

Veranzio's book, *Machinae Novae*, contains 49 drawings of various machines and devices. The work was not distributed widely at the time of its publication because Veranzio published it at his own expense and was not able to print many copies. However, the book became quite popular and left a significant impact after his death.

The work was probably inspired by Renaissance engineers like Francesco di Giorgio and Taccola and maybe even by Leonardo himself. In its pages, Veranzio described many innovative ideas such as aerial lifting (folio 36), a universal clock (folio 7), a life belt (which we will discuss in greater detail in Chap. 6), several types of mills and rotary printing, and different innovative bridges, including one utilizing suspension cables.

Many of his "inventions" appear somehow naïve and not particularly original, and the suspicion that he largely drew from previous sources, particularly the Sienese engineers, is more than a hypothesis.[12] Similar to the Renaissance engineers, Veranzio combined military and civilian projects and works of hydraulics and mechanics without much descriptive detail. Indeed, in a sense, his work represented a step back compared to Leonardo, appearing much closer to the

[9] Original text: "Buttarse giù d'una Torre, et non farsi male."

[10] The concept of *Mitteleuropa*, as intended by the Italian authors such as Claudio Magris, refers to a spiritual and cultural area which corresponds to the cosmopolitan Austro-Hungarian Empire.

[11] *Episcopus Csanadiensis in partibus.*

[12] Grmek, M.D. (2008) Voice on Faustus Verantius. www.encyclopedia.com. "Verantius became friendly in Rome with Giovanni Ambrogio Mazenta, a Barnabite like himself and, from 1611, general of the Congregation of St. Paul. Very possibly it was Mazenta who interested Verantius in the construction of machines and in architectural problems. Verantius undoubtedly had an opportunity to see many of Leonardo da Vinci's technical drawings, of which Mazenta had prepared a list about 1587. During his stay at Rome, Verantius had drawn and engraved a series of *new machines*."

Manuscript of the Hussite Wars and the *Bellifortis* of Konrad Kyeser of the German school, manuscripts in which innovative ideas were represented with great elegance but simplistic planning.

The illustrations in *Machinae Novae* are refined and charming but above all function as evidence that the dissemination of knowledge is a discontinuous process, slow and bumpy. If Leonardo's work on mechanics had been published and widely distributed during his lifetime or soon after his death, Veranzio's book would have seemed redundant.

It is difficult to determine with certainty whether Veranzio saw Leonardo's notes. Regardless, he certainly shared Leonardo's Renaissance spirit. They were both self-taught lovers of learning across different fields, including philosophy and history. Like Leonardo, Veranzio also harbored a passion for hydraulics and spent 2 years in Rome trying to regulate the floods of the Tiber. Back home in Venice, he maintained the wells and water supply of that city.

Veranzio died in Venice in 1617, a few years after the publication of *Machinae Novae*. He asked to be buried on the island of Pervicchio (modern Prvić in Croatia) in sight of his birthplace Sebenico. As mentioned above, we will encounter Fausto Veranzio again in Chap. 6, which discusses ships and lifeboats.

All these early parachutes were based on distinct geometrical shapes, such as cones and pyramids. This is not by chance but is in fact an expression of Renaissance thinking, a time in which geometry governed the world and delineated the horizon.

For Leonardo, the obsession with perfect proportions and geometry was a source of inspiration not only for architectural projects but also for his machines and inventions. The tendency during his time was to couple a geometrical shape with a specific use. This thinking would only be reversed in the modern age, and nowadays we generally accept that it is function that governs shape and not vice versa.

Brave Parachutists

The story of Leonardo's parachute does not end here, however. If anything, the curiosity to know whether Leonardo's parachute could work only increased over time. Meanwhile, technology has evolved such that modern parachutes can be made of lightweight and very resistant materials.

The first real parachute jump was accomplished by the Frenchman André-Jacques Garnerin (1769–1823) on 22 October 1797 at the Parc Monceau in Paris, when he used a basket attached to a silk parachute to jump from a hot air balloon at a height of 1000 m. Garnerin managed to safely reach the ground despite frightful wobbling due to the lack of stabilization of the parachute, which lacked an outlet for air. This problem was common to all the early attempts.

It was, however, necessary to wait until the year 2000 to finally see if Leonardo had been right in his parachute design. In that year, Adrian Nicholas, a famous English skydiver, jumped from a height of 3000 m employing an accurate reconstruction of Leonardo's parachute. Nicholas utilized a similar technique to that of the first parachutist, Garnerin. He remained attached to the hot air balloon until it reached the required height, then came off using the Leonardo-style parachute.

Nicholas's pyramidal parachute was made based on Leonardo's specifications and only used materials and tools that would have been available in his time. There was some concern regarding the stability of the parachute because the lack of a central hole could cause it to swing violently, as had happened to Garnerin. However, Nicholas finally decided to trust Leonardo.

Nicholas's descent using the pyramidal parachute was surprisingly stable, even better than with modern ones, according to Nicholas. The natural porosity of the material used for the sail allowed air to pass through, stabilizing the parachute. At a height of 600 m, Nicholas detached from Leonardo's system and landed using a modern parachute; the real danger was in fact landing with a heavy apparatus, weighing approximately 85 kg.

The reconstruction of the parachute and the attempt to fly with it were filmed and can be viewed in the beautiful BBC documentary *Leonardo—The Man Who Wanted To Know Everything*.[13] Adrian Nicholas's smiling face as he hangs from Leonardo's parachute is unforgettable.

Unfortunately, only a few years later, in 2005, during a high-speed jump, Nicholas died due to an accident while opening his parachute. The image of his happy expression during his extraordinarily courageous descent with Leonardo's carefully reconstructed parachute will remain in our memories.

Nicholas's challenge was taken up by another extreme skydiver, Olivier Vietti-Teppa of Switzerland. In 2008, Vietti-Teppa used a modified version of the pyramidal parachute that lacked the wooden square base and used modern materials for the sail. With this parachute, he made a complete (though slightly rough) descent to the ground at the airport in Payerne near Geneva in Switzerland. This attempt proved conclusively that Leonardo's pyramidal parachute could actually work.

[13] BBC documentary (2003) *Leonardo—The Man Who Wanted To Know Everything*.

Other Flying Machines

In addition to his parachute, Leonardo's passion for flying led him to imagine and design several machines and devices that are among his most impressive inventions. These machines are the ones that have most fiercely ignited the imaginations of so many people inspired by his work. Leonardo's studies of flight are usually divided into two periods. The first period began in the 1490s in Milan, where he largely devoted himself to designing machines with swinging wings. These were mostly "airships" that could be moved by human propulsion. A second period in Florence (from 1501 to 1508, with several quite long stays in different places in the meanwhile) was instead dedicated to observing the flight of birds, which allowed him to understand the fundamentals of flight itself. The result of these observations was the idea of gliding flight, to be carried out with the aid of simple machines featuring a fixed wing to support the weight of a man.

A Flying Monk

Leonardo da Vinci was not, however, the first to hit upon the idea of gliding flight. In fact, someone else had already tried to realize the dream of flying in this way, albeit with some unpleasant side effects.

In the Latin chronicle *Gesta regum Anglorum* (*The history of the English kings*), William of Malmesbury,[14] an English Benedictine monk, described in detail the flight of Eilmer of Malmesbury, a monk of the same monastery[15]: "He was a man learned for those times, of ripe old age, and in his early youth had hazarded a deed of remarkable boldness. He had by some means, I scarcely know what, fastened wings to his hands and feet so that, mistaking fable for truth, he might fly like Daedalus, and, collecting the breeze upon the summit of a tower, flew for more than a furlong [201 meters]. But agitated by the violence of the wind and the swirling of air, as well as by the awareness of his rash attempt, he fell, broke both his legs and was lame ever after. He used to relate as the cause of his failure, his forgetting to provide himself a tail."[16]

[14] The Abbey of Malmesbury is located in Wiltshire and is part of the Diocese of Bristol in Britain. The Abbey was founded in 671 by the monk Adelmo and had over time various vicissitudes that have damaged the original structure. The tower of the flight of Eilmer collapsed around 1500 due to a hurricane.

[15] White LJ (1961) Eilmer of Malmesbury: an Eleventh Century Aviator: A Case Study of Technology. Technology and Culture, 2: 97–111.

[16] William of Malmesbury, *Gesta regum Anglorum (The history of the English kings)*. Edited and translated by R. A. B. Mynors, R. M. Thomson, and M. Winterbottom, 2 vols., Oxford Medieval Texts (1998–9).

The attempt would have taken place between 1000 and 1010, while the first edition of the chronicle of William of Malmesbury was written in 1120, almost 100 years later. William is, however, considered a rather reliable source and, because he lived in the same Abbey, was likely able to have collected relatively trustworthy oral accounts of Eilmer's flight.

The glider used by Eilmer must have been of a considerable size to support the weight of the fearless young monk. By calculating the possible trajectory from the Abbey tower, we find that the flight should not have lasted over

Fig. 4.4 Stained glass window dedicated to Eilmer in the Malmesbury Abbey (1920). (This work is in the public domain in its country of origin and other countries and areas where the copyright term is the author's life plus 70 years or less. Wikimedia Commons. CC-BY-3.0. Author: Radicalrobbo. https://upload.wikimedia.org/wikipedia/commons/5/5d/Elmer_flying_monk.jpg)

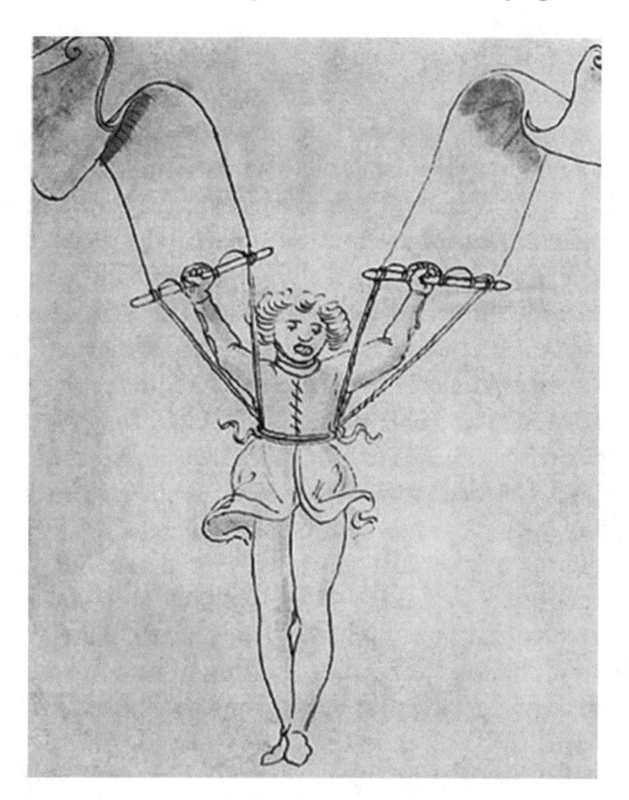

Fig. 4.5 Man with flying wings. Anonymous Sienese engineers. Libro di macchine, Additional Manuscript 34, 113, folio 189v. BL

15 s—nonetheless long enough to ensure Eilmer's primacy as the first flying man in history. This much can be deduced from the documentation.

In 1920, a stained glass window depicting the young Eilmer holding a glider was placed in the Abbey of Malmesbury. The glider looks much like one of the flying machines Leonardo was already famous for at the time (Fig. 4.4).

Returning to the time of Leonardo and the anonymous Sienese engineers, in the same manuscript that carries the first parachute design, it is possible to find yet another interesting drawing (Fig. 4.5). This time, the elegantly dressed flying man uses a rudimentary pair of wings. It is not clear if this system was simply meant to cushion his fall or if it was supposed to allow the man to fly.

It is hard to know if Leonardo ever saw this flying man or the one with the parachute from the same manuscript. Regardless, the existence of this design is a clear indication that the theme of flight had begun to attract the attention of the Renaissance engineers well before Leonardo began his investigations in this area.[17]

[17] Laurenza D (2004) Leonardo: On Flight. Giunti Editore.

Leonardo's Glider

As we have mentioned, Leonardo planned two distinct types of flying machines: the first group mechanically operated and man powered, and the second for free flight. Among these machines there is one model that has attracted particular attention because of its simplicity and seeming modernity.

At the bottom of folio 64r of *Codex Madrid I*, there is a drawing depicting a kind of giant kite with a hanging support system for a man (Fig. 4.6). Some critical points are specifically labeled: *a* and *b* mark the position of the pilot's torso, while *m* is where he would place his feet. Control of the glide[18] is made possible by the presence of two ropes, which are attached at both ends of the glider at the points labeled *s* and *r*. The pilot can change the direction of his flight by pulling these ropes in a controlled manner.

Leonardo did not dwell much on construction details but did care to give instructions about how to use the glider in a little note under the drawing. "The man holds his feet in m and his upper body in a and b. If the wind, which is along the line h, sometimes entering from below… could raise h more than necessary, then the man throws the rope s below. And so he should

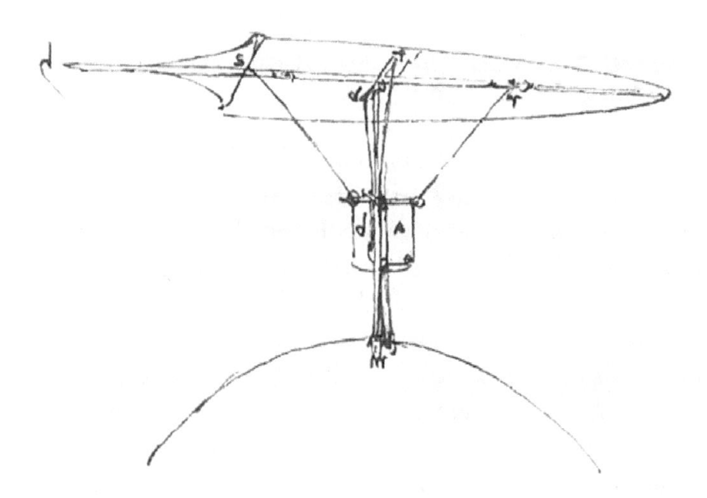

Fig. 4.6 Study for a "hang glider." Leonardo da Vinci. *Codex Madrid I*, folio 64r (detail). BNE

[18] Modern hang gliders use a unique delta-shaped wing which was invented and patented by the American Francis Melvin Rogallo in 1951. It is only since the late 1960s, however, that the Rogallo wing has been used for the construction of hang gliders for gliding free flight. The models of glider from Lilienthal and Leonardo did not have the form of a delta, so they are not, strictly speaking, true hang gliders, but we can, for convenience, give them that appellation.

do with the other strings where necessary, and when he wants to descend pull the rope s to direct toward earth."[19]

This project is simple but also very realistic, and in 2002, the University of Liverpool's Research Group on Science and Technology realized a flying scale model of Leonardo's glider that was carefully tested in a wind tunnel by Robbie Whittall, a famous hang gliding champion.[20] Thanks to the experience acquired through these simulations, the group then manufactured a glider that Whittall tested in Wiltshire, UK, in November 2002, filmed by BBC television (see footnote 14).

The first tests showed the need to add a vertical stabilizer (in the form of a vertical tail) to avoid catastrophic falls, as had already been observed by Eilmer of Malmesbury. This addition is clearly visible in the photograph of the updated model, though it did not, of course, appear in Leonardo's original design. In addition, in the glider used by Whittall, even the system that sustains the pilot is taken from modern hang gliders. These additions, however, do not change the basic idea of the project in which Leonardo dramatically anticipated the idea of free flight.

This is not the only model of a glider that Leonardo designed, but it is perhaps the most modern and advanced in its streamlined simplicity. In his other designs of flying machines, he was unable to move away from the idea of creating systems that somehow mimic the flight of birds. Leonardo perhaps remained trapped by the charm and elegance of this natural model of flight with swinging wings, despite the fact that he clearly understood the principles of flight, as evidenced by his in-depth study of birds, which he carefully recorded in one of his notebooks. Leonardo dedicated much time to the design of artificial wings that reproduced somehow birds' wing structure. These observations and studies cover many pages of Leonardo's notebooks and are part of the ideal of flight rooted in the archetype of Daedalus and Icarus. On the other hand, Leonardo was able to surpass this "primitive" model of flight and displayed incredible intuition in his vision of such projects as the parachute and the gliders—all of which show his capability to imagine the future as very few other people could.

Leonardo recorded a second example of a glider prototype in the *Codex Atlanticus*, this one based on the design of a bird's wing, whose skeleton should be covered with some fabric to allow for gliding free flight.

[19] Original text: "E ll'omo tieni i piedi in m e 'l bussto in a b. E sse 'l vento, che viene per la linia h, qualche volta entrando sotto, a uso di femina colla sua vite, potrebe qualche volta alzare h più che 'l dovere, allora l'omo tiri la corda S in basso. E così facci co' l'altre corde dove bisognia, e quando vol calare, tiri la corda S e dirizzerassi a terra."

[20] http://pcwww.liv.ac.uk/eweb/fst/davinci.html.

A complete project of this glider cannot be found in Leonardo's notebooks. However, in conjunction with a Channel 4 documentary on Leonardo's machines,[21] Martin Kemp, a leading Leonardo expert, used different elements appearing in the codices to create a working model of a glider with a structure based on Leonardo's bird's wing design.[22] The glider was built by a specialized British company using period materials. The flight attempt took place in October 2002 between the hills of the Sussex Downs, just before the one by Whittall discussed above, with another English champion of hang gliding, Judy Ledden, piloting the glider.[23] The hang glider was able to ascend 9 m and to travel a distance of approximately 30 m. The weakness of the model was its handling. According to Ledden, it was a bit like driving a car with brakes and an accelerator, but without the steering wheel. So we see that Leonardo's glider is able to fly without problems, but does not turn easily. By making some small changes and giving the pilot a bit of training, this glider illustrated that Leonardo's idea of flight, at least his intuitions about free flight, are a lot closer to our modern conceptions than we could have ever imagined.

Necessary Sacrifices

After Eilmer, other attempts to jump from towers and campaniles to fulfill the dream of flight were periodically reported over the years, among them that of Giambattista Danti, who reportedly managed to glide from Perugia to Lake Trasimeno in Italy in 1503. The sources recording these attempts are generally unreliable, but they all share a similar description of how the flight ended—usually with a pair of broken legs. At the beginning of the 1800s, the attempts become more realistic, and a short flight was successfully performed by Sir George Cayley,[24] who devised a prototype of a paraglider in 1804.

Leonardo, as we have seen, progressed over time from the idea of a flying machine with swinging wings to that of gliding flight. Several later sources cite attempts at flight performed by Leonardo in Tuscany. However, there is no reliable evidence of attempts to test his ideas practically, and his flights on the hills of Fiesole are probably just legends created over time.

---

[21] Channel 4 documentary (2004) Leonardo's machines.

[22] Martin Kemp, Leonardo da Vinci. Experience, experiment and design. V & A Publications 2011.

[23] http://archiviostorico.corriere.it/2003/gennaio/27/macchina_Leonardo_vola_Dopo_500_co_0_0301272277.shtml.

[24] Ackroyd JAD, Cayley G, (2002) The Father of Aeronautics Part 1. The Invention of the Aeroplane, Notes Rec R Soc Lond 56:167–181 The Royal Society.

It was not until the middle of the nineteenth century that a successful hang glider allowing human flight in a controlled and repeatable way was built. This success belongs to a German inventor from Pomerania, Karl Wilhelm Otto Lilienthal (1848–1896). Much like Leonardo, Lilienthal patiently and accurately studied the flight of birds in order to gain inspiration for an aerodynamic model.

The result of this study is an important book in the history of aviation: *The flight of birds as a base for aviation*,[25] published in 1889. Its similarity to Leonardo's *Codex on the Flight of Birds* is extraordinary.

In Lilienthal's time, lighter than air flying machines, such as hot air balloons, were already being successfully developed, but the idea that it might be possible to fly with something heavier than air was still a subject of doubt. Displaying tenacity and inventiveness, Lilienthal developed different models of "flying kites," which could be steered by shifting one's body weight in a manner similar to modern hang gliders (e.g., Fig. 4.7). These models had some technical limitations, tended to beat down, and were difficult to operate. Compared to modern hang gliders, in which the body hangs from the

Fig. 4.7 Picture of Otto Lilienthal during his flight (around 1895) with his hang glider. (This work is in the public domain in its country of origin and other countries and areas where the copyright term is the author's life plus 70 years or less. Wikimedia Commons. The image has been modified to reduce the noise. https://upload.wikimedia.org/wikipedia/commons/b/b8/Otto_Lilienthal_gliding_experiment_ppmsca.02546.jpg)

[25] Der Vogelflug als Grundlage der Fliegekunst.

frame, in these older models, the glider leans on the shoulders of the pilot, restricting his ability to rotate his body and shift his weight, thereby reducing maneuverability. Despite these limitations, Otto Lilienthal succeeded in making more than 2000 small flights. The longest of these covered a distance of approximately 250 m, though none reached great heights. Nevertheless, these flights were enough to demonstrate the feasibility of planar flight and inspire the Wright brothers in the making of their first airplane.

During one of these flights, on 9 August 1896, a sudden gust of wind caused a wing to break, and Lilienthal dropped about 15 m, causing him to fracture his spine. He died just 36 h later, exclaiming a sentence which remains famous to this day: "Sacrifices must be made!" (*Opfer müssen gebracht werden!*). The Lilienthal Tegel Airport in Berlin is dedicated to his memory, commemorating his boldness and his important contribution to the history of flight. Incidentally, the main airport in Rome is dedicated to Leonardo da Vinci.

Lost Codices

We have seen that the first hang gliders were produced toward the end of the 1860s, but Leonardo had presented the idea of free flight with a fixed wing glider long before that. Is it possible that no one had noticed Leonardo's theory?

Leonardo's drawings on the subject of free flight were located in the *Codex Madrid I* and *II*. In keeping with the usual tangled history surrounding all of Leonardo's manuscripts, these codices apparently vanished into thin air even though there was a historical record of them. Once again, the story passes through Pompeo Leoni. When he moved to the court of King Philip II of Spain in Madrid, Pompeo brought with him the *Codex Atlanticus* as well as many other of Leonardo's manuscripts. Two of these manuscripts must have remained in Spain and, after Pompeo's death, were bought by Juan de Espina, a Spanish gentleman and art collector. Upon his death in 1642, the manuscripts were bequeathed to the King of Spain and housed in the library of the palace, which become part of the National Library of Spain in 1830. At this point, the exact location of the manuscripts was forgotten, and all attempts to locate them failed. Eventually they were considered lost. A trivial classification error hid the manuscripts among the extensive collection of the National Library. In 1964, convinced of the existence of the codices, a very stubborn French scholar by the name of André Corbeau decided that the time was ripe for another attempt. The director of the library was finally convinced to allow him to try, and the lost volumes were found on shelves Aa119 and Aa120, whereas the locations specified by the catalog had been Aa19 and As20. In 1967 the discovery was officially announced to astonishment and disbelief. A printed version of the manuscripts was published immediately thereafter.

Leonardo and the Flight of Birds

As we have noted, Leonardo documented his studies on the flight of birds in a small notebook, which was eventually named the *Codex on the Flight of Birds*. He undertook these studies during his return to Florence in 1501. The *Codex* is composed of 18 pages, *recto* and *verso* (front and back), containing the attentive observations and deductions derived by his studies. This notebook was unfortunately among those looted by Napoleon and taken to France. Here, our famous book thief, Guglielmo Libri, once again comes into play. Libri not only removed pages and pages from Leonardo's various manuscripts, but he also managed to steal the entire *Codex on the Flight of Birds*. Furthermore, he was not satisfied to sell it in bulk, but (probably to maximize his profit) tore off 5 pages—numbers 1, 2, 10, 17, and 18—which he sold separately in London to an English collector, Charles Fairfax Murray. The 13 pages that remained of the *Codex on the Flight of Birds* were sold instead to an Italian nobleman, Count Giacomo Manzoni. When the Count died, this part of the *Codex* passed from one hand to another and eventually made its way to Russia after being purchased in 1892 by a passionate scholar of the Italian Renaissance, Theodore Sabachnikoff. At this point, the greatest part of the damage had been done, leaving the manuscript divided into two parts—one in Russia and the other randomly dispersed in London. Sabachnikoff, however, had a mind to reunify the *Codex* and managed to buy page 18 from Murray. Immediately afterward he published the manuscript, as was his original plan. Once Sabachnikoff had achieved this goal, he made a gift of the *Codex* to Queen Margherita of Italy, who deposited it in the Royal Library of Turin in 1893. There were still four missing pages. Page 17 somehow returned to join the manuscript a dozen years later, while the final three pages were purchased by the Italian collector Enrico Fazio, who donated them to King Victor Emmanuel II of Italy. The manuscript was then reassembled in its entirety and placed in the Royal Library of Turin, where it still resides. The incredible ordeal of *Codex on the Flight of Birds* was then complete, but unfortunately too much time had passed before it could be studied in its entirety and its importance understood.

The travels of the *Codex on the Flight of Birds* do not end here, however. Another trip, no less adventurous, brought it in electronic form all the way to

Mars! It reached the surface of the red planet in 2012 aboard the rover Curiosity, which carries a scanned copy in its electronic memory. Surely, no better homage could have been made to Leonardo's dream of flight.

The Great Kite

The *Codex on the Flight of Birds* has its own peculiarity. It does not contain any explicit drawings of flying machines, such as those scattered across other codices, particularly the *Codex Atlanticus*. Or are we missing something?

Throughout the pages of the *Codex on the Flight of Birds*, there is no unified plan of a machine. This is not unusual: Many of Leonardo's machines designed on paper appear as little more than rough sketches, as in the cases of the parachute and the submarine. Other times, the main parts of a machine are drawn on different pages or even in different manuscripts, which make it very difficult to reconstruct the complex devices planned by Leonardo. Of course, in yet other cases, the description is self-consistent, and the project is described and designed with great detail.

Edoardo Zanon of the Leonardo3 Museum[26] is an expert at reconstructing Leonardo's machines, including complex ones, like the musical instrument *Clavi-Viola*. Zanon succeeded in identifying several sketches across pages of different codices that could be put together like a puzzle. This method enabled the reconstruction of one of the most complex of Leonardo's flying machines, the *Grande Nibbio* (Great Kite).[27] The name is not accidental; in fact, a kite is a large predatory bird uniquely adapted for gliding. However, the kite also holds a particular significance in Leonardo's memory. He recalled a kite from an episode that had occurred when he was still in the cradle: "This writing so distinctly on the Kite seems to be my fate, because in the first memories of my childhood, when I was in a cot, a kite came to me and opened my mouth with his tail and hit me many times with the tail inside my lips" (*Codex Atlanticus*, folio 186v).[28]

Zanon's machine is a glider, in line with Leonardo's final conclusions regarding flight. It is clearly impossible to rebuild everything exactly as it had been conceived by Leonardo, and a certain degree of freedom in reconstruc-

[26] www.leonardo3.net.

[27] Cinzia di Cianini. *Decollo con il Grande Nibbio. L'aliante segreto di Leonardo*. La Stampa, 11 marzo 2009.

[28] Original text: "Questo scriver si distintamente del nibbio par che sia mio destino, perchè nella prima ricordazione della mia infanzia e mi parea che, essendo io in culla, un nibbio venissi a me e mi aprissi la bocca colla sua coda e molte volte mi percotessi con tal coda dentro le labbra."

tion is necessary. Nevertheless, the overall result seems consistent and is very impressive—a great glider with two wings and a tail, operated by a pilot who is free to shift his center of gravity and to move the wings (Fig. 4.8). The dimensions suggested by Leonardo were 30 arms for the wingspan, (about 18 m) with the center of gravity positioned 4 arm's length below.

The wings of the kite are covered with canvas and mimic the wing structure of a bat. With this configuration, the machine reaches a wing area of 110 m^2, larger than in modern hang gliders, guaranteeing, in principle, an advantageous wing loading. This flying machine has never been tested, but the reconstruction was exhibited in 2009.

Fig. 4.8 The reconstruction of the "Grande Nibbio" (Great Kite) designed by Edoardo Zanon of Leonardo3. (This file is licensed under the Creative Commons 3.0 unreported. Courtesy of Edoardo Zanon. https://upload.wikimedia.org/wikipedia/commons/6/6e/Leonardo3-CodiceDelVolo-GrandeNibbio-GreatKite.jpg)

Fig. 4.9 A man covered with "air bags." Leonardo da Vinci. *Codex on the Flight of Birds*, folio 16r (detail). BRT

It is quite curious to observe Leonardo's attention to the rider's safety. Leonardo must have written his notes on flight with the conviction that they would be published and that someone would sooner or later try to build his flying machines by following his instructions. Therefore, he even noted a suggestion for a kind of air bag: "Wineskins through which a man falling from a height of 6 feet can avoid injury, whether he falls into the water or on the ground. These containers must be tied carefully all around the body."[29]

The note is accompanied by Leonardo's stylized drawing of a man covered in small wineskins inflated with air in the upper right corner of folio 16r of *Codex on the Flight of Birds* (Fig. 4.9). The small drawing is roughly sketched but well represents Leonardo's idea to protect those who fly from their own rashness.

Ornithopters

Human flight in imitation of birds is one of the most basic ideas of flight. It is seemingly the simplest model but probably the most difficult to translate into a working machine. The flight of birds has always held a profound fascination for man, and the first attempts to fly, beginning with the myth of Icarus, were inspired by birds. Naturally, the first machines for flight were designed with the movements of birds in mind. Machines with flapping wings that mimic, or

[29] Original text: "Otri di cuoio tramite i quali un uomo che cada dall'altezza di 6 piedi possa evitare di farsi male, sia che cada in acqua o sul suolo. Questi otri debbono essere legati con attenzione tutto intorno al corpo."

attempt to imitate, the flight of birds are defined as *ornithopters*, from the Greek *ornis* (bird) and *pteron* (wing). Long before Leonardo, another great European visionary erudite, Roger Bacon (c. 1220–c. 1292), predicted the *ornithopter*, among a variety of other visionary machines: "Ships will go without rowers and with only a single man to guide them. Carriages without horses will travel with incredible speed. Machines for flying can be made in which a man sits, and skillfully devised wings strike the air in the manner of a bird. Machines will raise infinitely great weights, and ingenious bridges will span rivers without support."[30] Some centuries later, Leonardo went beyond simple description and began to actually design concrete engineering projects with the goal of realizing machines such as those envisioned by Bacon.

Ornithopters and Other Strange Flying Machines

As we have discussed, during his first period dedicated to studies of flight, Leonardo focused on the design of *ornithopters*, envisioning possible flying machines with flapping wings that were propelled by human muscle power, which would enable them to hover in the air. At that time, he was also studying mechanics and human anatomy, so it was quite natural for him to transfer this knowledge to the designs for flight. Leonardo's *ornithopter* models were essentially based on two different configurations, which differed in the position of the human pilot: the *vertical ornithopter* and the *horizontal ornithopter*.

The vertical *ornithopter* is a practically impossible machine but still appears as one of the most fascinating and futuristic of Leonardo's intuitions. Looking at it very carefully reminds one of a drawing of a spaceship, with stairs to go up and down and a hull for hosting the pilot (Fig. 4.10). The *ornithopter* is operated by a man in vertical position by means of a complex mechanical system of power transmission. He must activate the two wings by moving his hands, feet, and even head. "The movement of the wings will be crossed in the manner of horses. For this reason, I believe that this method is better than any other,"[31] Leonardo explained.

As we know, Leonardo developed a strong theoretical background in flight thanks to his observation of birds. These studies initially convinced him of the feasibility of birdlike flying machines. His basic understanding of what would later be known as the third law of motion convinced him of the possibility of human winged flight. In the *Codex Atlanticus*, he clearly described the basic

[30] Bacon R, De secretis operibus artis et naturae IV.

[31] Original text: "Il movimento delle ali sarà incrociato alla maniera dei cavalli. Per questo motivo ritengo che questo metodo sia il migliore rispetto a qualunque altro." Manuscript B, folio 80r.

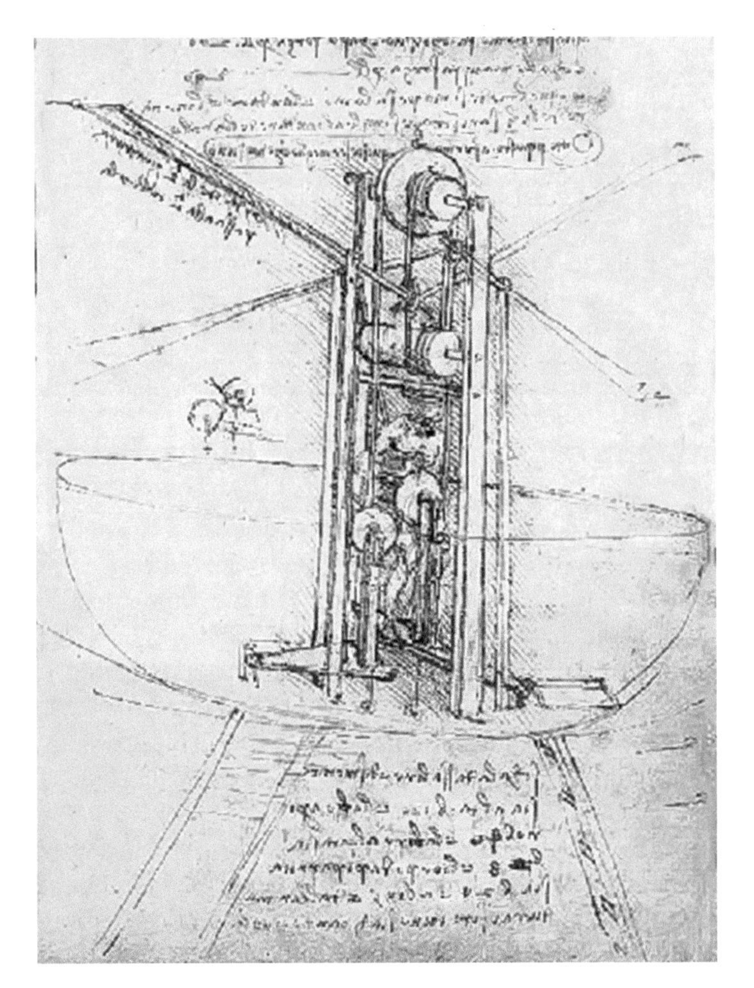

Fig. 4.10 Project for a vertical ornithopter. Leonardo da Vinci. Manuscript B, folio 80r, (detail). BIF

principle of flight based on the principle of action and reaction: "The same force that one object exerts on the air will be exerted in turn from the air against the object."[32]

Leonardo clearly had in mind that the downward movement of a bird's wing causes compression of the air, which in turn produces an upward push able to support the bird in flight. He well described what he assumed was the main enabler of bird flight: "Look how the wings, striking against the air, support the heavy eagle…. For these demonstrated reasons you can know the

[32] Original text: "Tanta forza si fa colla cosa in contro all'aria, quanto l'aria contro alla cosa." Codex Atlanticus, folio 1058v.

man who, with his large, contrived wings, presses on the air that resists, and winning such resistance, can subdue (the air) and fly above it."[33]

Leonardo's notes illustrate the great difference between him and those who had come before him, including the creative Sienese engineers. Leonardo's innovation was not limited to the machines themselves but also explained how those machines were envisioned and designed. In the case of flight and in many others, he developed a well-defined conceptual method for the observation of natural phenomena, their analysis in scientific terms, and the application of this data to engineering projects.

There is another illustration of one of Leonardo's flying machines that is very impressive, because it resembles so much the design of a spaceship as we might see in modern movies (Fig. 4.11). It is located on folio 89r of *Manuscript B*, where two overlapping sketches show a flying machine before and after taking off. The machine is clearly meant to takeoff vertically, but one is there-

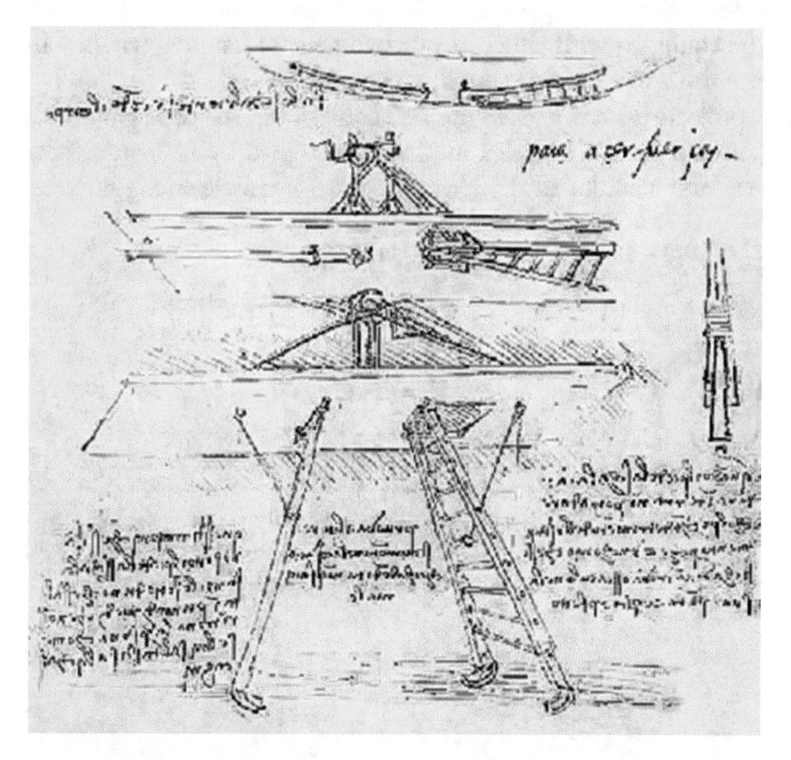

Fig. 4.11 Study of a mobile system for getting into the flying ship. Leonardo da Vinci. Manuscript B, folio 89r, (detail). BIF

[33] Original text: "Vedi l'alie percosse contro l'aria fa[r] sostenere la pesante aquila sulla suplema, sottile aria vicina all'elemento del fuoco. ....potrai conoscere l'uomo colle sue congegnate e grandi ale, facendo forza alla resistente aria e vincendo, poterla soggiogare e levarsi sopra di lei." Codex Atlanticus, folio 1058v.

fore faced with the problem of how to enter into the ship with a movable system. The solution is simple; a staircase is used to climb up to the platform which, immediately after takeoff, could be retracted with a rope, as explained in Leonardo's descriptive note: "And after taking off, pull up the stairs, as I show in the second upper figure."[34] Ultimately, Leonardo's system is not so different from the one used in modern airplanes.

This idea of a flying machine with a vertical structure was later abandoned in favor of the *horizontal ornithopter*. As discussed, in his studies on flight, Leonardo moved from complex models to simpler projects and finally to the idea of free gliding flight.

An example of the evolution of the idea of the *ornithopter* can be found on folio 79r of *Manuscript B*, where one can see a flying machine controlled by a man who is stretched out horizontally (Fig. 4.12). The pilot, through alternate movement of hands and feet, is able to transmit motion to the wings via a complex system of ropes and pulleys. Leonardo noted that this machine could be equipped with either one or two pairs of flapping wings: "This can be done with a pair of wings but also with two."

The projects for *ornithopters* are fascinating both for their graphical representation and for the complex mechanics designed by Leonardo. Leonardo, however, later abandoned the idea that a man's muscle strength alone could

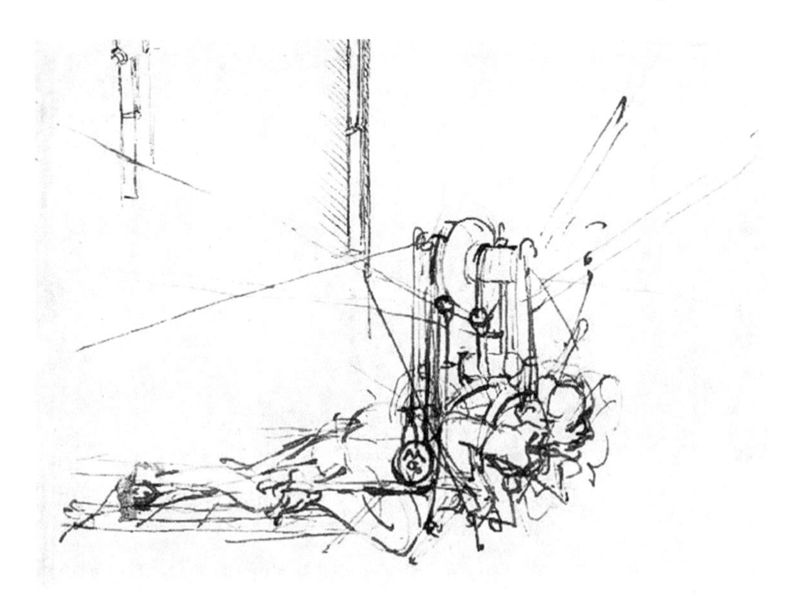

Fig. 4.12 Horizontal ornithopter. Leonardo da Vinci. Manuscript B, folio 79r (detail). BIF

[34] Original text: "E quando s'è levato, tira le scale in alto, com'è dimostro nella seconda fugura di sopra."

power the flight of a machine. He gradually realized that the elevated weight of the machine itself constituted an insurmountable obstacle to such a model.

Leonardo Was Right After All

From Leonardo's time onward, the conviction that human-powered flight was impossible became deeply rooted. However, the discovery in the 1960s of ever lighter and stronger materials opened up new possibilities.

In 1957 an American industrialist, Henry Kremer, officially opened a challenge and set out a series of prizes for those who were able to beat certain records for human-powered flight. It wasn't until 1977 that the first Kremer prize of £50,000 was awarded to Paul MacCready, whose Gossamer Condor was able to complete a figure-eight course covering a total of 1 mile, as established by the jury.[35] Notably, the human-powered aircraft was even capable of taking off by itself. MacCready's Condor was driven by a pedal system that moved a rear propeller able to exploit the huge wingspan (29 m) and low weight (32 kg) achieved through the use of lightweight materials. An amateur cyclist in a transparent gondola under the wings could move the aircraft with human muscle power alone.

After the success of the condor, which had flown for but a short distance and at a modest height, MacCready posed an even more ambitious goal, winning the Kremer Prize for a human-powered aircraft able to fly over the British Channel. On 12 June 1979, his new aircraft, the Gossamer Albatross, succeeded in traversing the 35.8 km channel in 2 h and 49 min, reaching a top speed of 29 km per hour at a height of 1.5 m (Fig. 4.13).

Leonardo's dream was finally realized. As in so many things, he had simply been too far ahead of his time, and the materials available to him did not allow a practical exploitation of his designs. Moreover, it seems as though he may not have been interested in building working models of his machines, all of which, to our knowledge, remained confined within his fertile imagination. Nevertheless, it is always surprising that, though Leonardo never tested any of his models, most of their replicas seem to work pretty well or to anticipate ideas that would later be transformed into real machines.

What about *ornithopters* with human-powered flapping wings? This seemed an almost impossible challenge. However, in 2006 a group of engineering students at the University of Toronto in Canada initiated a very ambitious

[35] Grosser M (1991) Gossamer Odyssey The Triumph of Human-Powered Flight. Riverwash Books. New York Dover.

Fig. 4.13 The NASA's Gossamer Penguin during the flight above Rogers Dry Lakebed at Edward in California. This version uses also a solar panel perpendicular to the wing to support the flight. July 25, 1979, NASA Photo Courtesy of Bob Rhine

Fig. 4.14 Aerovelo's Snowbird during a human-powered flight

project inspired by Leonardo's ideas. After few years, in 2010 they succeeded in constructing a working model of an *ornithopter* with flapping wings, nicknamed Snowbird.[36] It had a total span of 32 m and a weight of only 44.7 kg and could reach the speed of 25.6 km allowing a flight longer than 20 s (Fig. 4.14). The flexible wings, the low weight of the materials, and the aero-

[36] http://www.aerovelo.com/our-projects/#ornithopter-summary.

dynamic design were the main elements that contributed to the success. Even when he doubted himself, Leonardo was frequently right.

In the story of *ornithopters*, yet another amazing machine must be mentioned. It is the wing-beating, foot-propelled flying machine realized by the American aviator George R. White in the 1920s. White built his *ornithopter* with a chrome molybdenum frame covered with a nonflammable transparent cellulose fabric. After several attempts, he was able to cover a distance of almost a mile in 1928. The surprising part of this story is that if we look at a picture of his *ornithopter* (Fig. 4.15), we immediately notice its similarity to Leonardo's project shown in Fig. 4.10. George White was ignorant of Leonardo's design, but he had almost the same idea 400 years later. And, much like the Tuscan master, he could not stop pursuing his dream of flight.

Fig. 4.15 The ornithopter of George R. White, 1928. (This work is in the public domain in its country of origin and other countries and areas where the copyright term is the author's life plus 70 years or less. https://commons.wikimedia.org/wiki/File:Ornithopter_and_creator_George_R._White_at_St._Augustine.jpg)

The Flying Ball

Leonardo's flying machines do not even end here, however. In addition to the famous screw that somehow anticipated the idea of the helicopter, there is another system that is perhaps less well known but no less fascinating: the flying sphere (Fig. 4.16).

The design for this machine is on the same folio of the *Codex Madrid I* as the sketch and commentary on the glider. Leonardo briefly describes his strange idea: "Make a similar fan, as is presented here and is composed of sandalwood, rods and poles with a diameter of 20 braccia (see footnote 1) or more. In the center should be placed a perforated ball made of green circles of elm wood. And this ball with these circles works like the magnet of a compass, and there is a man in the middle of the ball. This instrument should be placed on a hill, in the wind, and it will follow the winds, and the man will always remain in a standing position."[37]

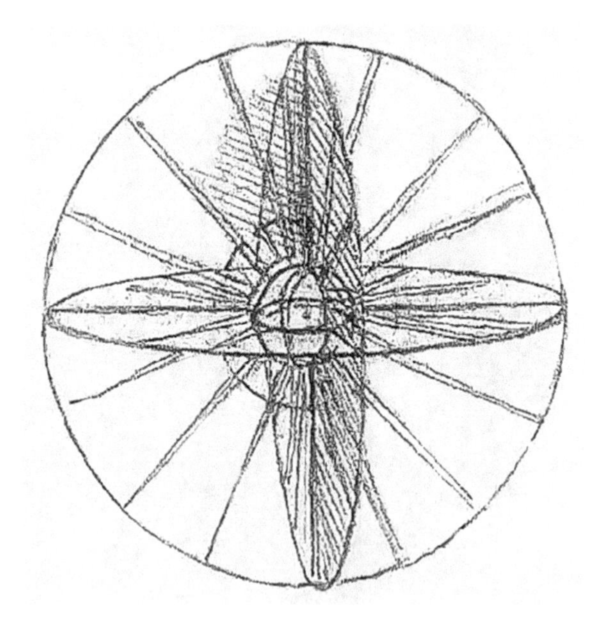

Fig. 4.16 The flying sphere. Leonardo da Vinci. *Codex Madrid I*, folio 64r (detail). BNE

[37] *Codex Madrid I*, folio 64 r. "Sia fatta una simile ventola, come qui è figurato, e ssia composta do zendado, corda, canne e asste, di diamitro di bracci 20 o più. Nel mezo della quale sia collocata una balla traforata, fatta di cierchi verdi, che ssieno d'olmo. E ssia detta balla con tali cierchi aconcia a uso di bussola di calamita, e nel mezo d'esa balia stia un olmo. E ssia tale strumento collocato sopra uno monte, al vento, e ttale strumento s'aconpagnierà dorso de venti, e ll'omo senpre starà in piedi."

The presence of this machine on the same page as the sailplane is probably not accidental, as it represents yet another possible method of free flight. The flying sphere, constructed in accordance with Leonardo's plan, would reach the impressive size of almost 12 m in diameter and would be made of three wooden circles with some spokes for support. A standing passenger is accommodated at the center of the system in a perforated sphere that acts much like a magnetic needle. Once in the air, this machine would fly while keeping the passenger always standing, as in a gyroscope. This flying device has not yet been tested, but simulations performed on a scale model have shown that there might be problems of instability.[38] Nevertheless, the aesthetic fascination exercised by a flying man at the center of a precisely defined geometric shape makes it one of the most intriguing of Leonardo's flying machines.

[38] Bartoli G, Borsani A, Borri C, Martelli A, Procino L, Vezzosi A (2009) Leonardo, the wind and the flying sphere. Proceedings of EACWE 5, Florence, Italy.

5

Flotation, Walking on Water, and Diving Under the Sea

The artist-engineers of the Renaissance, including Leonardo, had a special passion, even obsession, for the natural element of water. They dedicated a lot of time to hydraulic studies and to designing innovative machines using water as the driving force. This attraction to water should not come as a surprise. After all, aside from animal power, water was the only known continuous source of mechanical energy available at the time. Water pressure is capable of operating a variety of different types of mechanical devices, including water mills. Water mills were fundamental tools during the Middle Ages, and the survival of a community could be highly dependent on their ability to make such mills work properly. Water is also a natural element that must be carefully controlled, for instance, by dams and sluices, and it is a fundamental medium for the transportation of goods and people.

Water was an extraordinary source of inspiration for the design not only of innovative hydraulic machines but also for pleasure and fun. Leonardo was very much attracted by the endless potential of this natural element. He carefully observed storms, vortices, and the flow of water in rivers, filling his notebooks with numerous sketches and notes. At the same time, he studied the ships and hydraulic machines designed by the Sienese engineers.

Devices for Flotation

Water, however, could also be a very hostile element, and, at the time, sailing and fording a river could be very risky activities. Shipwrecks were quite frequent, and any unfortunate person who fell into the sea was likely to lose his or her life. Very few people were able to swim, so systems that allowed flotation

P. Innocenzi, *The Innovators Behind Leonardo*, https://doi.org/10.1007/978-3-319-90449-8_5

for as long as possible were necessary. This was the impetus for the first prototypes of the life preserver. Some possible solutions had already been envisioned before Leonardo's time, even if his life preserver is the most famous today. As with so many of Leonardo's so-called "inventions," his sketch is an almost exact copy of models already imagined by Taccola and Francesco di Giorgio.

Francesco di Giorgio, in particular, solved the problem of allowing a man to float in water by proposing a belt of light material. This idea is well represented in an elegant drawing (*Manuscript II.I.141*, BNF), where he sketched a man wearing a "life belt" (Fig. 5.1a). Leonardo reproduced exactly the same drawing (*Manuscript B*, folio 81v) and was later credited with its invention (Fig. 5.1b). The similarity between the two sketches is remarkable, leading to the conclusion that Leonardo's is a stylized copy. Even the body position

Fig. 5.1 Different types of flotation devices. (**a**) Swimmer with oar and life preserver (*top left*). Francesco di Giorgio. Ms. II.I.141, f. 196v. BNCF. (**b**) Swimmer with life preserver (*top right figure*). Leonardo da Vinci, Manuscript B, folio 81v (ca. 1487–1490). BIF. (**c**) Swimmers with life-preserving devices (*bottom left*). Francesco di Giorgio, Ms. 197.b.21 folio 49v. (BML). (**d**) Swimmers using different flotation devices (*bottom center*). Anonymous from Taccola and Francesco di Giorgio. Ms. Palat. 767, folio 10. BNCF. (**e**) Man with a life preserver and an oar who plays a cornet (*bottom right*). Paolo Santini (from Taccola) Ms. Lat. 7239. folio 82v. BNF

Fig. 5.2 Swimmer with life preserver, second version. Leonardo da Vinci. Codex Atlanticus, folio 784r. BAM

assumed by the two swimmers is identical, though Leonardo's sketch (likely a draft) appears much simpler.

This is not the only sketch of a man with a life preserver produced by Leonardo. An almost identical drawing, without any accompanying note, can be found on folio 748r of the *Codex Atlanticus*. It is also a rough sketch with the swimmer assuming exactly the same position, showing that Leonardo was very attracted to the idea of a life preserver (Fig. 5.2).

The purpose of Leonardo's life preserver is made clear by a small note above the drawing (folio 81v, *Manuscript B*), which briefly describes the object's function: "way to save oneself during a tempest." Interestingly, the note is written with a different ink and in the conventional way from left to right, rather than in Leonardo's mirror writing. It is reasonable, therefore, to assume that the note was added at a later date by someone other than Leonardo.

On the same page, just beneath the sketch of the swimmer with the life preserver, Leonardo described another system to save one's life during a shipwreck. As in several other cases, he used no sketches. However, the note describes invention in detail: "Way to save oneself during a tempest and a shipwreck. It is necessary to wear a leather dress with a double layer which has the thickness of a finger. This dress should be done with waterproof leather and should be double from the girdle to the knee. When it will be necessary to jump into the sea, you have to inflate the dress, then jump and let yourself be driven by the sea waves. Keep always in your mouth the pipe to inflate the air which goes to the dress, and when you cannot take breath because of the sea waves you can inhale the air from the dress".[1]

This description indicates that Leonardo also had in mind an alternative device—a leather "dress" with a waterproof cavity that could be inflated with air by blowing from the mouth. The dress could also be used as an air reserve when the sea waves were too strong to breathe regularly. Leonardo described something which is almost similar to our modern inflatable life jackets (sometimes termed "Mae West" style life preservers). However, it took another 400 years before a patent in 1928 described such a device.[2] This jacket became part of the essential standard equipment of British and US pilots during World War II.

However, neither the life preserver nor the inflatable life jacket was originally Leonardo's idea. While the life preserver, as we have already seen, was taken from Francesco di Giorgio, the idea for the jacket was borrowed from the description and illustrations in Roberto Valturio's *De Re Militari* (*On military things*), in particular from Liber (Book) XI, which is dedicated to naval warfare.

Roberto Valturio

Roberto Valturio (1405–1475) has a particularly important role among the "other" innovators because he was one of the main sources of inspiration for many engineers of the Renaissance, such as Francesco di Giorgio and Bonaccorso Ghiberti. Leonardo owned a personal copy of *De Re Militari*, from which he took several hints and ideas. Indeed, Leonardo had a rather rich collection of books that included important treatises of mathematics and geometry. Lists of his books are reported in *Codex Trivulzianus* 2162, folio 3r; *Codex Atlanticus*, folio 559r; and *Codex of Madrid II*, folios 2v-3r. In particular, the *Codex of Madrid II* list, probably compiled between 1490 and 1492, includes the name of Valturio's book, *De Re Militari*. It is listed under a heading entitled "memory of the books that I leave locked in the drawer."[3]

[1] Original text: "Modo di salvarsi in una tempesta e naufragio marittimo. Bisogna avere una vesta di corame (cuoiame) ch'abbi doppi i labbri del petto per ispazio di un dito, e così sia doppio dalla cintura insino a ginocchio; e sia di corame sicuro dello esalare (cuoiame impermeabile). E quando bisognassi saltare in mare, sgonfia (in realtà si intende gonfia in italiano moderno) per li labbri del petto le code del tuo vestito, e salta in mare e lasciati guidare all'onde, quando non vedi visina riva, né abbi notizia del mare. E tieni sempre in bocca la canna dell'aria che va nel vestito, e quando per una volta o 2 ti bisognasse torre dell'aria comune e la schiuma t'impedissi, tira per bocca di quella del vestito."

[2] Peter Markus (1885–1974) in 1928 received a patent describing an inflatable life preserver (US Patent 1,694,714).

[3] Original text: "Ricordo de' libr[i] ch'io lascio serati nel cassone." Codex of Madrid II, folio 559r.

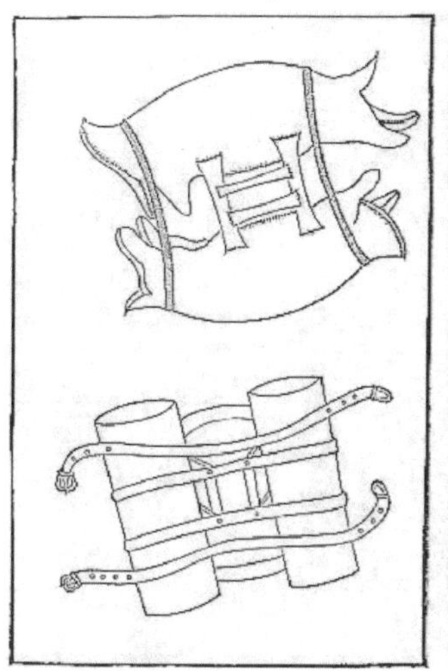

Fig. 5.3 Flotation devices (*left*), pg. 318; inflatable life jacket (*right*) pg. 319. Roberto Valturio, De Re Militari, Book XI. (1532). BSBM

As we have said, in *De Re Militari*, one can find a detailed drawing (Fig. 5.3) that perfectly matches Leonardo's own description of the inflatable jacket. The illustration of the system is very clear, and although Leonardo was not very good at Latin, he could have easily understood the purpose of the jacket from the explicit language and imagery used by Valturio.[4] The figure shows a soldier who, before crossing a river, blows air into an inflatable jacket strapped to his body. Two types of flotation devices appear in the figure. The first is simpler: two wineskins or cylinders inflated with air to allow for flotation. The other corresponds to the description of Leonardo's jacket.

Francesco di Giorgio, proposing alternative solutions, drew two naked swimming men, one using a life preserver and the other a pair of floaters (Fig. 5.1c, d). The same drawing, with some modifications, can be found in an anonymous copy of this manuscript (Ms. II.I.141) and in the colored version splendidly illustrated by Paolo Santini (Fig. 5.1e). In the latter copy in particular, the man wearing the life preserver is playing a cornet during his attempt to cross the river, while in Francesco di Giorgio's original copy, the

[4] The editio princeps (first edition) of Valturio's *De Re Militari* was published in Verona in Latin in 1472, followed by a second edition in 1482. The first printed edition in Italian was printed by Paolo Ramusio, again in Verona, in 1483. Other editions in Latin were printed in Paris in 1532 and 1534, and the copy translated in French was published in 1555 (translation of Loys Meigret). Leonardo da Vinci likely consulted at first the Latin version of the book and only later the Italian translation. The list of Leonardo's books in the *Codex Atlanticus* with the reference to *De Re Militari* was likely compiled between 1490 and 1492, when the Italian translation version was already available, but it is difficult to establish which version Leonardo had.

Fig. 5.4 Navicula axillaris of Fausto Veranzio, in De Machinis Novae. 1606

men are always naked and in a statuesque pose. Other than in Leonardo's sketch, all the men with life preservers use an oar, which suggests that most of these solutions were envisioned as providing support for swimming as opposed to a purely life-saving purpose.

It seems, looking at Francesco di Giorgio and Valturio's illustrations, that they had in mind another very practical problem—fording rivers. At the time, a river represented an almost insurmountable barrier, especially for armed soldiers. This is also quite clear in the illustration (Fig. 5.4) of a later inventor, Fausto Veranzio, whom we have already met in the previous chapter about the invention of the parachute.

Veranzio called his belt *navicula axillaris* (underarm small boat) which resembles very much a combination between the modern fisherman's boots and an inner tube. The proposed use of Veranzio's flotation device is indeed for river fording, if possible without getting wet, using the hands as oars to reach the opposite bank (Fig. 5.4).

Walking on Water

Leonardo borrowed more than just the concept of the life preserver from Francesco di Giorgio. There is a second such idea which is quite fantastical, at least from our modern point of view: that of walking on water. In this case as well, Leonardo's sketch seems to reproduce Francesco di Giorgio's original drawing with few modifications (Fig. 5.5).

Fig. 5.5 (**a**) Men walking on water (*left*). Francesco di Giorgio, Ms. 197.b.21 folio 55v (BML). "Way of walking on water". (**b**) Leonardo da Vinci. *Codex Atlanticus*, folio 26r (ca. 1480–1482), detail (*right*). BAM

Francesco di Giorgio imagined that, using barrels or other similar flotation devices, it could be possible in principle to move or walk on the water's surface without sinking (Fig. 5.5a). Leonardo liked this idea and made a slight modification by envisioning a kind of water skier with sticks and snowshoe-like footwear that could support a walk on water (Fig. 5.5b). Francesco di Giorgio's drawings are really beautiful, especially in comparison to Leonardo's simple sketch. The latter was perhaps just meant to be a memo of an idea that Leonardo intended to use later. So as not to leave any ambiguity about the meaning of the small drawing, Leonardo added the annotation: "way of walking on the water."[5]

Divers: Maritime Sabotage from Below

Besides floating and walking on water, it was also important to be able to dive in a safe and controlled manner for fishing and other underwater activities, such as covert attacks on enemy ships. This practical need started to be taken into consideration as early as the beginning of the fifteenth century, at least according to remaining documents (*Bellifortis* and the *Manuscript of the Hussite Wars*). A variety of solutions were proposed. They may now appear

[5] Original text: "modo di cammina[r] sop[r']acqua."

naive, but, as we will see, they contained the basic ideas that would later result in much more complex devices. Leonardo, starting from these early-stage models, used his own distinctive, rigorous, and technical approach to obtain working solutions.

Leonardo's famous surface-supplied diver (Fig. 5.14), which we will discuss in more detail below, is of particular interest because his unique approach clearly stands out in comparison to previous solutions. As in the previous example of the life preserver, Leonardo probably got the basic idea from Taccola and Francesco di Giorgio's manuscripts. Leonardo, however, was able to go beyond their efforts in this case, turning the idea into a complex design with an unprecedented level of technical detail and precision. This has allowed modern researchers and enthusiasts to realize and test an accurate reconstruction of the device based on his drawings. Nobody before Leonardo had reached the graphical ability to represent engineering projects in such detail.

The notion that a man could move and even work under water dates back to the Greeks and Romans. But it was not until the Middle Ages that the first, elementary practical tests and attempts started to appear.

These are represented in several picturesque illustrations in different manuscripts. One such example can be found in the *Bellifortis* of Konrad Kyeser and another in the so-called *Manuscript of the Hussite Wars* (Figs. 5.6, 5.7, and 5.8), two important treatises that describe technologies and machines for warfare, very much rooted in the techniques and traditions of the Middle Ages but somehow also foreshadowing the times to come.

In the context of the Renaissance artist-engineers' deep preoccupation with warfare, the possibility of diving and working under the sea was generally conceived as a military technology that could offer new possibilities, such as the sinking of enemy ships. This was exactly the approach taken much later from the nineteenth century, when the technology was finally developed, but, fortunately for us, today diving is primarily popular as a peaceful, leisure activity.

During the Renaissance, the idea of surface-supplied and scuba divers was therefore triggered by the need to secretly attack enemy ships. After proposing this possibility, however, Leonardo was somehow terrified by the idea of extensive use of his invention and explicitly wrote that he had no intention of revealing the details of his project: "It is not possible to stay under the water if you cannot breathe. Many can stay with this instrument under the water. How and why I will not write my way to stay under the water, as such as I (will not reveal) how long I can stay without eating. This I do not publish or

Fig. 5.6 Three different solutions for jackets to be used as air reserve. Konrad Kyeser. "Bellifortis" (1402–1405), folios 114, 115, and 116 from left to right. ULJCS

Fig. 5.7 Divers from "Bellifortis" of Konrad Kyeser (1402–1405), folios 123, e 124 from left to right. ULJCS

make known due to the bad nature of men. They will use [such knowledge] for killing under the sea, for breaking ships, and will let them sink with the men who are inside. Even if I teach other methods, they are not dangerous because the top of a cane will appear above the water, where they breathe, and

Fig. 5.8 Copy of the inflatable life preserver from Roberto Valturio, *De Re Militari*. Philipp Mönch. *Kriegsbuch*. Manuscript Pal. Germ. 126, Universitätsbibliothek Heidelberg

is put on wineskins or cork."[6] As we can see from this comment and others, though Leonardo designed many war machines, he had a peaceful attitude and probably never actualized any of his military projects.

As we have mentioned, one of the very first representations of a diving technology is found in the *Bellifortis* of Konrad Kyeser.[7]

[6] Codex Arundel. Folio 22v. Original text: "Come è non si po star sotto l'acque, se non quanto si po ritenere lo alitare. Come molti stieno con istrumento alquanto sotto l'acque. Come, e perché io non iscrivo il mio modo di star sotto l'acqua quanto i' posso star senza mangiare; questo non pubblico o divolgo per le male nature delli omini, li quali userebbono il assassinamenti ne[l] fondo de' mari, col. rompere i navili in fondo, e sommergerli insieme colli omini, che vi son dentro (per); e benchè io insegni delli altri, quelli non son di pericolo, perché disopra all'acqua apparisce la bocca della canna, onde alitano, posta sopra li otri o sughero."

[7] White L (1969) Kyeser's "Bellifortis": The First Technological Treatise of the Fifteenth Century. Technology and Culture 10: 436–441.

Konrad Kyeser

Kyeser was not in fact an engineer but a medical doctor who led a very adventurous life. He was born in Eichstatt in 1366 and went on to practice medicine for a time at the court of the Signore of Padua. Kyeser was later seduced by the call of adventure and joined a crusade against the Turks. The expedition ended in a military disaster with the battle of Nicopolis,[8] which the crusaders lost in 1396. Kyeser blamed this defeat on Sigismund, King of Hungary and, later, Holy Roman Emperor. This was not a wise choice because, soon after Sigismund managed to take over the region of Bohemia (where Kyeser had established himself after returning from the crusade), he began to search for any opponents and detractors. The same King Sigismund who was seen by Taccola as the savior of Siena represented for Kyeser the cause of his ruin.

Kyeser was condemned to exile in the mountains of Bohemia, where he dedicated himself to the task of writing his treatise on warfare. By sheer coincidence, some unemployed artists arrived in the village to which he was confined, likely to look for jobs. They are the likely illustrators of the beautiful colored sketches that can be found in the various copies of Kyeser's treatise. The drawings look simple, still rooted in a pre-Renaissance age, but they are fascinating in their naïve attempt to illustrate the new technologies and war machines conceived by Kyeser.

Kyeser died in exile in 1405 before finishing his manuscript, which nevertheless became very popular, especially in Germany where several copies were later made.

The charm of *Bellifortis* is its mixture of military technologies and techniques with elements of magic and astrology. In this way, the manuscript, which contains some interesting innovative ideas, is still very much rooted in the culture of the Middle Ages. It would be necessary to wait for the Renaissance to step away from this old world approach.

Among the many colored illustrations in the *Bellifortis* dedicated to war machines and warfare, there are some interesting sketches that show a life jacket with an air reservoir (Fig. 5.6). All of these end with a mouthpiece. The similarity to Valturio's system and to the one described in Leonardo's notebooks is clear.

However, this is not the only solution Kyeser developed for moving and breathing under water. Two additional drawings show a scuba diver with a mask and a small floating device from which to breathe just under the surface of the water (Fig. 5.7). The colored sketches do not leave any room for doubt: The devices were to be used to cross small rivers and reach an enemy castle unnoticed.

---

[8] The battle of Nicopolis was fought on 25 September 1396 between a European army, mainly composed by Germans, French, Hungarians, and Bulgarians, against the Turkish army. The battle finished with a heavy defeat of the European army and the end of the crusade against the Turks.

The connections between the German school and its protagonists, such as Konrad Kyeser, and the engineers of the Italian Renaissance are still largely to be explored. Several manuscripts show that the German school remained alive and well following Konrad Kyeser's contributions. Some of this original work, especially in the development and use of firearms, was of high quality. What is clear is that Valturio's work played a pivotal role in bridging the German and Italian schools, especially concerning the art of war. Leonardo was not the only person to study Valturio's treatise in detail and to borrow his ideas to reproduce or develop new machines and systems, such as the inflatable life preserver. The large diffusion of the printed versions of *De Re Militari* facilitated the spread of Valturio's ideas far beyond Italy's borders. In Germany, we can find an almost exact copy of Valturio's life preserver in Philipp Mönch's *Kriegsbuch*,[9] dated 1496, 24 years after the first printed edition of *De Re Militari* (Fig. 5.8). Mönch likely used his manuscript as a kind of catalogue of skills to offer his services to a powerful German patron.

As is evident from the figures we have seen, the illustrations clearly show the differences between the German and Italian schools in terms of graphical representation, technical skills, and creativity.

The Surface-Supplied Diver

It is in the so-called *Anonymous of the Hussite Wars*[10,11] that the idea of the surface-supplied diver is first documented. The drawing, which is quite simple, shows a man under the water's surface dressed in a waterproof suit. He breathes by inhaling air through a long and flexible pipe that extends above the top of the water (Fig. 5.9).

[9] Manuscript Pal. Germ. 126, Universitätsbibliothek Heidelberg, Germany. The digitalized version is available online at the site of the library.

[10] Munich Codex Latinus Monacensis 197.

[11] Hall BS (1979) The Technological Illustrations of the So-Called *Anonymous of the Hussite Wars*: Codex Latinus Monacensis 197,1. Ludwig Reichert Verlag, Wiesbaden.

The *Anonymous of the Hussite Wars*

Notwithstanding its name, this manuscript is unrelated to the Hussite Wars (1420–1436)[12] and was compiled by two distinct authors. The treatise is composed of two parts, which both primarily describe techniques for warfare. The title comes from the manuscript's two references to the Hussite Wars. The first part of the book was likely written around 1475 and the second between 1480 and the beginning of 1490.

This dating suggests that the two authors were born after Taccola and Giovanni Fontana. The style, however, is similar to that of these two Italian authors, although in reality, the authors were actually almost contemporaries of Francesco di Giorgio and Leonardo da Vinci.

A comparison between these manuscripts illustrates the level of innovation which was occurring in Italy as compared to the rest of Europe. The German school was not able to catch up to the ongoing technological revolution beyond the Alps.

Nonetheless, the popularity of the *Anonymous of the Hussite Wars* was enormous. This is a very critical point and provides a lesson to be learned. The diffusion of Leonardo and the Siena engineers' work remained rather limited. In this, we see that sometimes an idea's opportunities for diffusion can be even more influential than its originality and quality.[13]

Another difference between the German and Italian schools is that, both in *Bellifortis* and in the *Anonymous of the Hussite Wars*, even traditionally civilian machines, such as mills, are mostly depicted in the context of war, for instance, during a siege. This focus on military usage is another distinctive character of the German school.

In the minds of the Italian engineers, however, machines go beyond military purposes, becoming instead a way to emancipate mankind from drudgery while also providing an element of amusement and leisure.

Along with the anonymous authors of the *Anonymous of the Hussite Wars*, Francesco di Giorgio and Taccola also dedicated their attention to the project of diving. Several examples remain of these models, such as a drawing with two men concentrating on their underwater activity (Fig. 5.10, left). They are likely looking for sponges and use two different methods to breathe underwater: a pipe with a float and a strange folded contraption that resembles an accordion. Both use a waterproof diving suit with goggles that seal the opening for the eyes. If we compare these figures with the one that later appears in *Anonymous of the Hussite Wars* (Fig. 5.9), the resemblance is striking. It seems that sometimes ideas travelled quite fast, even four centuries ago.

[12] The term Hussite Wars was used to identify a series of religious wars which broke out in Bohemia between the years 1420 and 1436. The execution via burning at the stake of the preacher Jan Hus for heresy on 1415 incited the riots of his followers.

[13] Keller A (1984) Technology and Culture. 25:109–111.

Fig. 5.9 *Anonymous of the Hussite Wars*, the surface-supplied diver

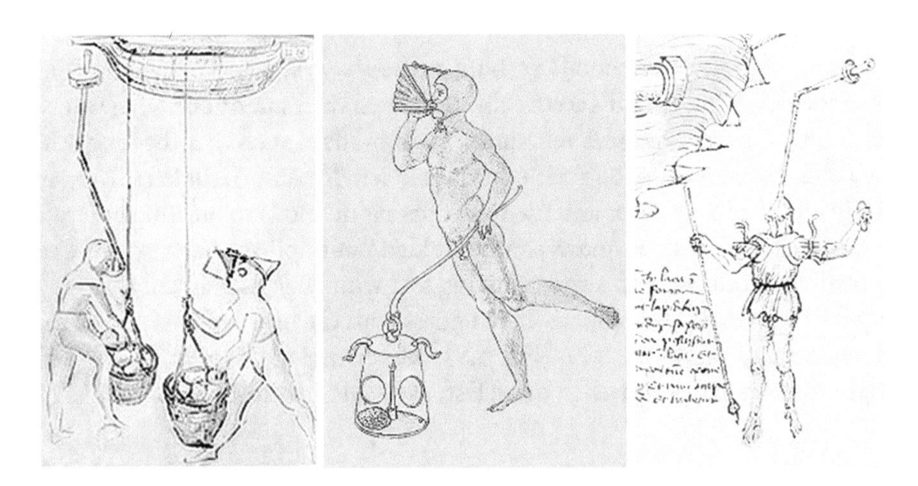

Fig. 5.10 Divers looking for sponges (*left*). Anonymous Siena Engineer (from Taccola and Francesco di Giorgio). Ms. Additional 34,113, folio 180v. BL. Diver with a lantern (*center*). Anonymous (from Francesco di Giorgio). Ms. Palatino 767, folio 9r. BNCF. Soldier diver (detail). Taccola, De Ingeneis (*right*). BSBM

The curious accordion-like contraption is also used by another diver, who appears in an illustration in the *Manuscript Palatinus 767* in the National Library of Florence (Fig. 5.10, middle). The idea clearly originated from Francesco di Giorgio's sketch, but the quality of this second drawing is very impressive. Some other details also catch the attention. It is possible to identify a swimming cap and a waterproof lantern. The latter is formed by a watertight bell jar containing a candle to illuminate the sea floor. On the other hand, the man is clearly naked.

Another diver—a soldier in full armor—was designed by Taccola in *De Ingeneis* (Fig. 5.10, right). The sketch shows a soldier with a rigid pipe, likely made out of bamboo, which is connected to the surface to breathe. These models, at least in comparison to the divers outlined in the *Anonymous of the Hussite Wars* and *Bellifortis*, represent an improvement. Solutions to some technical problems are well represented, such as the necessity of using a floating system for the air pipe. It was, however, only with Leonardo that a truly implementable system to breathe underwater would be designed.

Before moving on to Leonardo's project, however, we should pause to consider Taccola and Francesco di Giorgio's systems for snorkeling. A couple of Taccola's drawings display a diver's mask and snorkel that are surprisingly similar to our modern equivalents (Fig. 5.11, left and center). Another diver with a snorkel, from Francesco di Giorgio, is clearly a copy of Taccola's (Fig. 5.11, right).

As we have mentioned, Leonardo also faced the problem of the diving mask. His solution (Fig. 5.12) can be found on the same page (folio 26r, *Codex Atlanticus*) as the small sketch of the man walking on water which we encountered earlier in this chapter. The diver is furnished with sophisticated swimming goggles, another incredible anticipation of times to come. In a second drawing on the same page, a flexible pipe with a float at the water's

Fig. 5.11 Diving mask with snorkel. Taccola. De Ingeneis, II Cod. Lat. Monacensis 197, folio 57r (*left*) and folio 12r (*center*), details. BSBM. Copy of Taccola's mask. Francesco di Giorgio. Manuscript Vat. Urb. Lat. 1757, folio 13r (*right*), detail. BAV

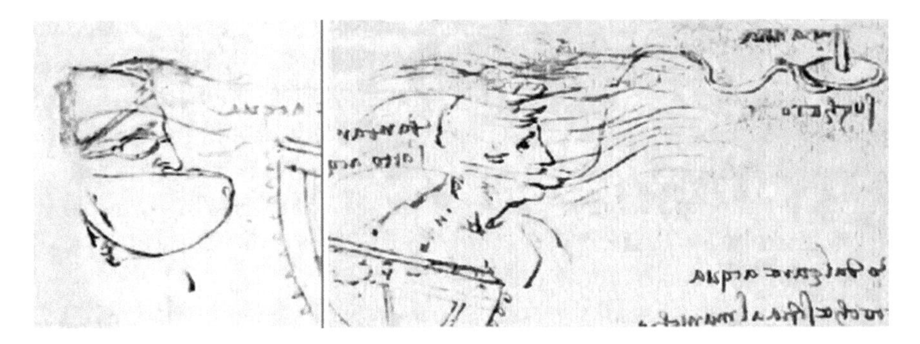

Fig. 5.12 Diver with mask and swimming goggles (*left, detail*). Diver with a pipe system to breathe from the surface (*right, detail*). Leonardo da Vinci. Codex Atlanticus, folio 26r (ca. 1480–1482). BAM

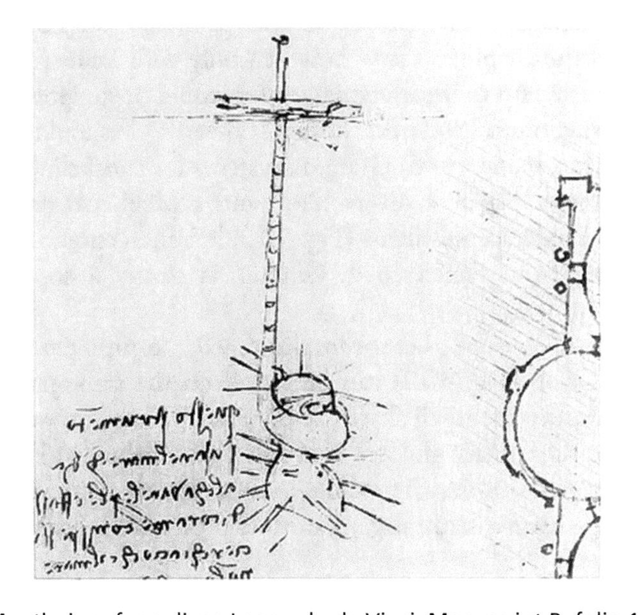

Fig. 5.13 Mouthpiece for a diver. Leonardo da Vinci. Manuscript B, folio 18r. BIF

surface represents an alternative, smaller device for diving just under the water's surface.

On the other hand, a small sketch in *Manuscript B* (folio 18r) showing a mouthpiece tells us a different story about the possible origin of the idea of surface-supplied diving (Fig. 5.13). Leonardo explains the origins of this system, which could support respiration under the sea surface: "This tool is used in India for fishing sea pearls. It is constructed with thick circles of cordage so that the sea cannot close (the pipe). A companion remains in the boat waiting while the other one gathers pearls and corals and wears glass goggles like those used for snow and armor with spikes for fish."[14] The metallic spikes around the bust and the head of the diver are, as described by Leonardo, a protection from dangerous sea creatures, such as sharks. This is particularly interesting because, in the Mediterranean Sea, there are no dangerous fish to justify the use of such protection. This means that Leonardo seems to be truly reporting something he must have heard from travelers coming from Asia, a further demonstration of how ideas could travel quite far at the time.

As we have seen, the idea of the surface-supplied diver and, more generally, a system to breathe underwater was envisioned in various ways, some quite naive and others more realistic. Leonardo borrowed from these various concepts to develop his own detailed project. He elaborated on and represented the idea in a series of sketches, which became progressively more complex until he arrived at a design for the equipment of a surface-supplied diver, complete with technical details.

In a sketch that can be found in the left margin of folio 1069r of the *Codex Atlanticus*, a simple inhalation system is depicted (Fig. 5.14, left). This is likely a re-elaborated version of Francesco di Giorgio's concept; in particular, the float appears very similar to that found in di Giorgio's sketches. Leonardo did not extensively explain this simple system and instead wrote just two words: "*cork-cane.*"[15] This breathing system has several drawbacks, however, as Leonardo quickly realized. A rigid pipe limits the movements of the diver, while the float on the surface does not prevent water from entering the cane. The major problem, however, is represented by the foul air that can become stagnant if only one pipe is used for both inhalation and exhalation. Leonardo solved this problem by designing a breathing system with two pipes for air—one for breathing in clean air and the second for breathing out (Fig. 5.14, center and right). The two ducts have valves that alternately open and close.

The structure of this system, as can be deduced from the drawing, was designed with extreme care. Several joints with springs are used to guard against water pressure. A large bell-like float makes it possible to control air from the outside. Leonardo also gave a detailed explanation of the device,

[14] Manuscript B, folio 18r. Original text: "Questo strumento s'usa nel mare d'India al cavar la perla, e fassi di corame con ispessi cerchi, a ciò che il mare non la richiugga; e sta di sopra il compagno colla barca [a] aspettare. E questo pesca perle e coralla e ha occhiali di vetro da neve, e corazza di spuntoni per pesci."

[15] Original text: "sughero-canna".

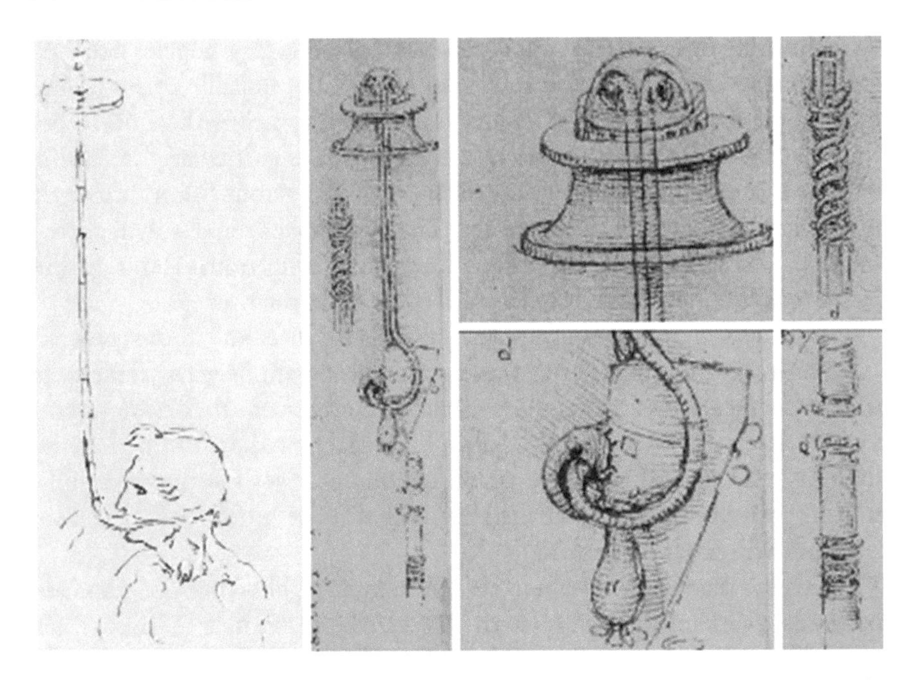

Fig. 5.14 System to breathe under water (detail), *left*. Leonardo da Vinci. Codex Atlanticus, folio 1069r. BAM. Apparatus for breathing when immersed [in water], the full device and enlarged details, *right*. Codex Arundel MS 263, folio 24v. BL (1507–1508)

describing how to join the canes, ensure waterproof cladding, and construct the float (Fig. 5.15).[16]

As with Leonardo's glider and parachute, this surface-supplied system has been reproduced today using period materials and technologies to test its feasibility. Pig leather was used for the diving suit and cane hoses for the breathing pipes. The joints were reinforced with steel spirals to avoid their collapse under water pressure. The apparatus was then tested by famous scuba diver Jacquie

[16] Original text: "a e b è in che modo li pezi delle canne si congiungano insieme; e questi fili hanno a essere come una mezana costa di coltello; e debbono essere tenperati, acciò che nel piegarsi tucta la canna, esso ferro non riservi la sua piegatura. E questo tal filo è posto infra le giunture della canne dentro alli corami che legano tal canne insieme, acciò che la potenzia che spigne intorno a esso corame, no 'l venissi a richiudere, perché essa potenzia è grandissima. E questo corame che veste tal ferro, debba essere doppio, acciò che, se quel di fuori si ronpe, quel dentro non si consumi. Il saco n è messo in tal loco acciò che, se nulla cadessi di sopra, tal saco la possa ricevere di dentro sanza dare impedimento all'anima dentro; e poi che tal saco n sarà pieno di materia, esso si potrà votare dalla parte di socto sanza dare alcun inpedimento alla sua anima. a, b sono le forme e vicinità delle fronti delle canne, e come esse si debbono fermare col. filo del ferro.

Per provarla, fa due cerbottane di carta inpastata fatte 'n una over sopra la medesima aste equale in grossezza e lunga quanto puoi. E se l'aste non fussi lunga, appica li pezi de la cerbotana nella lunghezza che a te piace, e cava la cerbottana d'in sulla forma col. taliarla per lo lungo, e po' rinsalda con carta inpastata. Ma intanti che allunghi la cerbottana, prova co' pezzi corti, e riuscendotu, allungali poi quanto vuoi e puoi isperimentare il tutto." Leonardo da Vinci, folio 24v of Codex Arundel MS 263, British Library, Londra.

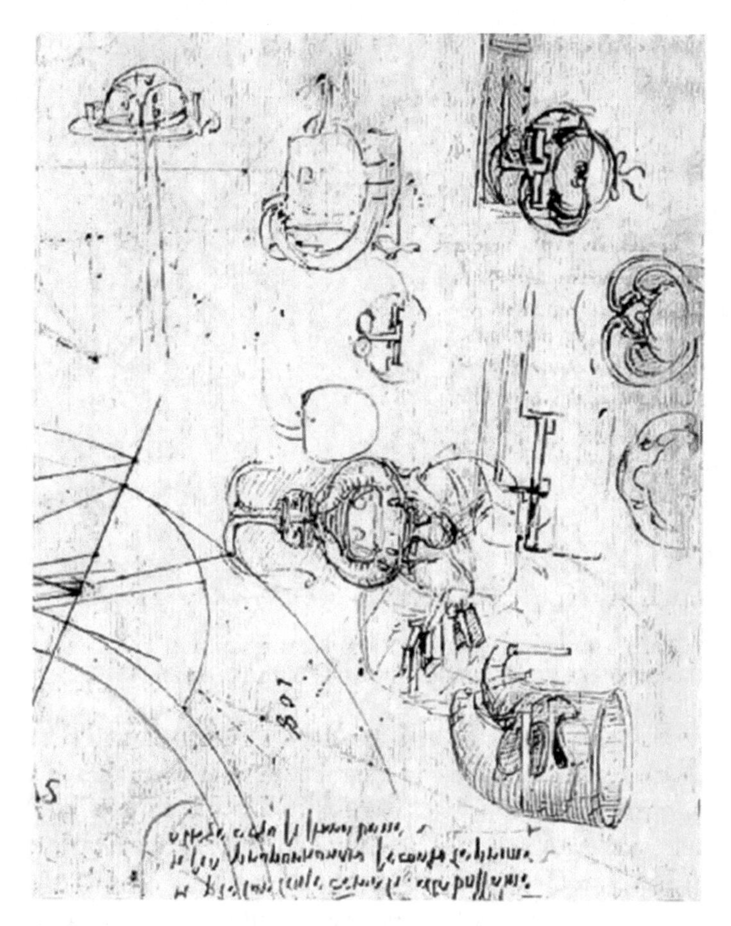

Fig. 5.15 Study for a diver's mask with valves. Leonardo da Vinci. Codex Atlanticus, folio 647v (detail). BAM

Cozens. It proved to work quite well in shallow water, although in deeper water, breathing problems generated by the increased water pressure emerged.[17]

Not long after Leonardo described his surface-supplied diving system and life preservers, in 1521, a lesser known author, Giovanni Battista della Valle (c. 1470–1550),[18] published a book dedicated to warfare, *Il Vallo* (also known as *The Book of the Captains*). This book is largely unknown to lay people, but it is considered the first modern military handbook and enjoyed enormous popularity at the time. It was originally written in Italian but had several

[17] The reconstruction and testing of the Leonardo's scuba diver suit was done for the BBC program Leonardo: The man who wanted to know everything (2003).

[18] Not so much is known about Giovanni Battista della Valle (Venafro, c. 1470 – c. 1550). He was likely working for some warlords as captain, and this was the opportunity to acquire the experience that he transferred in his war handbook.

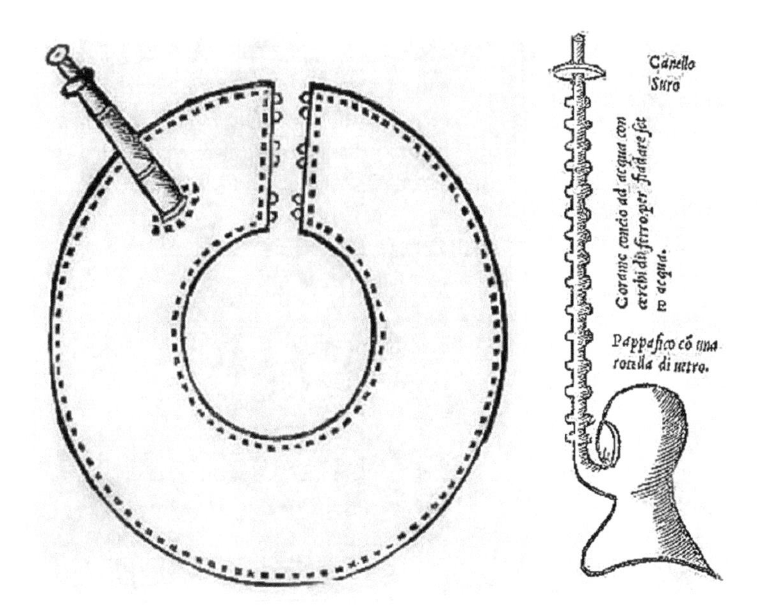

Fig. 5.16 The life preserver (*left*) and the surface-supplied diving system (*right*), (details). Giovanni Battista della Valle. The Book of the Captains (1520)

reprinted versions, some translated into French, German, and Spanish. The book is divided into four parts and describes warfare methods and technologies, mostly concerning the art of fortification, and the use of artillery, fireworks, and machines for assaults. In addition to more general instructions for a captain, such as troop placement, ethics, and dueling, some technical devices that might aid soldiers are also described, among them a life preserver for soldiers fording rivers (chapter XXXVI, book III) and a diving system (chapter XXXVII, book III, Fig. 5.16).[19]

The common source of these ideas is also quite clear, Valturio's *De Re Militari*. Many, if not most, of the military technologies that appeared in the different works of the time originated from this book. This is why we find the same device in so many sources with few changes or modifications. These are reported, however, without any mention of the original. Our modern-day citation mania was yet to come, and at the time, it seemed quite natural to use the ideas of other people as a source of inspiration if not indeed for a copycat work.

[19] Giovanni Battista della Valle. *Vallo libro appertinente continente a Capitanii, retenere e fortificare una Citta con bastioni, con novi artificii de fuoco aggionti, come nella tabola appare, e de diverse sorte polvere, e de espugnare una Citta con ponti scale, argani, tromba, trenciere, artegliarie, cave, dare avisamenti senza messo allo amico, fare ordinanze, battaglioni et ponti de disfida con lo pingere, opera molto utile con la esperienza de l'arte militare.* (Book about the captains. How to defend and fortify a city with ramparts, with new fireworks as shown in the tables. How to conquer a city with bridges, staircases, winches, trenches, artillery, mines. How to send messages to friends without using messengers, send orders, and make battalions.... Work very useful for the art of war). Naples, 1531.

Giovanni Battista della Valle's life preserver is therefore similar to that of Valturio, which is, in turn, similar to the model described by Leonardo. More novel, however, is the "pappafico" shown in the same image, an archaic Italian word that refers to the cap used by a horseback rider to protect his head from the wind. The description of making a sealed cap with goggles is quite accurate, but it is technically inconsistent when compared to Leonardo's version. Even though Leonardo was too afraid to divulge some of the technical details of his inventions for fear that they might be used for violent or nefarious purposes, others, as we can see, were not nearly so scrupulous.

Despite the many iterations of diving equipment that we have seen, the idea of the surface-supplied diver that originated in the late Middle Ages and Renaissance did not become a practical reality until several centuries later.

In the 1820s, Charles and John Deane of Whitstable, in the English region of Kent, created a copper helmet with the purpose of protecting firefighters from toxic smoke. A leather hose attached to the back of the helmet provided an air supply, and the helmet was connected to the firefighter's garment via a flexible collar. Lack of funds prevented full development of the invention, and the first smoke helmet was finally built in 1827 by another engineer, Augustus Siebe. A year later, the Deane brothers converted the system into a diving suit (Fig. 5.17). This suit was tested for diving more than 300 years after the first

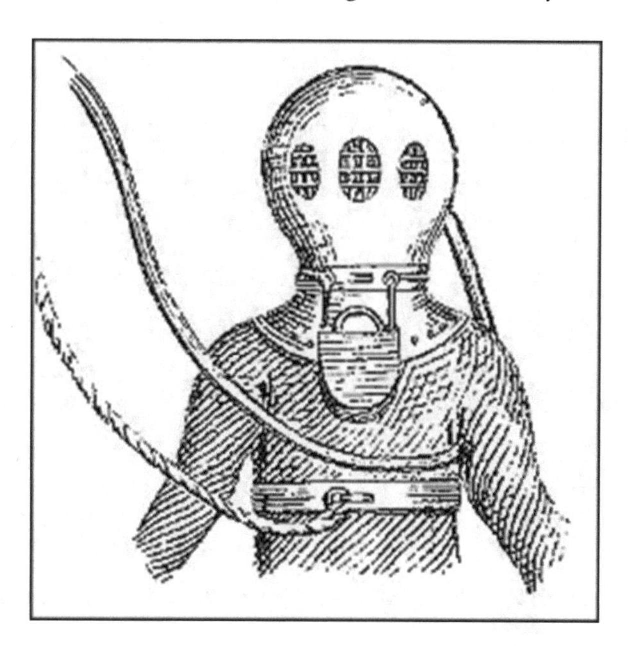

Fig. 5.17 The Deane helmet and suit, from an illustration in the Magazine of Science in 1842. (This work is in the public domain in its country of origin and other countries and areas where the copyright term is the author's life plus 70 years or less. https://upload.wikimedia.org/wikipedia/commons/f/f1/Deane_helmet_1842.jpg)

description of the basic idea. It is important to stress that a fundamental technical improvement for the practical use of the system was the application of a valve in the helmet, exactly as Leonardo had foreseen.

Scuba Diving: Beyond Surface-Based Air Supply

Leonardo did not limit himself to the idea of a surface-based air supply, however. In yet another drawing, he envisaged the possibility of a waterproof bag to fasten to the chest as an air reserve, allowing a diver to swim and move under the water without any ties to the surface (Fig. 5.18).

The same folio also bears the sketch of another strange diver. He uses two bags connected to a mask as his air reserve.[20] It is interesting that both these drawings are on the same page, meaning that Leonardo was exploring different solutions for breathing underwater. The two possible solutions are only outlined; they represent a vision likely inspired by previous works. They are also a personal elaboration and exploration of more sophisticated solutions. Comparing these drawings with the drawings of divers from earlier authors demonstrates the difference in Leonardo's work. He may not always have used original ideas, but he had the extraordinary ability to go beyond the general

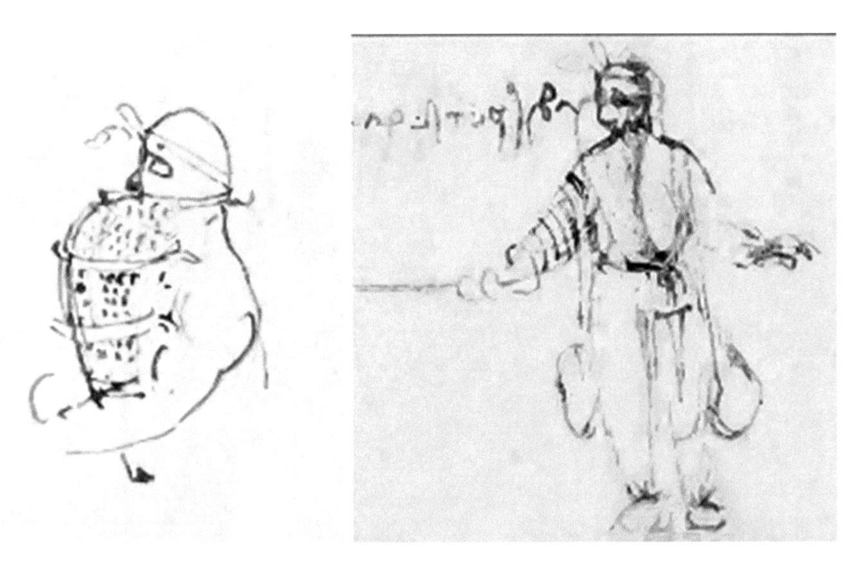

Fig. 5.18 Divers with air sack reserves. Leonardo da Vinci. Codex Atlanticus, folio 909v (details). 1485–1487. BAM

[20] Original text: "disperse da la vesta, se bisognassi romperlo."

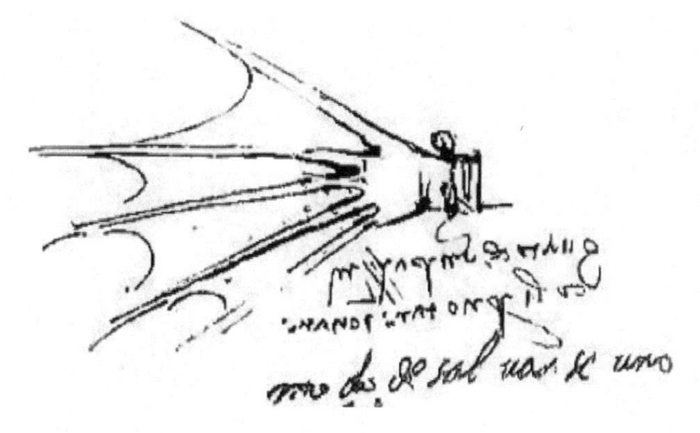

Fig. 5.19 Webbed gloves for swimming and diving. Manuscript B, folio 81v (detail). BIF

representation of a concept to design detailed working projects that we are still able to reproduce and test today.

Leonardo's investigations of swimming and diving do not stop here. On the same page where he sketched a man wearing a life preserver (folio 81v, *Manuscript B*), he also designed a new swimming tool, the webbed glove (Fig. 5.19). The glove is formed by five needles with a membrane stretched over them. Leonardo's description calls it "a glove with small fabrics to swim in the sea"[21] which, as with the wings of his flying machines, is clearly inspired by his observation of nature. He noticed that animals with webbed hands and feet moved faster in water. This glove can also be seen as a precursor to the flippers that are now popular among swimmers and divers.

As we can see, Leonardo devised a whole range of swimming and diving equipment. Of them, the only one yet to be realized in our modern age is the ability to walk on water.

Machines of Marine Sabotage and Warfare

Leonardo certainly designed a variety of different sophisticated diving systems, but to what end? Taccola and Francesco di Giorgio likely envisioned peaceful uses of the invention, such as collecting sponges or hunting for lost treasures under the sea. Leonardo's concerns were likely not as peaceful.

On folio 909v of the *Codex Atlanticus*, there is a drawing of a simple but dangerous device to be used for breaking the hull of a ship (Fig. 5.20). Perhaps by chance (though likely not), this system is drawn near the two divers we

[21] Original text: "guanto con pannicoli per notare in mare."

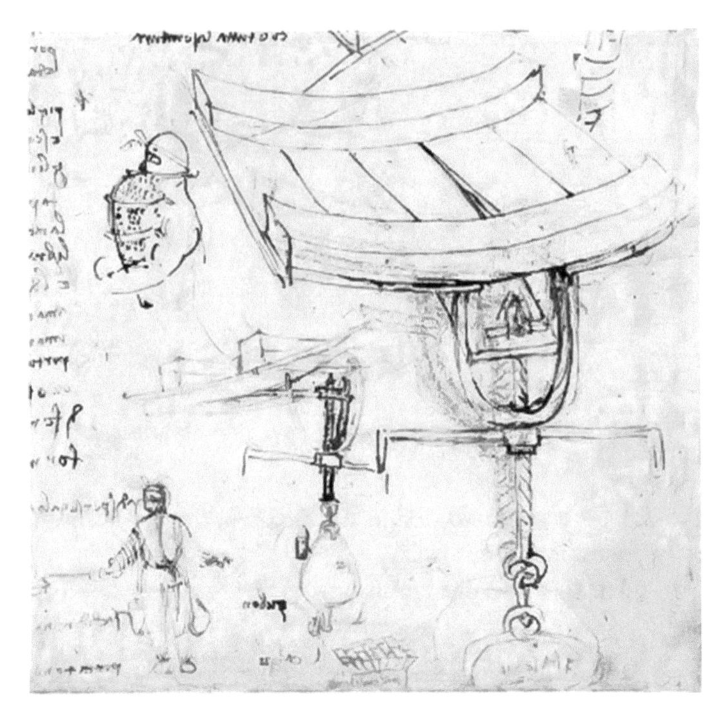

Fig. 5.20 Device for breaking a ship's hull. Leonardo da Vinci, Codex Atlanticus, folio 909v. BAM

encountered in Fig. 5.18. The tool is shaped like an inverted U and equipped with a central screw. To produce a crack on the bottom of a ship, the two arms of the U tool must be affixed to two non-contiguous boards. Once the central screw is firmly attached to the hull, the second crank is used to put pressure on the two arms, causing breakage of the hull and the eventual sinking of the ship.

On the other hand, Leonardo was not the only one to develop machines for maritime warfare. Despite their peaceful drawings of diving systems, Taccola and Francesco di Giorgio also designed a large number of hull-breaking devices. Some examples can be found on folio 125r of Francesco di Giorgio's *Codicetto* ("small codex", Fig. 5.21). The second device in particular seems very similar to that of Leonardo, even if the latter was designed to be used by a diver operating under the sea. This is another good example of how, at the time, the public dissemination of military innovations was considered as normal as that of civilian applications.

There is one more question to answer before closing this chapter dedicated to diving and swimming devices. Did these ideas spread, or did they go unnoticed? We know that Francesco di Giorgio and Taccola's manuscripts were copied and studied for a fairly long time before falling into obscurity until their recent revival in modern times among scholarly circles. As discussed in

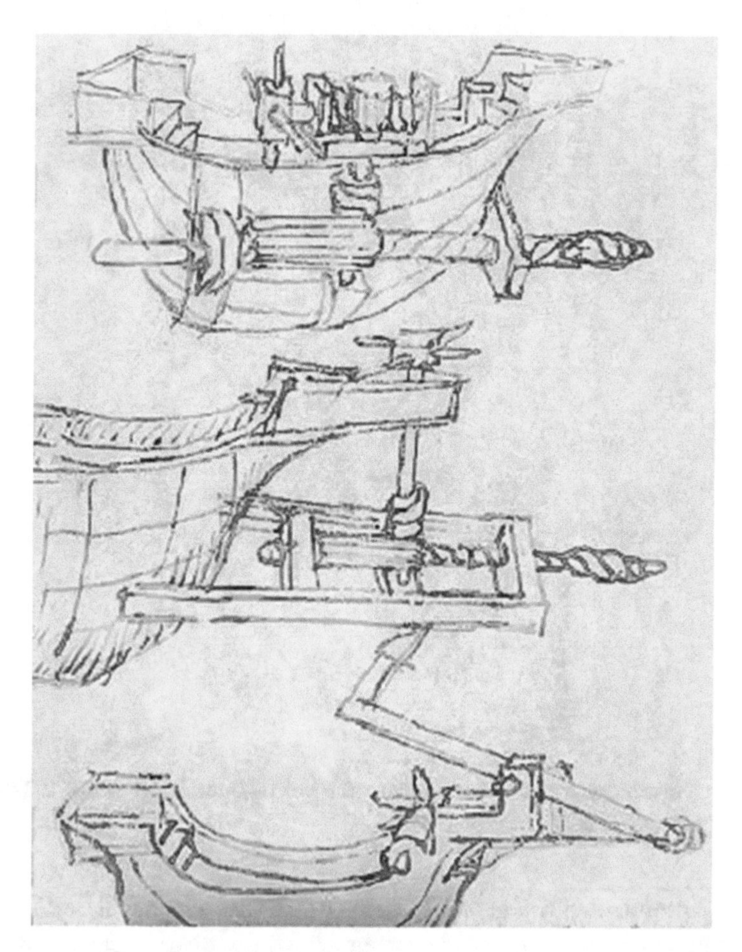

Fig. 5.21 Devices to be mounted on ships to break hulls. Francesco di Giorgio. Codicetto. Ms Lat. Urbinate 1757, folio 125r. BAV

previous chapters, Leonardo's work faced an even unhappier fate, remaining hidden, lost, or unavailable in some cases, for centuries.

Do any traces of Leonardo's ideas remain in later works? There is no clear answer to this question because some of Leonardo's notebooks were likely read and seen by several people, even if their dispersion made systematic consultation difficult. In other cases, the original idea is not from Leonardo's work but from other architect-engineers of the time who had similar ideas and devices. Some connections have been hypothesized[22] between Leonardo's work and that of Giovanni Alfonso Borelli (1608 Naples–1679 Rome). Borelli's *De Motu Animalium* (*On the Movement of Animals*),[23] which is considered his masterpiece,

[22] Rosheim M (2009) Leonardo's lost Robots. Springer Berlin Heidelberg.

[23] Borelli G (2015) *Borelli's On the Movement of Animals – On the Force of Percussion,* translated by Paul Maquet. Springer.

Fig. 5.22 Giovanni Alfonso Borelli's diver (De Motu Animalium). Book 1, Table XIV, Figure 8

contains, among other things, a submarine prototype, which will be discussed in more detail in a later chapter, and a system allowing a diver to move and breathe underwater (Fig. 5.22).

As we can see, the illustration shows a diver using a closed circuit for breathing (rebreather), which is an original and interesting idea. The exhaled air is cooled by sea water as it passes through copper tubing. In addition, the diver is also using two other strange and novel devices. One is a tube with a piston that helps the diver to reach a hydrostatic equilibrium by controlling the amount of air inside, and the other is a pair of unusual swim fins worn on the feet that aid in movement. The fins appear to have claws of some sort, which are likely meant to assist in walking along the muddy sea bottom, as opposed to swimming.

How much can the inspiration behind this diver be credited to Leonardo? It is hard to be certain, but the inspiration here, at least, seems to owe more to Borelli's own original studies of animal movement in water than to Leonardo's influence in the field of mechanical contraptions. In any case, as can be seen, the challenge of harnessing and mastering water was a consuming interest from Leonardo's time up until the present day.

6

Paddle Boats, Submarines, and Other Sea Vessels

During the Renaissance, engineering techniques for shipbuilding were of enormous civilian and military interest. At the time, ships were still powered by traditional systems of sails and oars, but new possibilities started to be proposed. These were mostly exploratory in nature and would only come to fruition centuries later, with the emergence of the steam engine.

It should not surprise us that naval technology exerted a special pull on the Renaissance engineers because of its key military and economic importance. Many pages of their manuscripts, including those of Leonardo, are full of schemes and projects for ships and maritime warfare techniques. These include several innovative models and ideas, such as primitive submarines and double hulls, together with numerous variations on traditional ships and naval strategy.

Paddle Boats

As is well known, during the Renaissance, knowledge of the repertory of Roman technologies slowly reemerged, thanks to renewed interest in and translation of surviving classical manuscripts. Among these were a range of Roman warfare treatises, such as Vegetius's *Epitome de Rei Militari*, written between the fourth and fifth centuries, and the anonymous *De Rebus Bellicis*.[1] Among other topics, the latter text described a variety of Roman war machines, including a particular type of ship, a *liburna* or liburnian, which, in some

[1] Anonymous *Le cose della guerra*. (Introduction, Text, Translation and Commentary: Giardina A) (1989) Mondadori, Fondazione Lorenzo Valla

P. Innocenzi, *The Innovators Behind Leonardo*, https://doi.org/10.1007/978-3-319-90449-8_6

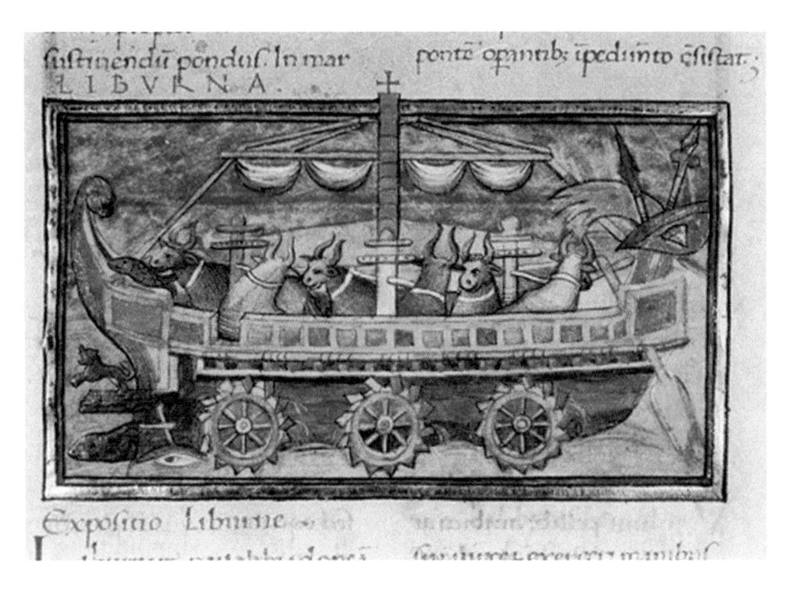

Fig. 6.1 Miniature of an oxen-powered liburna in a fifteenth-century copy of De Rebus Bellicis. Codex Oxoniensis Canonicianus class, lat. Misc. 378, IV sec. A.D., Bodleian Library, Oxford

versions, was outfitted with paddle wheels[2] (Fig. 6.1). The name derives from the people of Liburnia, who resided on the coast of modern-day Croatia.

The Roman *liburna* was particularly fast and maneuverable and, for this reason, was also used by the Roman navy for quick transport of supplies and soldiers.[3] The ship was narrow and could bear up to six orders of oars. It was used by the Romans for the first time in the huge naval battle of Azio in 31 BC, which settled the civil war between Octavian and Mark Antony.

---

[2] The description of the paddle wheel version of the liburna is reported in Chapter XVII. *Expositio liburnae. 1. Liburnam navalibus idoneam bellis, quam pro magnitudine sui virorum exerceri manibus quodammodo imbecillitas humana prohibebat, quocumque utilitas vocet ad facilitatem cursus, ingenii ope subnixa animalium virtus impellit. 2. In cuius alveo vel capacitate bini boves machinis adiuncti adhaerentes rotas navis lateribus volvunt, quarum supra ambitum vel rotunditatem extantes radii, currentibus iisdem rotis, in modum remorum aquam conatibus elidentes miro quodam artis effectu operantur, impetu parturiente discursum. 3. Haec eadem tamen liburna pro mole sui proque machinis in semet operantibus tanto virium fremitu pugnam capescit, ut omnes adversarias liburnas comminus venientes facili attritu comminuat.*

Description of a liburna. 1. The liburna is a vessel suitable for sea battles, but of such size that the strength of the men was insufficient to move it and ride it where the need recalled: with the help of intelligence, has been used the energy of animals for an easier navigation. 2. In her womb, therefore and in the interior space two oxen are introduced, they are linked to the systems that spin the wheels fitted to the hips (outside) of the boat. Above the band of these or the roundness of their circumference there are ledges as spokes, which rotate with the same wheels, and cleaving the water the same way the blows of oars and working for the wonderful effect of art, they open with their impetus the surface of the water. 3. This same liburna, however, because of its size and the machines that operate within it, is used to attack in the battle with such a tremor of forces that easily destroys all opposing ships that come on its way.

[3] Morrison JS,Coates JF (1996) *Greek and Roman Warships 399–30 B.C.* Oxbow Books. Oxford

The ship seen in Fig. 6.1 is equipped with three paddles for each side. These are moved by three pairs of oxen who rotate the capstans. The advantage of this system is evident. Instead of human oarsmen, the ship is propelled by animal power. This represents a major innovation in naval technology, and it enjoyed relatively successful diffusion from the Middle Ages up to the Renaissance. Illustrations of the paddle wheel *liburna* can be found in the manuscripts of several Renaissance engineers, such as the nearly identical copies that are reported in the work of Giuliano da Sangallo[4] (written between 1464 and 1516) and on folio 67r of the *Codex Escurialensis*[5] (c. 1491–1514), which is composed of a sketchbook and additional drawings likely from his workshop.

One of the first Renaissance sketches of a paddle wheel ship is reported in Konrad Kyeser's *Bellifortis* (1402–1405). This relatively detailed illustration shows a small boat outfitted with a system of internal and external toothed wheels with a man-operated handle (Fig. 6.2). The system was designed with some sophistication, featuring unpaired teeth instead of simple paddles to allow stronger propulsion in the water.

In another manuscript version of *Bellifortis* (Fig. 6.3), the movement to the external wheels is transmitted via a crank placed on a platform structure in the

Fig. 6.2 Paddle boat. Konrad Keyser. Bellifortis. BSBM

[4] Giuliano da Sangallo, *Codice Vaticano Barberiano Latino 4424*, Barb. lat. 4424, Biblioteca Apostolica Vaticana, Roma, f. 35r. Giuliano da Sangallo. A reproduction of the codex has been realized with the title: *Il libro di Giuliano da Sangallo: Codice Vaticano Barberiniano Latino 4424 (The book of Giuliano da Sangallo)*, Città del Vaticano, Biblioteca Apostolica Vaticana, 1984.

[5] *Codex Escurialensis*, Ms. 28-I-11. Real Biblioteca del Monasterio de El Escorial, Madrid, Spain.

Fig. 6.3 Small boat moved by two paddle wheels. Konrad Keyser. Bellifortis. Cod. Ms. philos. 63, folio 54v., beginning of fifteenth century. Universitätsbibliothek, Göttingen. (Repr. facs.: K. Kyeser aus Eichstätt, Bellifortis, Umschrift und Übersetzung von Götz Quarg, Düsseldorf, VDI-Verlag, 1967)

center of the boat. The man or men operating the boat could reverse the direction of movement by a simple change in the crank rotation. Several other similar versions of paddle boats outfitted with winch mechanisms can be found in other copies of the *Bellifortis*.[6]

Another nice illustration of a small paddle boat can be found in the *Manuscript of the Hussite Wars* (Fig. 6.4). The mechanism is much simpler than that in *De Rebus Bellis* and *Bellifortis*: two men sit back-to-back and directly rotate pairs of external wheels. Some basic coordination during operation was necessary, at a minimum to avoid turning the wheels in opposite directions.

Yet another model of the paddle boat is described and illustrated in *De Re Militari* (*About military matters*) of Giovanni Valturio (Rimini, 1405–1475),[7] whom we have already encountered as the original source of Leonardo's life preserver.

Giovanni Valturio

Valturio's work inspired many generations of engineers and military experts, and copies of his techniques and machines are found in most of the subsequent works dedicated to technological solutions for warfare.

[6] At least 12 manuscript copies of Konrad Kyeser's work are known; they reproduce, generally in a modified version, the original work.

[7] The *De Re Militari* of Roberto Valturio was published in Verona in Latin in 1472. In 1483 the first translated copy in Italian was printed. Leonardo used this printed edition.

Valturio was a member of a cultured family of notaries and high-ranking administrative officials. He knew both Greek and Latin fairly well and wrote his *De Re Militari* in the latter language. For some years, from 1427 to 1437, he taught rhetoric and poetry at the Studio of Bologna (later renamed the University of Bologna). After a period in Rome, where he worked as an official at the Vatican administration, he returned to Rimini to become a member of the private council (consiglio privato) of Sigismondo Pandolfo Malatesta, one of the most famous warlords of the time and chief commander of the Venetian Army in the war against the Ottoman Empire. Upon Valturio's death in 1475, Roberto Malatesta, Sigismondo's son, decided to bury him among the illustrious men in the Pantheon of Rimini, the *Tempio Malatestiano*, as a tribute to his loyalty to the Malatesta family and the importance of his work.

Valturio exerted great influence on the Renaissance engineers, including Leonardo, who carefully stored his copy of *De Re Militari*.

Valturio dedicated the *De Re Militari*, a treatise in 12 books on the art of war, to his patron Sigismondo Malatesta. Though much of the work consists of a compilation of classical sources, it nonetheless had an enormous influence, in part thanks to its vivid illustrations of war machines and military techniques.[8] As soon as Valturio completed the work in 1460, Sigismondo Malatesta sent copies to each of the major European courts. This allowed for tremendous diffusion of the work. The subsequent success of *De Re Militari* attests to the great interest in warfare during this tumultuous period in Europe.

While many of the war machines described in the *De Re Militari* were derived from Roman or medieval traditions, a number appear original, and some contain innovative ideas. A watertight boat similar to the modern-day submarine, two models of paddle wheel ships, a boat that could be disassembled, and the life preserver discussed in Chap. 5 are particularly notable.

As can be seen in Fig. 6.5, Valturio's paddle boat was reproduced exactly by Bonaccorso Ghiberti in his compilation *Zibaldone*. Valturio's system is similar to the one we encountered in the *Manuscript of the Hussite Wars*. In the single-wheel version, the boat is equipped with a helm for steering, while in the five-paddle design, a central shaft ensures that the motion of the wheels is coordinated.

Valturio's paddle boat was also used as a model by both Taccola and Francesco di Giorgio, with particularly interesting and creative versions to be found in the latter's *Codicetto Vaticano*.

Among the variations explored by Francesco di Giorgio were changes in the position of the wheels and in the method of movement transmission (folio 18v of *Codicetto Vaticano*, Fig. 6.6). One example features two boats with a treadwheel in the middle (Fig. 6.6, left). Another (Fig. 6.6, right) is outfitted

[8] The illustrations were not done by Roberto Valturio but very likely by Matteo de' Pasti, an artist working at the Malatesta court in Rimini.

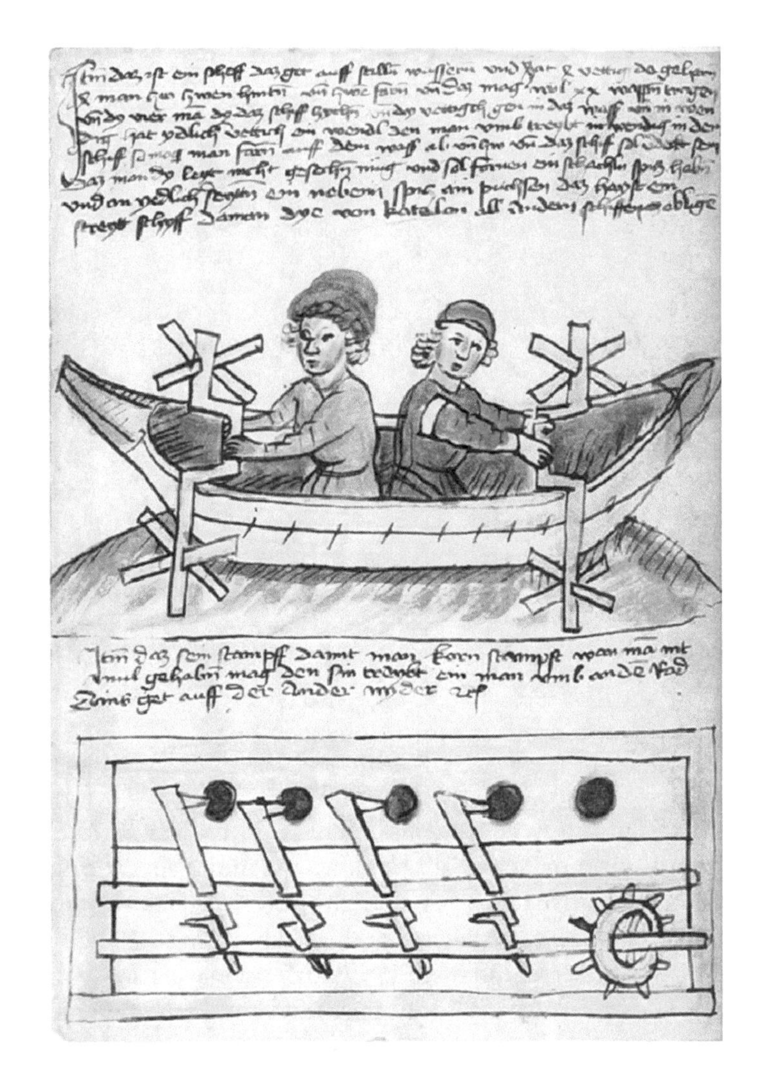

Fig. 6.4 Small paddle boat moved by a pair of men. Manuscript of the Hussite Wars. Clm 197, Part 1, Folio 17v Supra (Detail). Osterreichische Nationalbibliothek, Vienna

Fig. 6.5 Models of small paddle wheel boats of Valturio (*left*) from *De Re Militari* and copies from Bonaccorso Ghiberti in the *Zibaldone* (*right*), manuscript BR228, folio 216v. BNCF

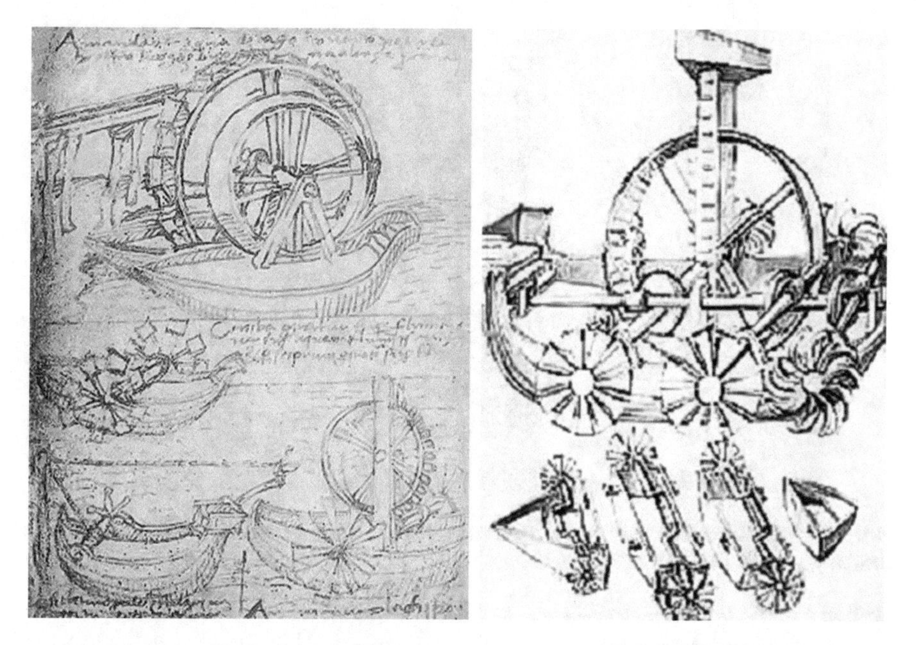

Fig. 6.6 Paddle wheel boats. Francesco di Giorgio. Codicetto Vaticano folio 18v (Vat. Urb. Lat. 1757), *left*. BAV. Manuscript 197.b.21, folio 43v (detail), *right*. BL

with external paddle wheels, as seen in the models by Valturio and Konrad Kyeser but with a central mechanism of much greater complexity. In this second model (found in *Manuscript 197.b.21*, BLL), the motion is generated by a man walking inside the rim of a treadwheel placed in a structure in the middle of the boat. The treadwheel is connected to six side wheels via a long shaft parallel to the deck. While these sketches are intriguing, they lack the level of mechanical detail so characteristic of Leonardo's approach.

Treadwheels

It should be noted that the use of treadwheels dates back to the Romans, who used such systems to create "walking cranes" for the purpose of raising stones during the construction of buildings and monuments. Similar machines were also utilized for theater performances. The Romans used both horizontal and vertical configurations, and their treadwheels were powered by both men and animals. Such devices were widely employed in Europe during the Middle Ages for the construction of castles and cathedrals.

Antonio da Sangallo's (1484–1546) elegant example of a ship moved by two side wheels (Fig. 6.7) was likely inspired by Francesco di Giorgio's model. The big central wheel is again moved by means of a treadwheel, and the motion is transmitted to the external paddles via a cogwheel. It is exciting to

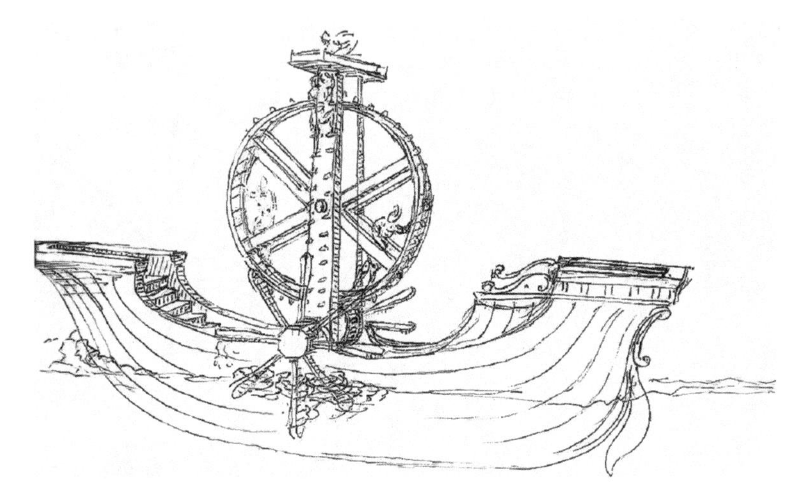

Fig. 6.7 Vessel moved by a treadwheel. Antonio da Sangallo the Younger, Architectural drawings, folio 1438r . GDSU

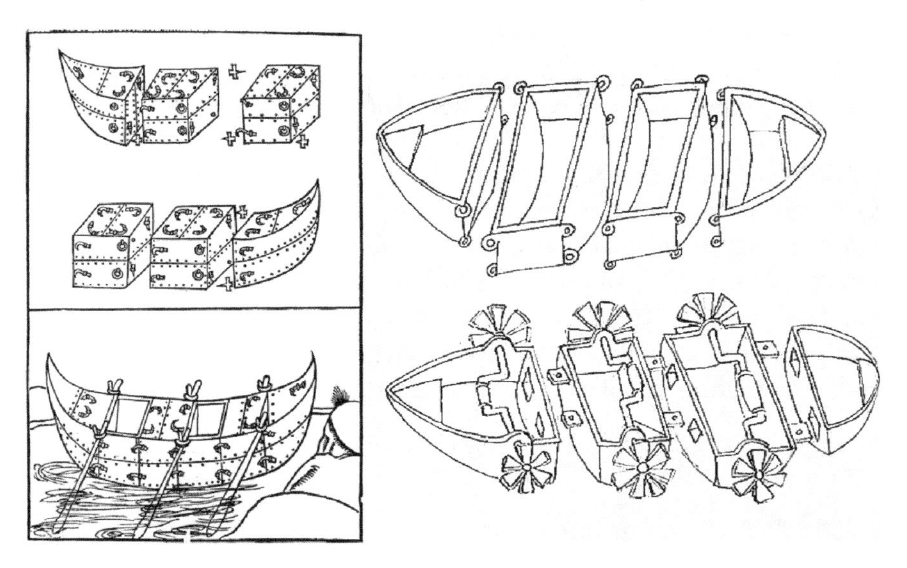

Fig. 6.8 *Left*: Modular boat. Valturio. De Re Militari. *Right*: Two models of modular boats, with and without paddle wheels. Francesco di Giorgio. Trattato di architettura. BNCF

see the speed with which ideas such as these were spreading within the circle of this new generation of architect-engineers. The rediscovery of Roman technologies, especially via Vitruvius, stimulated this age of innovation.

Let us now return to the right-hand image in Fig. 6.6, Francesco di Giorgio's large paddle wheel ship. The bottom of the page shows an intriguing sketch of the boat divided into four sections. This represents a slightly modified version of Valturio's portable boat, formed by modular caissons to be used by an army to ford small rivers (Fig. 6.8).

The many copies of this design underscore its appeal. For example, an almost identical sketch can be found in Bonaccorso Ghiberti's *Zibaldone*. Several other versions can be found among Francesco di Giorgio's works (Figs. 6.6 and 6.8 right-hand image). The latter's innovation in this instance was to combine two of Valturio's ideas: the paddle wheel (Fig. 6.5) and the modular boat. The means of joining the various sections of the portable boat is also modified, with the introduction of external rings to be wedged through a pivot.

Taccola also explored the idea of using side paddle wheels for traction. In Fig. 6.9 we can see his ingenious solution for propelling a boat upstream via a system of ropes.

Fig. 6.9 System for propelling a paddle boat upstream. Mariano di Jacopo (Taccola), De rebus militaribus (De machinis 1449), codex Santini, folio 87r, between 1449 and 1475, BNP. (Knobloch E (1992) L'art de la guerre: machines et stratagèmes de Taccola, ingénieur de la Renaissance. Gallimard, Paris)

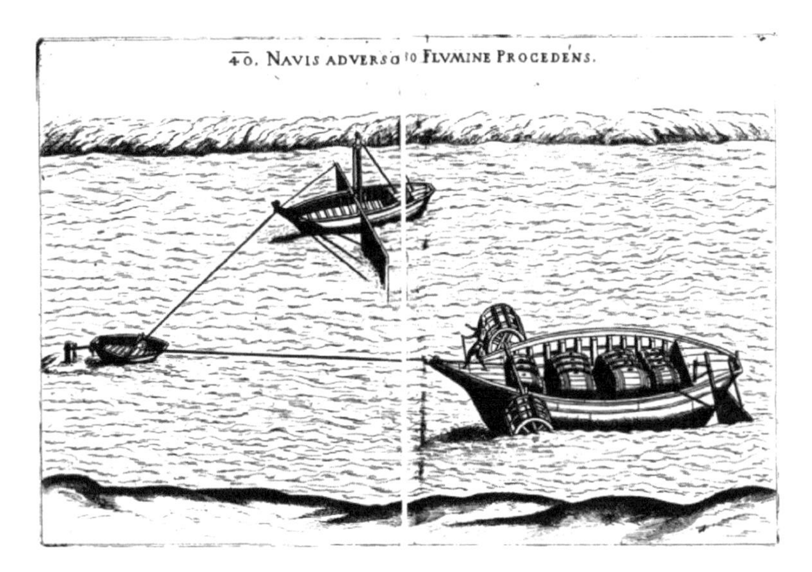

Fig. 6.10 Fausto Veranzio. System to navigate upstream. Illustration 40. Machinae Novae. 1606

During a time in which the discovery of other sources of transport was far in the future, such a mechanism represented a significant improvement and was replicated by later inventors such as Fausto Veranzio (c.1551–1617) who described a "*Navis adverso flumine procedens*" (*a boat which proceeds upstream*) in his *Machinae Novae* (Fig. 6.10). Comparison to Fig. 6.9 leaves few doubts as to Veranzio's source of inspiration.

Unsurprisingly, given his great interest in any possible mechanical innovation, paddle ships also attracted Leonardo's attention. As can be seen from the mechanics and general structure, Taccola, Francesco di Giorgio, and Valturio's models served as his starting point. Nevertheless, as we will see, he managed to achieve a much higher level of detail and sophistication. This is not to say, however, that all of his designs were technically correct and capable of implementation.

A small, poorly outlined sketch on folio 83r of *Manuscript B* (Fig. 6.11) depicts a small boat with a man moving a pair of paddle wheels using a crank. This is likely just a small draft taken as a memo.

While this boat is of quite rudimentary design, Leonardo would go on to produce an extremely sophisticated paddle boat design featuring a complex system of gears (Fig. 6.12) on folio 945r of the *Codex Atlanticus*. More extensive studies illustrating the details of the project can be found on folio 1063r of *Codex Atlanticus*.

Leonardo's design for the transmission of motion represented a true innovation, not simply an improved or modified version of Taccola and Francesco

Fig. 6.11 Sketch of a small paddle boat. Leonardo da Vinci. Manuscript B, folio 83r. BIF

Fig. 6.12 Plan of a paddle wheel boat. Leonardo da Vinci. Codex Atlanticus, folio 945r, detail. BAM

di Giorgio's models. The rotating paddles are moved by a system of large pedals which could transform the up-and-down motion of the pedals into uniform rotational motion of the paddle wheels. This is the same system of using vertically moving foot pedals to power a wheel that was addressed in our discussion of the Dobertec bicycle in Chap. 3.

Further Innovations After Leonardo: Captain Francesco Ramelli and the Amphibious Armored Tank

As we know, the flurry of creative activity of the Renaissance engineers does not end with Leonardo. One notable example is Captain Francesco Ramelli (Ponte Tresa, 1531–1608), whose 1558 treatise "*Le diverse et artificiose machine del Capitano Agostino Ramelli*" (*The various and ingenious machines of Captain Agostino Ramelli*) contains 195 chapters, each illustrating a different device with commentary in Italian and French. Most of these systems were hydraulic machines for mills, pumping water, and Archimedes screws. The treatise contains also few examples of warfare instruments and construction devices. This nicely illustrated treatise had a significant influence on the development of machines in Europe.

One of Ramelli's inventions which merits our attention here is an amphibious armored "tank" moved by paddle wheels (Fig. 6.13).

The tank was designed to allow an army to ford a river without being shot by enemies. On land, it rests on four large wheels and is drawn by horses. Once at the bank of the river, the large wheels are changed for smaller ones that allow the tank to be pushed into the water. Inside, the tank is fitted with small chambers from which the crew of four arquebusiers could shoot at the enemy while crossing the river. In the center of the tank, a standing man uses a crank to rotate a pair of paddle wheels that propel the tank through the water.

Fig. 6.13 An amphibious armored tank with paddle wheels for fording a river. Agostino Ramelli. Le diverse et artificiose machine del Capitano Agostino Ramelli. Figure CLII, pag. 250. 1558

Steamboats and Paddle Wheels

The various models of paddle boats envisaged by Renaissance engineers such as Leonardo and his precursors reflected a real need to move beyond the traditional maritime propulsion systems of oars and sails. However, it wasn't until several centuries later, with the invention of the steam engine during the industrial revolution, that a significant, viable alternative to these systems would become reality. This was also the destiny of the paddle boats, and the first working model was fabricated almost three centuries after Leonardo's projects.

The first paddle boat to be reported was the *Palmipede*, a 13-m-long steamboat built in France in 1774 by the Marquis Claude de Jouffroy. The ship was able to navigate for a relatively long distance in the Doubs River. A new and improved version built by the Marquis was the *Pyroscaphe*, which in 1783 navigated for 15 min along the Saone river before having a mechanical failure.

The beginning of French revolution interrupted these first attempts to develop the technology of paddle ships moved by steam engines.

The next attempt was realized in Scotland. After some initial experimental forays, William Symington and Patrick Miller of Dalswinton built the *Charlotte Dundas*, a commercial ship designed to navigate in a canal, for the Forth and Clyde Canal Company. Despite the ship's ability to navigate even against a strong headwind, no other boats of this design were ordered because of the fear that the movement of the paddles would somehow damage the canal.

The true commercial success of the paddle wheel ship arrived finally in the United States, where the *Clermont* built by Robert Fulton entered regular service in 1807, serving a route from New York to Albany. Initially, these paddle steamers were used for navigation in internal waters—rivers and lakes—but later, some models capable of facing the more demanding routes of the open sea were also developed. At the beginning, these were employed only for coastal navigation, but finally in 1819 the paddle steamer *SS Savannah* successfully crossed the Atlantic Ocean, the first such crossing by a steam engine ship. Kyeser, Leonardo, Di Giorgio, and Valturio had all been right about the potential of paddle wheels; they had simply lacked the proper engine to make their visions a reality.[9]

The triumph of paddle boats was, however, quite short-lived, as the invention of the screw propeller caused them to fall out of favor. They were never completely withdrawn from service, however. Their good maneuverability and small draft are a major advantage in shallow waters, and several such ships are still in service today on many rivers and lakes across the globe (Fig. 6.14).

---

[9] The SS Savannah was also equipped with sails.

Fig. 6.14 Illustration of a Chinese paddle wheel boat from a Qing Dynasty encyclopedia published in 1726 (This work is in the public domain in its country of origin and other countries and areas where the copyright term is the author's life plus 70 years or less. https://upload.wikimedia.org/wikipedia/commons/7/7f/Radpaddelsch.jpg)

While we have been focusing on the Western World, it is important to remember that the use of paddle ships is well documented in China as well, with the first report coming about a century later than its Roman counterpart. It is difficult to assess if this was an independent invention or one inspired by Western sources. In contrast to the West, in China, the use and development of paddle ships was never halted, and ships capable of carrying hundreds of people and equipped with more than 20 wheels were eventually built during the Song dynasty (960–1279).[10] During the First Opium War (1839–1842),

[10] Needham J (1986). Science and Civilization in China: Volume 4, Part 3, Civil Engineering and Nautics. Taipei, Caves Books.

there was an encounter between different fleets of paddle boats—the British ironclad paddle wheel steamships and the Chinese man-powered, wooden ones. Notwithstanding the valor of the Chinese sailors, the superiority of the British naval technology ensured the latter's triumph.

This does not, however, represent the final word on paddle boats. There is another interesting story that should be told about the development of boats powered by animals. Some of these, using the same system used in in the Roman *liburna* and described in the *De Rebus Bellicis*, worked perfectly and even operated a regular service.

Animals harnessed in this fashion were a common source of power for pumping water, milling, and grinding. The idea of using this system for the propulsion of a ship thus may appear as a rather obvious extension. The application of a vertical treadwheel to vessel propulsion is also a logical development of a widely used system. Today, these ideas may appear quite odd, but if we think back to the time in which they were conceived, they appear as a sign of those innovative times. As we have mentioned, there are no records of any attempt to actualize these models. This is true especially for Leonardo, who confined most of his work on machines within the pages of his notebooks.

The *horse-powered boats* (so-called team boats) which were built at the beginning of nineteenth century in the United States[11] represent an impressive case. Horse boats became popular as ferries, and several designs for using horses as power sources were developed. These solutions basically reproduced the Roman and Renaissance systems from Italian engineers even if the inventors, who patented their boats, were likely totally unaware of the previous models. One type of team boat was powered by horses moving in a circle on the deck (Fig. 6.15) and revolving a large wheel in a manner similar to the *liburna* system. The movement was transmitted to the side paddle wheels via a rotating gear. Another model employed treadwheels moved by horses, with the paddle wheels turned via gears connected to the shaft.

Notwithstanding the competition from steamboats (Fig. 6.16), horse-powered ferries, which could reach a speed of nearly ten knots per hour (compared to the eight to ten knots of the steamboats of the time), remained in service until the 1920s.

[11] Crisman KJ, Cohn AB (1998) When Horses Walked On Water: Horse-Powered Ferries in Nineteenth-Century America. Smithsonian Institution Press, USA

Fig. 6.15 "The Peninsula Packet, a horse boat which ferried passengers between Toronto and the Toronto Island" (Depiction: 1840s, Publication of image: 1896). The first version of the boat initially worked with only two horses; the modified version in the image was able to operate with five horses walking on a circular track on the deck. Volume 2, page 763 of Robertson's Landmarks of Toronto by J. Ross Robertson, Toronto (1893–1914) (This work is in the public domain in its country of origin and other countries and areas where the copyright term is the author's life plus 70 years or less. The image has been modified with respect to the original. http://static.torontopubliclibrary.ca/da/images/LC/d3-21b.jpg)

Fig. 6.16 The 1909 replica of the North River Steamboat. The steamboat carried passengers between New York City and Albany, and it was the first one to have a commercial success (Detroit Publishing Co. collection at the Library of Congress, USA. This work is in the public domain in its country of origin and other countries and areas where the copyright term is the author's life plus 70 years or less. The image has been modified with respect to the original to reduce the noise. http://www.loc.gov/pictures/resource/det.4a16094/)

Submarines

One of Leonardo's purported "inventions" which has greatly attracted the present-day imagination is the submarine. Together with the helicopter and the tank, the submarine has become a virtually ubiquitous icon with which to represent Leonardo's incredible capability of anticipating future developments.

In his *Manuscript B*, folio 11r (Fig. 6.19), there is a small and simple sketch which, at least to our modern eyes, looks like the profile of a submarine. However, the idea of a watertight boat capable of navigating underwater can be traced to at least two others prior to Leonardo: Guido da Vigevano and, again, Valturio.

Guido da Vigevano

Guido da Vigevano[12] was born in Pavia[13] and may be considered the first of the Italian architect-engineer innovators. He was born around 1280, well before the start of the Renaissance, and died in Paris in 1349, likely a victim of the big plague which devastated Europe at the time. His seminal Latin essay the *Texaurus Regis Francie*[14,15] presents the first ideas of totally new machines, including a self-propelling cart and a submarine, which probably looked quite strange at the time.

Guido da Vigevano[16] was trained as a medical doctor and practiced medicine in his native city of Pavia. However, he was eventually forced to leave because of his involvement in the fierce political struggle between supporters of the Pope (known as Guelphs) and supporters of the Holy Roman Emperor (known as Ghibellines). He then moved to France where he became the personal physician of Queen Joan of Burgundy (1293–1349). His well-rounded interests gave him a

[12] Settia A (2004) Entry Guido da Vigevano. In: Dizionario Biografico degli Italiani. Treccani. Volume

[13] Vigevano is a small city close to Milan. In Italy it was common to add to the first name the place of birth, as in the case of Leonardo, who is known as da Vinci because the village of Vinci is his birth place. In the case of Guido da Vigevano, notwithstanding the name, it seems that his original place of birth is another city close to Milan, Pavia. In several documents, in fact, he is indicated as "de Papia," from Pavia.

[14] The manuscript of Guido da Vigevano is: *Texaurus regis Francie acquisicionis Terre sancte de ultra mare necnon sanitatis corporis eius et vite ipsius prolongacionis ac etiam* cum *custodia propter venenum*. Fonds lat. 11015 of the Bibliothèque Nationale of Paris. Another manuscript, which is a partial copy of the original in Paris, is conserved at the British and Art Center, rare books and manuscript of the Yale Library, Paul Mellon Collection, Military Mss., 1375 (http://hdl.handle.net/1079/bibid/4510397). This copy contains only the second the part of the Texaurus which regards the siege warcraft and was owned by Guglielmo Libri.

[15] Hall A R (1976) *"Guido's Texaurus, 1335."* In: *Humana Civilitas: Sources and Studies Relating to the Middle Ages and the Renaissance. Volume 1: On Pre-Modern Technology and Science. A Volume of Studies in Honor of Lynn White, Jr.*, eds. Bert S. Hall and West DC, Malibu, Undena Publications, p. 10–52.

[16] Ostuni G (1993) *Le macchine del re. Il "Texaurus regis Francie" di Guido da Vigevano; trascrizione, traduzione e commento del codice Lat. 11015 della Biblioteque Nationale di Parigi*. Diakronia, Vigevano.

lot in common with Leonardo. He was curious and devoted not only to the practice of medicine (writing a medical treatise containing a translation of the work of the Greek physician Galen of Pergamon (129–201) as well as 18 illustrated tables of anatomy[17]) but also to mechanics and military warfare. We do not know if Leonardo was aware of this work and used it as a source for his anatomical studies.

Da Vigevano's *Texaurus* was written with a practical purpose. The French King Philip VI the Fortunate (1293–1350) had plans (never realized) to organize a large crusade to reconquer the Holy Land.[18] Guido da Vigevano gave his personal support to the crusade by writing this treatise containing medical practices, warfare strategies, and machines. The *Texaurus* is divided into two parts: one dedicated to medicine[19] and a second with a more general character,[20] containing 13 chapters with several illustrative drawings. This section is essentially a military treatise with descriptions of many war machines, some of them dedicated to naval warfare.

These warfare techniques and instruments were probably not Guido da Vigevano's original ideas. They were most likely inspired by the machines used during the wars between the communes in the north of Italy. It is well documented that as far back as the eleventh century, the Italian communes, such as Milan and Bologna, employed sophisticated war carts in several battles. Da Vigevano's illustrations are simple, lacking perspective and three-dimensionality, and are very much rooted in the medieval conception of technology. Nonetheless, as we have indicated above, some significant innovations are also included. It is these elements that make the *Texaurus* one of the true predecessors of Leonardo's work.

Guido da Vigevano did not have the opportunity to see the practical exploitation of his inventions in the intended crusade because Philip VI meanwhile turned his bellicose initiatives to Europe and instead of a crusade had the better idea of starting the Hundred Years' War (1337–1453).

Among the illustrations in the *Texaurus* can be found a sketch (Fig. 6.17) of what is sometimes credited as the first historically documented trace of a "submarine." The sketch, which appears in Chap. 8, is accompanied by a richly detailed description.[21] The ship, which has a round hull, appears to be designed to navigate in several different sea conditions by means of side paddle wheels operated by strong men in the interior of a waterproof hull. As the

[17] *Liber notabilium illustrissimi principis Philippi septimi, Francorum regis, a libris Galieni per me Guidonem de Papia, medicum suprascripti regis atque consortis eius inclite Iohanne regine, extractus, anno Domini 1345, papa vivente sexto Clemente*. The codex is conserved in the Condé Museum of Chantilly in France (ms. 569).

[18] The Fall of Acre in 1291, the last stronghold in Palestine, marked the end of the Christian presence in the Holy Land; from that time, there were several attempts to prepare a new crusade, such as that of Philip VI.

[19] *Liber conservacionis sanitatis senis*

[20] *Modus acquisicionis Terre sancte, Christi nomine invocato, regi Francie intitulatio*

[21] *Capitulum Viii. De Modo Faciendi Naves Per Omnes Aquas Navigantes /. Et Super. Equo Involutas Aportare*

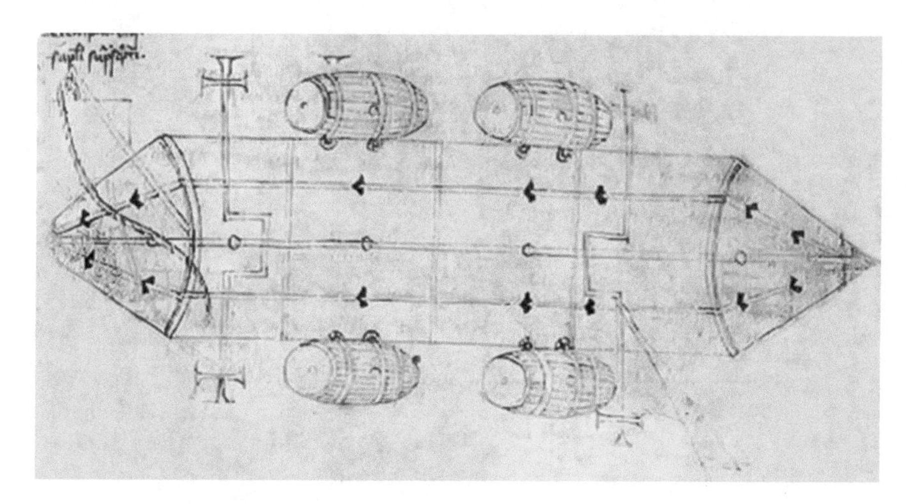

Fig. 6.17 A model of a "submarine". Guido da Vigevano, Texaurus Regis Francies, 1335. Manuscript Latin, 11015. BNF

descriptive heading *On the way of making boats sailing upon all sorts of waters and carrying them folded up on horseback* makes clear, the ship was designed to be both portable and unsinkable.

The text goes on to describe how this is to be achieved: "And so that the boat may be made unsinkable let them attach to its sides four wooden casks each a yard and a half long and a span wide. And thus is completed a good and safe boat for four horses, which can be got ready in an hour and put into the water either by night or by day; a single horse will carry one (or three horses can carry two) of them folded up separately. And so that the boat may be strongly propelled through the water by fewer men, let them make paddles in this way, namely let them take two logs as thick as a man's arm and two yards long each, according to the judgment of the makers, and on their ends let them fit two paddles crosswise, a foot broad and a yard long more or less, as the boatmen shall judge best for propelling it through the water. And those two timbers are joined together by an iron [crank] handle and placed one at each end of the boat upon two polished pieces of wood as high above the sides as the makers think fit. And so they are turned like a [hand-] mill."

Despite its structural similarity to a submarine, da Vigevano's boat was actually designed for navigating only on the water's surface. Indeed, it lacks any specific air reservoir or device for getting air from the surface. The external barrels serve to guard against uncontrolled rotation of the hull, which would otherwise be difficult to avoid because of the boat's shape.

Nonetheless, it is possible that the intriguing form of this vessel may have inspired Valturio, as it is possible to find a similar boat in *De Re Militari* (Fig. 6.18). This vessel was clearly designed to be completely sealed, and when

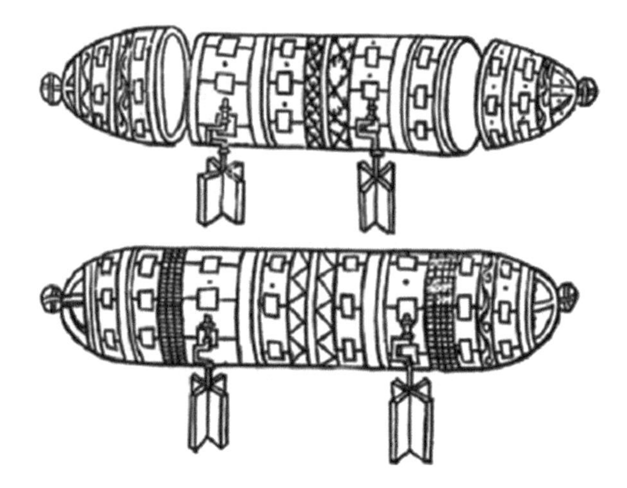

Fig. 6.18 Model of "submarine". Valturio, De Re Militari

we compare the two prototypes, the similarity is striking. Valturio's boat was to be covered with wax-treated linen to make it waterproof and propelled via a relatively sophisticated system of external paddle wheels. In all other aspects, the ship recalls da Vigevano's model, barring, of course, the absence of the stabilizing external barrels and the rear tiller.

We now come to Leonardo's design. Its great fame arises from its incredible similarity to modern boats, with its rounded hull and upper turret (Fig. 6.19). The short, cryptic note that accompanies the small sketch is, however, not so clear: "Ship to be used to break down boats with the instrument you know."[22] The text that follows is (likely intentionally) even more obscure. What is clear is that the vessel was intended to be employed to attack and destroy enemy ships.

Less clear is whether Leonardo truly intended an underwater vessel or simply wanted to represent a ship which, due to its shape, could impact and break enemy hulls, a common naval warfare technique already well developed by the Greeks and Romans.

Another drawing in the same folio (Fig. 6.20) addresses this challenge from the opposite angle: that of how to build an unsinkable ship. This drawing shows a double hull, which could increase the ship's stability and the resistance to collisions, be they inflicted by man or nature. The inner hull forms a second barrier in case of damage to the outer wood layer. It is interesting to observe that Leonardo was considering at the same time how to sink a ship and how to protect it.

[22] Original text (Manuscript B, folio 11br): "*Nave da usare a sfondare navili collo strumento [che] sai. E fassi quel rilievo in mezzo, perchè ti schifi il frugare colle lance dallo schifoRico[r]dati [ch]e entri e che serri, di mandare fori la l t, e che ripigli della quantità del vacuo.*"

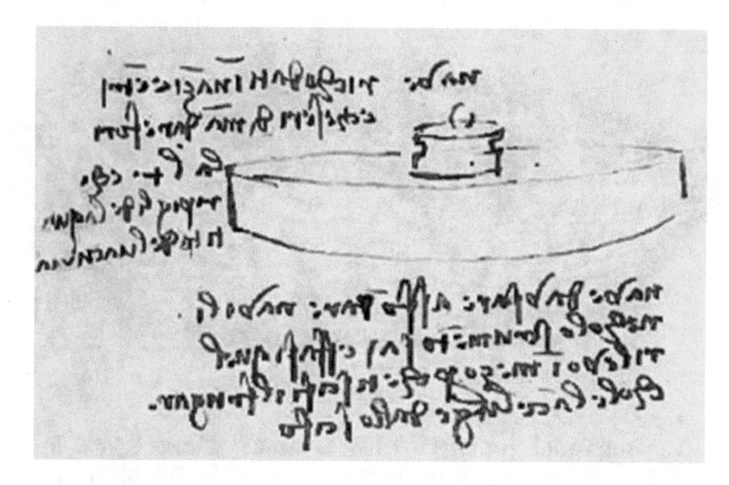

Fig. 6.19 Drawing of a "Ship to be used to break down boats with the instrument you know." Leonardo da Vinci. Manuscript B, folio 11r, detail. BIF

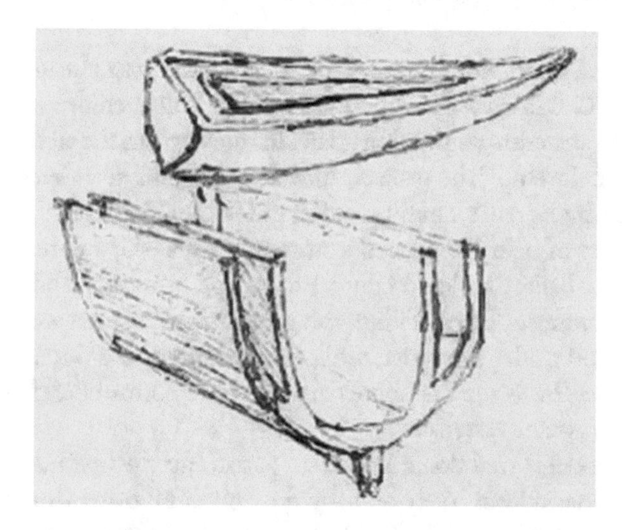

Fig. 6.20 Sketch of double hull. Leonardo da Vinci. Manuscript B, folio 11r, detail. BIF. (1487–1490)

After these first simple and naive prototypes of submarines which appeared, as we have seen, even well before the Renaissance, it was necessary to wait over 100 years before the emergence of a new project for an underwater or diving ship. We have already encountered this innovator, Giovanni Alfonso Borelli, once before, as the designer of a diving device discussed in Chap. 5. In Table XIV of Book I of his *De motu animalium* (*About motion of animals*), just to the right of the illustration of the diver (Figure 9 in da Vigevano's book), we find the design of what is, without any doubt, a submarine (Fig. 6.21). At first

Fig. 6.21 The submarine prototype from Giovanni Alfonso Borelli in De Motu Animalium (1680), Table XIV and Figure 9 (This work is in the public domain in its country of origin and other countries and areas where the copyright term is the author's life plus 70 years or less. https://upload.wikimedia.org/wikipedia/commons/9/98/Giovanni_Alfonso_Borelli_De_Motu_Animalium_1680.jpg)

sight, this vessel does not appear to be such a great step ahead compared to Valturio and Guido da Vigevano's models. The hull is composed of a watertight wooden structure containing goatskins and oars in the stern to guide the movement of the ship. The project, however, contains an interesting innovation inspired by Borelli's careful studies, guided by his medical training, of a fish's capability of being immersed in water. The goatskins were conceived, in fact, as simple ballast tanks to control the ship's buoyancy and the stages of submersion and resurfacing. Filling the goatskins with water would cause the boat to descend under the water, while squeezing out the water should cause it to resurface. This is the first reported attempt to control the diving capability of an underwater vessel.

Borelli's model is well documented on paper, but the first navigable prototype of a submarine was likely built by the Dutch inventor Cornelis Drebbel (1572–1633), whose ship in 1620 was able to navigate submersed in the Thames river, albeit for a short distance. No illustrations or accurate descriptions remain, however, of Drebbel's submarine.

After this, nearly two centuries would pass before a new impulse for the idea of building an underwater ship would emerge. Returning to Leonardo, as we have seen in other contexts, in the case of submarine warfare as well, he appeared very much afraid that his inventions would be used for "the killing of men" (*assassinamenti de li omini*). For this reason, he recorded only undetailed sketches of most of his system, accompanied with descriptions that were largely obscure and extremely difficult to understand. He wrote: "How and why I do not describe my way to stay under water… this I do not publish or

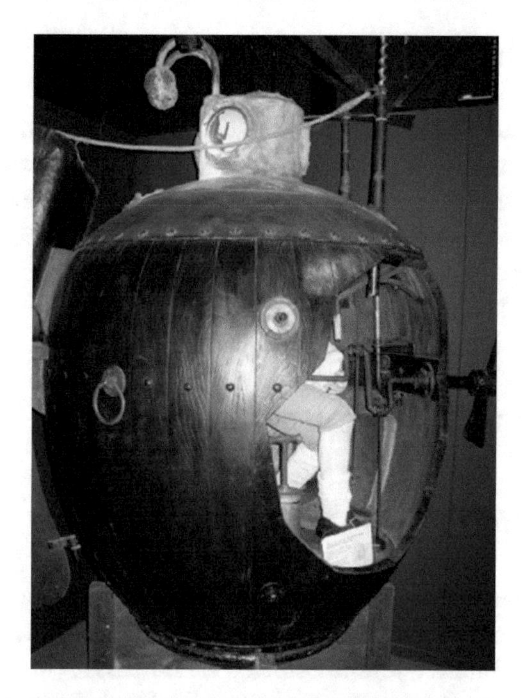

Fig. 6.22 Picture of a full-size replica of the American Turtle at the Royal Navy Submarine Museum in Gosport, UK (Permission is granted to copy, distribute, and/or modify this document under the terms of the GNU Free Documentation License. Courtesy of Geni under GFDL CC-BY-SA. Version 1.2 or any later version published by the Free Software Foundation,; with no Invariant Sections, no Front-Cover Texts, and no Back-Cover Texts. A copy of the license is included in the section entitled *GNU Free Documentation License* (https://commons.wikimedia.org/wiki/Commons:GNU_Free_Documentation_License,_version_1.2). This file is licensed under the Creative Commons Attribution-Share Alike 4.0 International (https://creativecommons.org/licenses/by-sa/4.0), 3.0 Unported (https://creativecommons.org/licenses/by-sa/3.0/deed.en), 2.5 Generic (https://creativecommons.org/licenses/by-sa/2.5/deed.en), 2.0 Generic (https://creativecommons.org/licenses/by-sa/2.0/deed.en) and 1.0 Generic (https://creativecommons.org/licenses/by-sa/1.0/deed.en) license.)

make known because of the bad nature of men, who will use [my inventions] to kill [people] in the depth of the sea, to break the ships and let them sink with all the people inside."[23] Leonardo's predictions in this regard were, of course, spot on. By chance—or not—one of the first modern submarines, the *American Turtle*, built in 1775 by David Bushnell, was designed to attach explosive charges under English ships (Fig. 6.22). In practice, the attempts failed, but this was only the beginning of a new era of devastating war machines, exactly as foreseen by Leonardo.

[23] Original text: "*Come e perchè io non iscrivo il mio modo di star sotto l'acqua quanto io posso star senza mangiare; e questo io non poblico e divolgo, per le male nature de li omini, li quali userebbero li assassinamento ne' i fondi de' mari, col. rompere i navili in fondo e sommergerli insieme colli omini che vi son dentro.*"

The Amphibious Ship

As we have already seen, many precursor models of modern ships can be traced back to the late Middle Ages. There is another such idea which we should mention in passing for completeness—that of amphibious boats along the same lines as Captain Agostino Ramelli's amphibious armored tank (Fig. 6.13). In his *Trattato di Architettura*, Francesco di Giorgio shows, among several other models of ships, a small boat with several side wheels (Fig. 6.23). These wheels, which are manually moved by a pair of gears, are most likely designed to get a better grip in muddy ground. Despite the basic design, the similarity to the modern-day car is quite impressive.

The Mud Dredge

We have seen that Francesco di Giorgio's manuscripts are particularly rich in studies devoted to innovative methods for the propulsion of ships. Warships were his primary interest, and his designs encompass an incredible collection of bizarre and somewhat outdated devices for breaking hulls, such as rams, hooks, and more. These were clearly rooted in the tradition of Roman naval warfare, but they also indicate a need or desire to go beyond the known solutions. The drawings incorporate a wide variety of firearms, from bombards to the more advanced cannons. Most of the drawings are clearly not designed for practical use but rather act as a proof of concept.

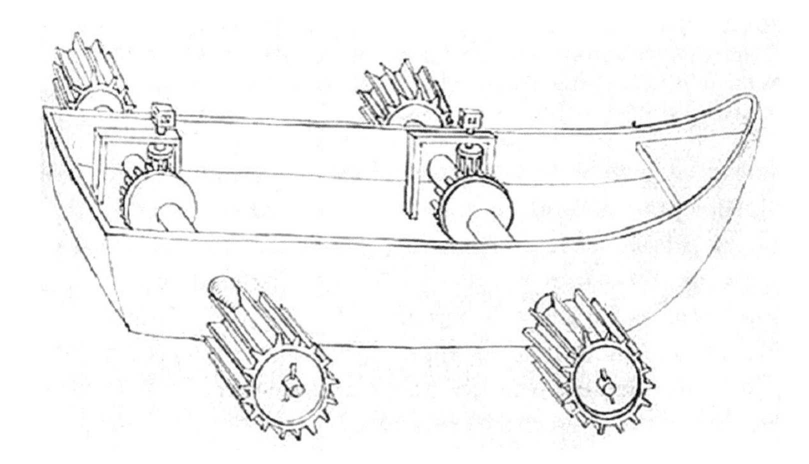

Fig. 6.23 Project of an amphibious boat. Francesco di Giorgio. Trattato di architettura. BNCF

Fig. 6.24 *Left*: Mud dredge from Francesco di Giorgio (Manoscritto Saluzziano 148, f. 64v. Biblioteca Reale di Torino). *Right*: Copy realized by Leonardo (Manuscript E, folio 75v). BIF

However, Francesco di Giorgio was also very attracted to the possibility of constructing specialized ships such as floating mills, mud dredges, and pontoons. One such interesting example, which was copied with a small change by Leonardo, is a design for a mud dredge (folio 64v of *Manoscritto Saluzziano 148,* Fig. 6.24, left).

The mud dredge was designed to clean the beds of canals, lagoons, and harbor areas. As can be seen from the figure, it was to be formed by a pair of boats supporting a central wheel with four grabbers, which collect mud from the depths. In Francesco di Giorgio's design, the mechanism was to be manually moved by a crank. In Leonardo's sketch, which most likely dates to 1513–1514, this crank is moved to the wheel hub.

Leonardo labels his dredger an "instrument to take out soil."[24] As in so many instances, even if his model of a dredger is largely a copy, Leonardo introduced small variations to improve its performance. Indeed, Leonardo's crank serves a double purpose: It not only directly rotates the wheel removing the mud but also moves the dredger forward using a sheet anchored to the shore. In this way, the dredger slowly advances as each section of the canal bed is cleaned of mud. This innovative mechanism is described in a detailed note by Leonardo.[25]

An interesting question arises: Why was Leonardo thinking about a mud dredger during this period of his life? At the time he drew the sketch, Leonardo was in Rome, where Pope Leo X (1475–1521) was trying to reclaim the big swamplands surrounding the city. With this project on the forefront of his

[24] Original text: "*strumento per cavare terra.*"

[25] Original text: "*Il voltare del mani[c]o n volta una rocchetta, e questa rocchetta volta la rota dentata f, e questa rota f è congiunta colla croce delle casse, portatrice della terra del pantano, che si scarica sopra le barche. Ma le due corde m f e m b s'avvoltano al polo f e fa[n] camminare lo strumento colle 2 barche contro al m; e queste corde per tale uffizio sono utilissime, ecc.*"

mind, Leonardo most likely remembered Francesco di Giorgio's mud dredger. Unfortunately, however, the work never came to fruition, as the Pope died soon after and the project was abandoned.

The examples of paddle wheel boats and submarines that we have seen in this chapter clearly show how the new ideas may a somewhat linear time development but are also very much dependent on the level of technology available at the time in which they are conceived. Leonardo's projects and the Sienese engineers' ideas were very sophisticated from a technological point of view, but they had very few or any chances to be practically exploited. On the other hand, there is a common cultural approach which links these first innovators through the centuries, from Guido da Vigevano to Leonardo and beyond. They were interested more in the concept, the basic idea behind the innovation. They considered the vision of the future to be more important than the present, and this may be the true reason they fascinate us so much.

7

Wind Chariots and "Automobiles"

The Renaissance engineers devoted special attention to a particular kind of war instrument: the war chariots. The result of their efforts is an impressive list of mobile systems for warfare: scythed chariots, wagons loaded with soldiers, armored vehicles with protection from external attacks, mobile towers built to break through defensive walls, and more. They are all part of an almost inexhaustible catalogue of war tools described in great detail and with extraordinary inventiveness. Despite their wide variety, these systems are still part of the legacy of a distant past and mostly rely on Roman sources still well known in the Middle Ages.

The wide variety of war chariots found, in particular, in the work of Taccola and Francesco di Giorgio, addressed the need for innovative warfare solutions in a world experiencing a rapid technological evolution. It is interesting that the attention of the Sienese engineers was directed not only to the structure and design of the war wagons but also to the means of generating and transmitted motion in vehicles. The case of the *liburna* paddle warships moved by pairs of oxen or by treadwheels which we encountered in the previous chapter is just such an example of the several innovative solutions that were beginning to spread in Europe. Although these initial plans were in large part naïve and embryonic and were not translated immediately into real machines, they represent the beginning of a long process leading to our modern cars. It is also worth noting that, despite limits on communication in the period, they enjoyed considerable circulation and dissemination through various channels.

P. Innocenzi, *The Innovators Behind Leonardo*, https://doi.org/10.1007/978-3-319-90449-8_7

The First Wind Chariots

The chief limitation for war chariots was the challenge of generating sufficient traction force. Oxen, horses, or men were commonly employed for this purpose. A creative alternative which we shall explore here is wind power.

Based on the extant sources, Guido da Vigevano was the first to imagine a chariot driven by the wind (Fig. 7.1, left). It appears as a kind of self-propelled wheeled mill equipped with a wind-driven propeller. The propeller transmits the motion to the wheels via a system of gears. The idea is clearly derived from the operation of the mills, with the thinking going that, if wind is able to move a mill, it is not a large logical step to presume that it could rotate the wheels of a wagon through a similar technology.

The wagon's structure is rather complex, more closely resembling a kind of self-propelled castle than a means of human transport. In another illustration of the same source, there appears a virtually identical wagon with the exception of the upper part (Fig. 7.1, right), which confirms the most probable intended use of the wagon: on the summit we find a tower full of soldiers with their spears pointed outward. It is a kind of tank for attacking castles and fortified towns.

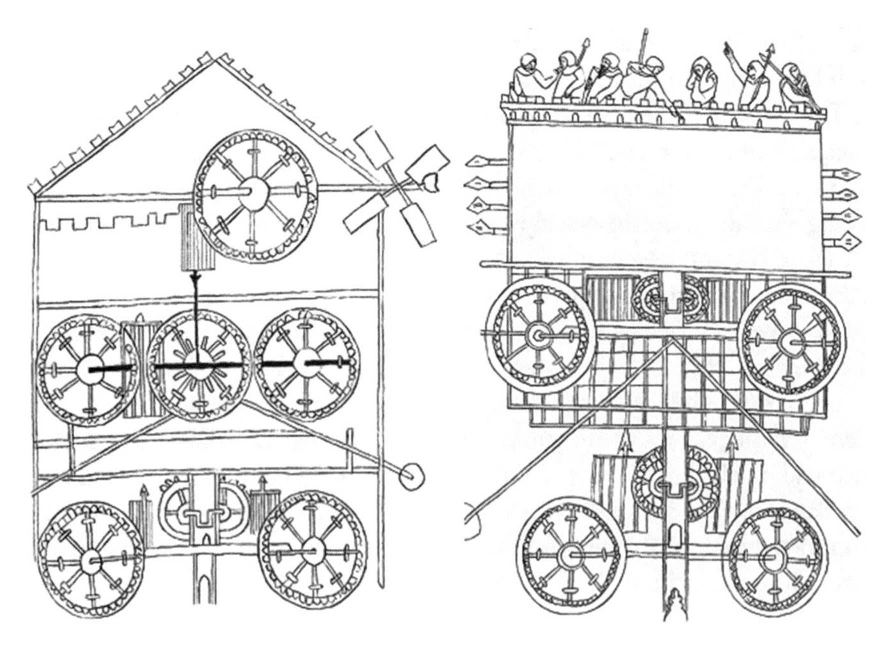

Fig. 7.1 A wind chariot folio 52v (*left*) and a battle wagon, folio 51r (*right*). Guido da Vigevano, Texaurus Regis Francie, 1335. Manuscript Latin, 11015. BNF

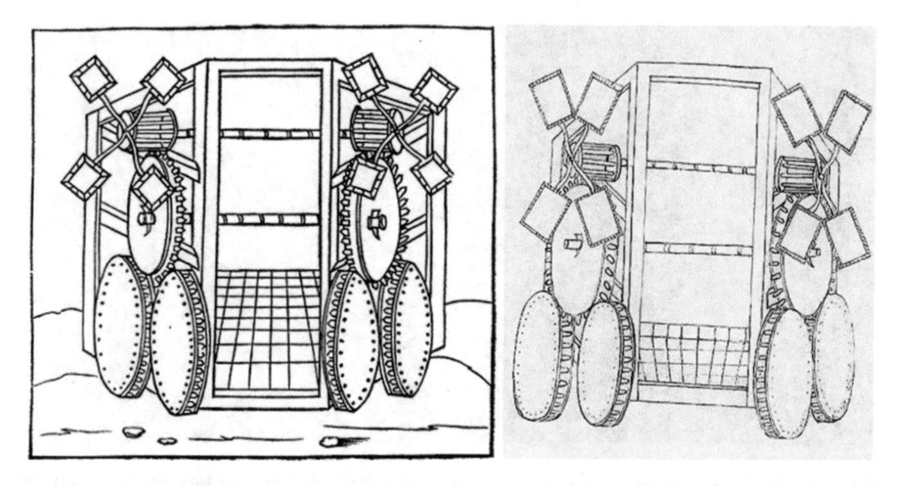

Fig. 7.2 Wind chariot (*left image*). Valturio. De Re Militari, liber X, page 232. Paris 1534. BSBM. Copy of Valturio's wind chariot (*right image*). Bonaccorso Ghiberti. Zibaldone, pag. 192. BNF

Guido da Vigevano's wind-propelled wagon probably opened the way for Valturio's chariot (Fig. 7.2, left), which is also the more famous of the two due to the wider diffusion of the latter's printed work. Valturio's wind chariot is very similar, and it seems likely that he had encountered da Vigevano's earlier design.

The structure of Valturio's wagon is simpler and easier to understand, even if the lack of perspective in the drawing doesn't help. The cart has two pairs of wheels on the sides that are moved via a toothed wheel driven by wind blades located on opposite ends of the upper frame. This design attracted the interest and imagination of many Renaissance engineers, and similar versions of the wind wagon are reported in a number of other texts, such as Bonaccorso Ghiberti's *Zibaldone*, which includes a nearly exact copy (Fig. 7.2, right).

Wind-Driven Amphibious Vehicles

Among the inventors of self-propelled wagons, we cannot leave out Taccola, who designed a multipurpose amphibious tank, perhaps the first of its kind, which can be found on folios 27v, 28r, and 28v of *Liber tertius de ingeneis ac edifitiis non usitatis* (Fig. 7.3, left). The tank consists of a cart with a mast in the middle outfitted with a sail and rudder. However, it has the distinction of also featuring four wheels to allow for movement on land as well as by sea. In both instances, the force of the wind would aid the movement.

Fig. 7.3 Wind-driven amphibious chariot (detail), *left*. Mariano di Jacopo (Taccola). De Ingeneis, BNF. Models of amphibious chariots, (detail), *right*. Francesco di Giorgio. Codex Ashburnham 361. BMLF

In his text, Taccola listed, at great length, the advantages of a multipurpose transport vehicle such as the one he had designed. The bottom wheels, which consist of two disks connected by pins, made it possible for the wagon to be dragged by buffalos on dry land even in the presence of mud. The amphibious wagon could also move on mixed terrains with both water and mud or in large rivers, where the chariot could float and be pulled by water buffalos also capable of moving along the bottom of the river due to their great strength. Finally, with the animals decoupled, the same wagon could navigate on open water using the sail and rudder.

Taccola also provided brief instructions on the reasons for the wagon's particular construction. Instead of being flat, the bottom should be slightly concave to ensure better navigability, while the caisson should be thoroughly treated with pitch to ensure a watertight seal.

Although not a major innovation, Taccola's chariot clearly represents an attempt to address the practical need for a multipurpose transport system capable of moving in shallow water such as when fording rivers or traversing wetlands. As with so many of the inventions of the Renaissance artist-engineers, its primary purpose was most likely military.

Unsurprisingly, Francesco di Giorgio could not fail to follow his master and be inspired by his wind chariot. di Giorgio designed two amphibious ships, both more sophisticated than Taccola's stocky chariot, but designed to be towed on land rather than being propelled by the wind (Fig. 7.3, right). Konrad Kyeser had already imagined something similar (c. 1405), a self-propelled cart for launching bridges to ford rivers. Both di Giorgio and

Fig. 7.4 Amphibious wind-propelled ship. Giovanni Fontana. Bellicorum Instrumentorum Liber, folio 98r. BSBM

Kyeser's vehicles were designed for specific, primarily military needs, and, appropriately, they are the precursors of our modern military technology.

There was another individual, however, who combined several of these ideas to design an amphibious warship: Giovanni Fontana (c. 1395–c. 1454). His ship (Fig. 7.4) featured four toothed wheels to allow for movement in muddy soil as well as on land. The addition of these wheels represented Fontana's single stroke of innovation. However, because he did not include a system to enable changes in direction, it is very unlikely that this system could have effectively worked on land.

To our knowledge, Taccola, Valturio, and Giovanni Fontana's designs for wind chariots remained confined to their notebooks, and no attempts were made to test the inventions. This was a common approach which Leonardo also shared, and in most cases his incredible inventiveness failed to be transformed into working devices. Regarding wind chariots, Leonardo for sure

Fig. 7.5 The wind chariot or "land yacht" of Stevinus for Prince Maurice d'Orange. Engraving by Jacques de Gheyn (This work is in the public domain in its country of origin and other countries and areas where the copyright term is the author's life plus 70 years or less. The image has been modified with respect to the original to reduce the noise. https://upload.wikimedia.org/wikipedia/commons/9/96/Simon_Stevins_zeilwagen_voor_Prins_Maurits_1649.jpg)

knew at least Valturio's project, but it seems he was not particularly attracted by the idea and did not record designs for any such chariots in his extant manuscripts.

The idea of a wind chariot, however, became reality not much later. In Northern Europe, in fact, another very talented inventor, the Flemish scientist Simon Stevin (1548–1620), also known as Stevinus, realized a carriage propelled only by the force of the wind (Fig. 7.5). Stevinus received patents for several windmill-related innovations, and he was likely inspired by mills to use wind as a force for propelling a cart. In contrast to the vehicles that we have encountered thus far in this chapter, Stevinus's wind wagon was designed not for war but, rather, for fun. It was a land yacht to be used for traversing the flat windy beaches of Northern Europe. Chronicles of the time reported that in 1600 Stevinus's vehicle covered the distance between Scheveningen in Belgium and Petten in Holland (around 100 km) in 2 h solely propelled by the wind.[1]

[1] Duyvendak J. J. L (1942) Simon Stevin's "Sailing-Chariot," *T'oung Pao*, Second Series, Vol. 36, Livr. 3/5 (1942), pp. 401–407.

Land sailing or land yachting is still a very popular sport, primarily practiced in windy flat areas and with international competitions, such as the Landyachting World Championship, regularly organized all over the world.

The First "Cars"

The projects which we have just seen, though very fascinating in their own right, can hardly be directly connected to the modern automobile. They were, perhaps, the forerunners of our modern-day trucks and reflected the need to transport people and goods using innovative traction modes beyond animal power. However, these innovations were inspired by extant technologies such as the windmills which are clearly the basis for Taccola and Valturio's models.

However, in a manuscript of the time we can also find an illustration that appears remarkably similar to the modern car, at least in its general appearance (Fig. 7.6). The manuscript in question is the Latin *Bellicorum Instrumentorum Liber* of Giovanni Fontana, whose amphibious ship we have just discussed.

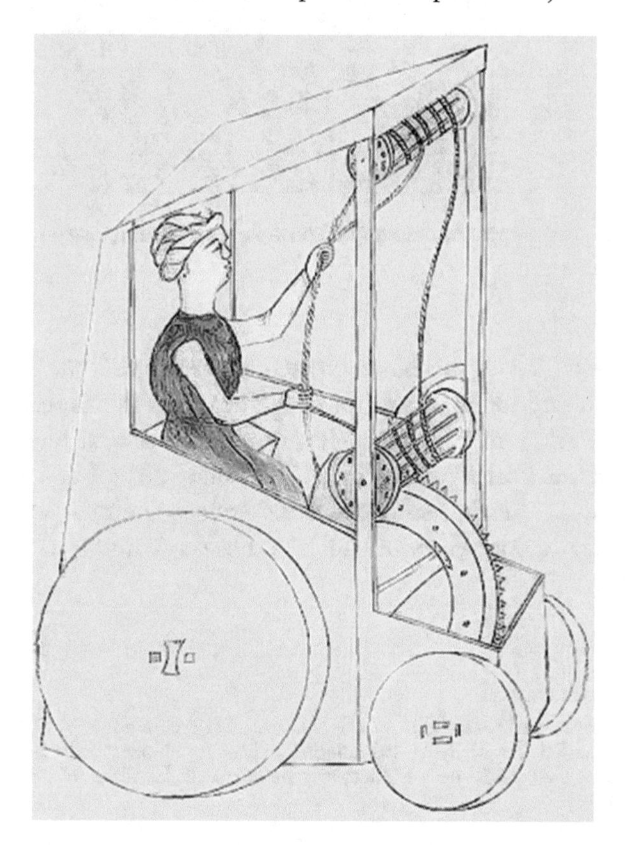

Fig. 7.6 "Catredam deambulatoriam." Giovanni Fontana. Bellicorum Instrumentorum Liber, folio 18r. BSBM

Fig. 7.7 A fantastical figure. Giovanni Fontana. Bellicorum Instrumentorum Liber, folio 60r. BSBM

As the title indicates, this book, written between 1420 and 1430, is dedicated to the instruments of war, but together with the expected drawings and illustrations of military strategies, it contains several inventions alternated with fantastical figures (Fig. 7.7). Much of the book is written in cipher, the first time that such a writing style was employed outside the context of secret communications such as those containing sensitive matters of diplomacy.[2]

[2] Battisti E, Saccaro Battisti G (1984) Le macchine cifrate di Giovanni Fontana. Con la riproduzione del Cod. Icon. 242 della Bayerische StaatsBibliothek di Monaco di baviera e la descrizione di esso e del Cod. Lat. Nouv. Acq. 635 della Bibliothèque nationale di Parigi. Arcadia Edizioni Cinisello Balsamo, Milano

Giovanni Fontana

Giovanni Fontana was born in Venice probably around 1390, and, as was also the case for Konrad Kyeser and then Veranzio, he studied at the University of Padua, completing his doctorate in the arts in 1418. He was a contemporary of Taccola's, although, as far as we can tell, the two ignored each other's works.

Fontana continued his studies to become a medical doctor and practiced his profession in Udine, a small city in the Republic of Venice in northeast Italy. Despite the heavy burdens of his medical practice, he used much of his free time to cultivate his passion for mechanics, astronomy, and trigonometry. Not much is known of his life. He probably made a series of trips as far as Rome and Crete and died in an unknown location just after 1494.

After his death, his manuscripts were dispersed in a number of European libraries without being published and, with the exception of *Bellicorum Instrumentorum Liber* which was printed about a hundred years after his death, his work remained in obscurity until about 1700. Ten of Fontana's treatises have been preserved. Five others mentioned in his other works are considered lost.

Fontana still is not well known outside of specialists circles, but he represents one of the most versatile and talented personalities of the early Renaissance. With his humanistic culture and passion for science and technology, he embodies the figure of the new man that would mark the transition from the Middle Ages to modernity.

Fontana used a variety of different historical sources for his books but then took these further with his own independent ideas. His works contain, among other designs, plans for a mechanical organ, fountains, magic castles, theatrical effects, stoves and locks, the first description of a magic lantern, and a series of automata. Fontana's memory and encryption machines are among his most original designs and can be found in his *Secretum de experimentorum imaginationis thesauro hominum* ("The secret of the treasure of the human imagination and experimentation").[3]

Among the wide variety of machines and projects to be found in the *Bellicorum* is the one mentioned above for a "*catredam deambulatoriam*" or "moving chair" (Fig. 7.6), whose shape undoubtedly very closely resembles that of a modern-day car. This "car" stands out with regard to both design and general conception from the self-propelled armed military wagons we have previously seen. In contrast to those vehicles, this one is designed to transport an individual comfortably seated in a "chair" with wheels. The car is advanced via hand-operated system of ropes attached to the wheels. This represents perhaps the first depiction of a car for the transport of individuals in the modern sense.

[3] Paris, Bibliothèque Nationale, Cod. Lat. Nouv. Acq. 635.

Fontana's description of its use and operation is very detailed: "I called [this thing] that I invented for your sake mobile chair. It starts with such force to overcome a running horse and sometimes it is hard to restrain, if you drive on a flat surface or downhill. The innovation introduced in this object is to place a chair on four wheels; the rear however must be larger than the front ones, as in wagons. Then those who rise [on the chair] must roll a cylinder, which turns with strength the iron gears, and then the four wheels."

Maybe Giovanni Fontana somewhat exaggerated the performance of his moving chair, especially when comparing its speed to that of a horse. Putting this aside, for its level of ingenuity and inventiveness, the model was certainly worthy of praise and attention.

Francesco di Giorgio's "Automobile"

Neither Francesco di Giorgio nor Leonardo himself, who also developed projects for self-propelled machines, would come as close as Fontana to envisioning automobiles as a means of transporting people. Nonetheless, it was di Giorgio who first used the term *automobile*. This was the result of several studies whose traces remain in many of his manuscripts.

On folio 25v of the *Opusculum de Architectura*, there can be found a beautiful drawing of a machine with four wheels fitted with an equal number of winches located on the roof (Fig. 7.8). The drawing appears in perspective, and the mechanics of the machine can easily be understood. To move the car forward, it is necessary to rotate the winches, presumably by hand. The wheels are decoupled, and it is possible, thanks to a simple pivoting system, to turn the wagon via a shaft controlled by a central winch also located on the roof. From the roof, the pilot of the cart could easily look at the route and coordinate the work of the men moving the car.

We can easily imagine that the mechanisms for wind and water mills were the basis for these models.

Additional studies of self-propelled wagons of various shapes and sizes are to be found in di Giorgio's *Codex Ashburnham*, folios 46v and 47r (Fig. 7.9). The drawings show detailed and technically advanced projects that present several possible mechanical solutions for the transmission and control of movement. These sketches demonstrate a sophisticated understanding of the mechanics for building self-propelled carts, a subject which clearly attracted so much of di Giorgio's interest. The power source is still human traction. However, the main interest here is the mechanical systems of gears and winches rather than the question of an appropriate power source, and the

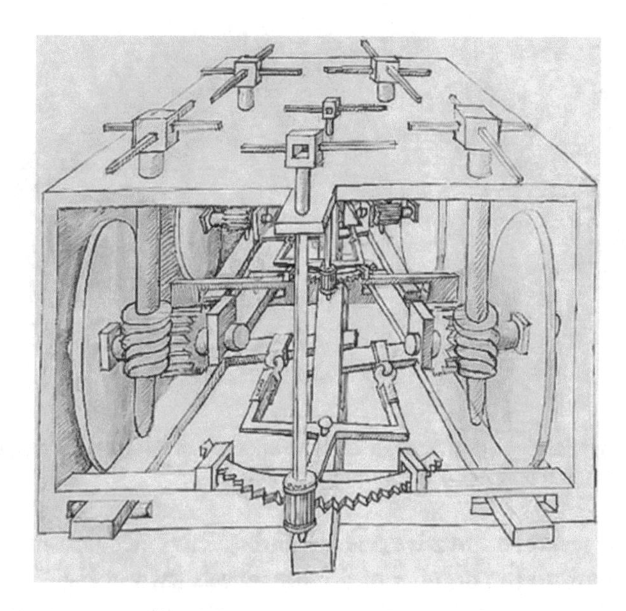

Fig. 7.8 Project for an "automobile." Francesco di Giorgio. Opusculum de Architectura, manuscript 197.b.21, folio 25v. BML

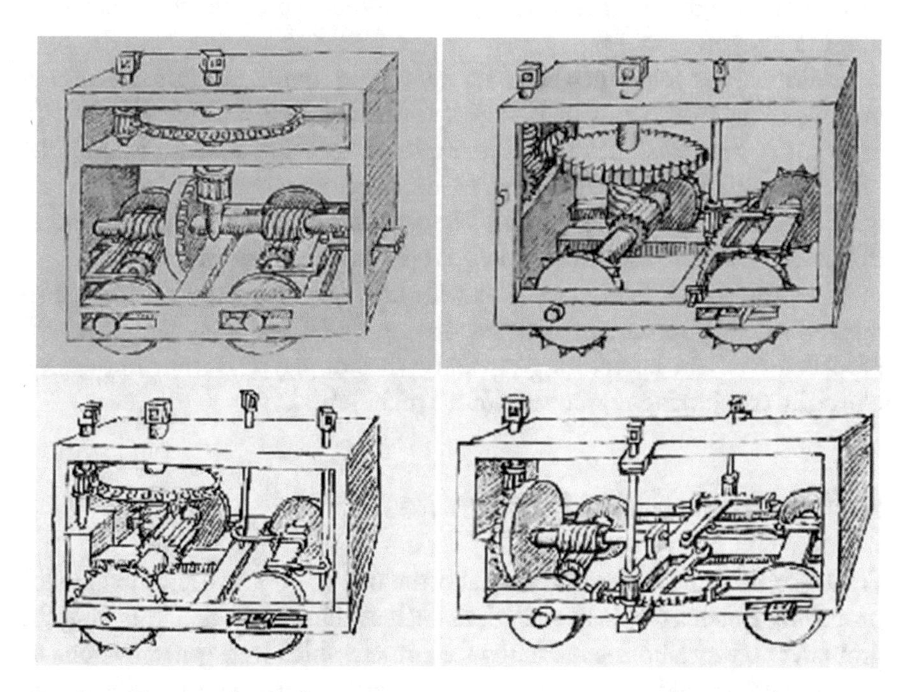

Fig. 7.9 Projects for self-propelled carts. Francesco di Giorgio. Codex Ashburnham 361, folio 46v (details). BMLF

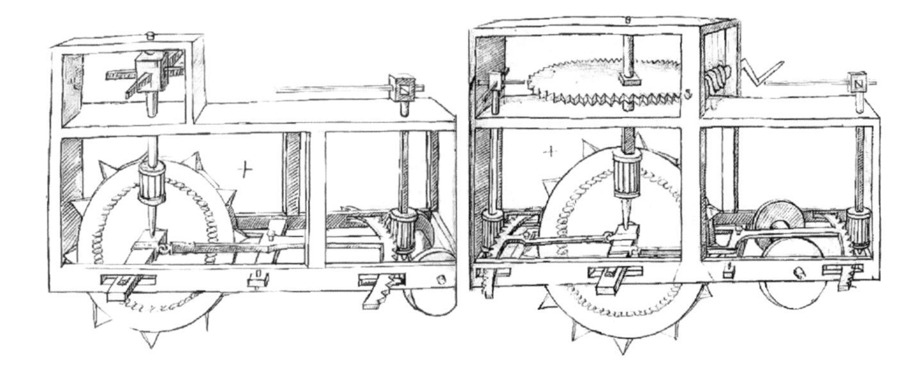

Fig. 7.10 Projects for self-propelled carts. Francesco di Giorgio. Codex Ashburnham 361, folios 23v (*left*) and 24r (*right*), details. BMLF

designs show several innovative turning and steering mechanisms. The models vary with regard to the number of driving wheels (two in Fig. 7.9 top left and four in Fig. 7.9 top right) and the type of steering system (e.g., the front and lateral systems found in Fig. 7.9 bottom left and bottom right, respectively). As can be seen, the wheels of these chariots could also vary with regard to size and surface design, for example, including toothed wheels for additional traction in muddy terrain.

Additional designs, which make for interesting comparison, are to be found on folios 23v and 24r of di Giorgio's *Codex Ashburnham 361* (Fig. 7.10). These are two rather similar models to explore once again different possibilities of mechanical design. The left-hand model has three wheels—a driving wheel with two auxiliary steering wheels. The right-hand design differs in that, in this case, the larger rear driving wheel is also capable of steering.

As we can see, di Giorgio made a wide and systematic study of the transmission of motion and of coupled driving and steering systems for self-propelled cars. The engineering capability thus displayed is really impressive, especially considering that it was not his main job….

Self-Propelled Work Machines

Francesco di Giorgio was probably also the first to conceive and design self-propelling systems for daily operations such as tilling the soil and hoeing the ground, two very laborious activities for farmers which were generally done by hand or with the aid of oxen. Within di Giorgio's notebooks is possible to find several projects for such machines, in particular, a mechanical hoe and a tractor for tilling (see Figs. 1.6 and 7.11, respectively). Both are man-powered and

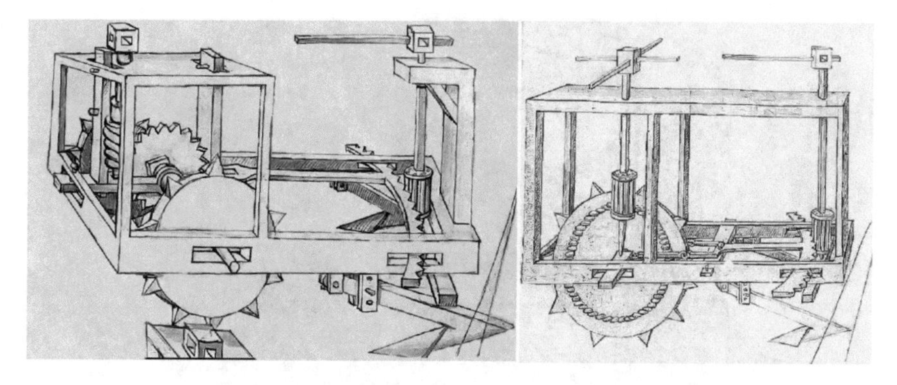

Fig. 7.11 Projects for a tractor for tilling the soil. Francesco di Giorgio. Opusculum de Architectura, MS 197.b.21, folios 22v (*left*) and 23r (*right*), details. BML

feature a design derived from the "automobile" projects that we have seen above. The mechanical hoe has a steering system derived from the "automobile" with a pair of hoes moved by a man via a crank. The crank puts in motion a toothed wheel which, in turn, transmits the movement to the four wheels of the cart, allowing for a slow forward motion. The same toothed wheel alternatively lifts and lowers the hoes. As we have mentioned, di Giorgio designed a second machine to aid work in the fields: a tractor for tilling the soil. He realized two models with one and two traction wheels (Fig. 7.11, right and left, respectively). The wheels have pyramidal teeth to ensure the proper grip in muddy soils, and the motion is obtained via man-powered shafts connected to the wheels. Another shaft is used for controlling direction.

di Giorgio's designs for the tractor and the hoe show that he appreciated the importance of the systems for both transmission of movement and control of direction in a self-propelling wagon, which required both accurate knowledge and careful design.

Leonardo, in a small sketch in the *Codex Atlanticus*, folio 17v (Fig. 7.12), made his contribution to the mechanics of "cars," suggesting a simple system, similar to a differential gear, for transmitting motion to the cart axle: When the horizontal gear transmits the motion to one of the wheels via a pinion on the axle, the other moves at a different speed, allowing for rotation.

The Treadwheel-Powered Car

The "automobile" models we have encountered thus far centered around non-animal sources of motion such as wind and human arm strength aided by structured mechanical systems.

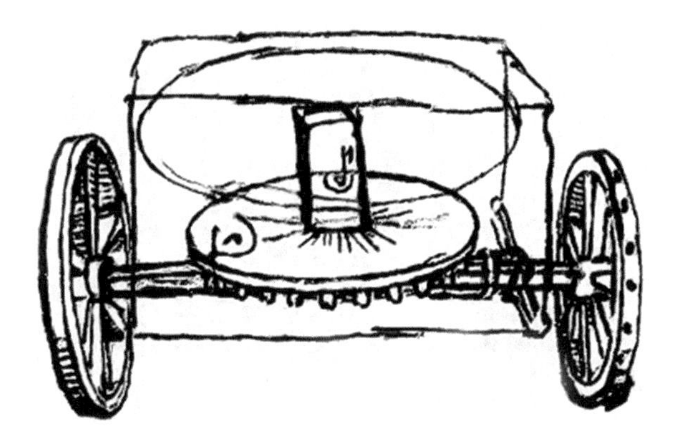

Fig. 7.12 Differential gear for a wagon. Leonardo da Vinci. Codex Atlanticus, folio 17v, detail. BAM. 1478–1480 about

Another well-known and widely used technique for generating motion using man rather than animal power was the treadwheel, which, as we have seen, was largely used in construction technologies. With some creativity, at least in the fantasy sketches of the Renaissance engineers, its use was extended to other systems, such as ships and, as we will see here, even for a self-propelled "car" designed by Antonio da Sangallo (Fig. 7.13). His rough sketch shows an "automobile" propelled via a pair of parallel treadwheels.

The treadwheel puts in motion a system of gears which transmit the movement to the axle and the wheels. Although the sketch itself is not very detailed, it epitomizes the Renaissance artist-engineers' passion for exploring new designs and mechanical systems for realizing self-propelling systems. It is fitting that the only word written on the page is the Italian "sognio," or "dream." No better possible comment could be imagined.

Among the approximately 1200 drawings left by Antonio da Sangallo and his studio, there is another sketch which can be related to a project of car (Fig. 7.14), this one clearly inspired by Francesco di Giorgio's work in this area. da Sangallo's version appears as a simplification of the more refined original model of the Sienese artist. da Sangallo's description states that it was designed for military purposes, which may explain the heavy and labor-intensive mechanism of the two man-powered gears at the top.

This drawing seems a step back with respect to the sophisticated design of Francesco di Giorgio. Thus we see that, although Antonio da Sangallo still shared with the previous generation of Renaissance artists the multifaceted approach mixing art, science, and engineering, he is clearly losing the inventiveness and freshness which characterized the work of Leonardo and Sienese engineers. As we will see in numerous instances, it is a sign of a fading time: the increasing complexity of science would come to require specialization, and the unique season of these polymaths would soon vanish.

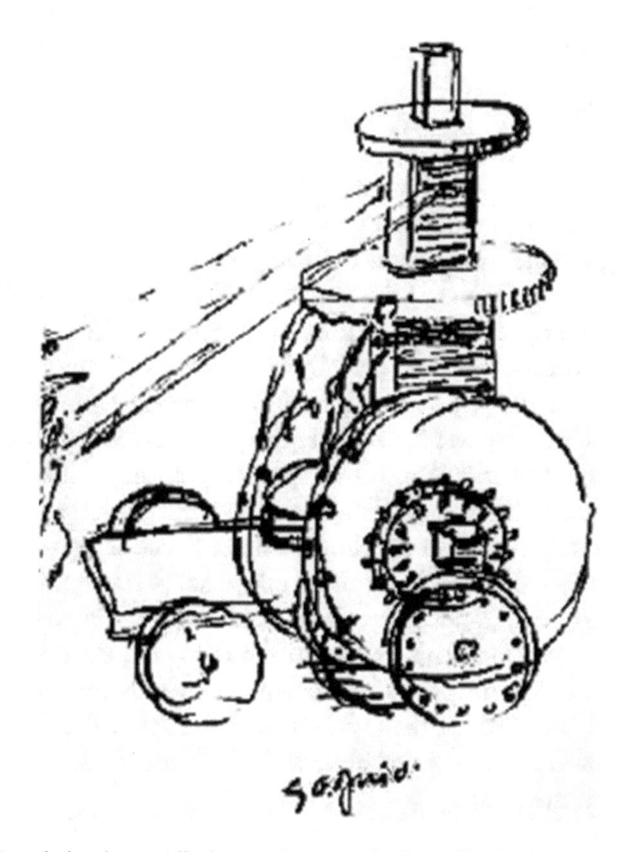

Fig. 7.13 Treadwheel-propelled cart. Antonio da Sangallo the Young. GDSU. (about 1526, +/– 20)

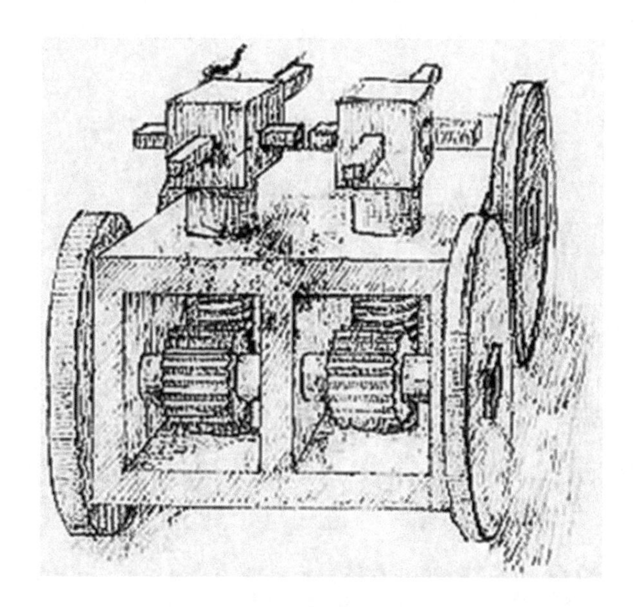

Fig. 7.14 Model of car. Antonio da Sangallo the Young. folio 1450ra, (about 1526, +/– 20). GDSU

Leonardo's Car

At this point, it remains to us to finally introduce the most famous "car" of the Renaissance, which is known as Leonardo's car (Figs. 7.15 and 7.16). Paradoxically, despite its reputation of being the first project of a car, it is, in fact, probably the only one of the cars we have encountered in this chapter that was not intended as such, at least in its function. The wind-propelled chariots and the self-propelled "cars" by Francesco di Giorgio and Giovanni Fontana were designed for the transportation of people, animals, and goods, but Leonardo's machine was instead intended to amaze for its ability to move without any external source of power, probably during a theatrical performance. In practice, therefore, it is more akin to a mechanical automaton than a car and was likely capable of traveling for short distances, up to 40 m, thanks to a spring, which, after compression, is able to act as traction force.

Leonardo's project (folio 812r, *Codex Atlanticus*) was probably drawn in 1478. No notes or comments appear in the folio, and for a long time, the details of the automaton's operation remained rather obscure. The drawing is not particularly detailed. At the top of the folio (Fig. 7.15) is shown a three-dimensional side view of the machine, with three wheels and a small driving wheel placed at the front.

In the same folio, immediately below, there is a top view of the same device, showing the working mechanical system (Fig. 7.16).

As we have seen, some vagueness in design details was quite common not only to Leonardo but also to his predecessors. In many cases it represented a

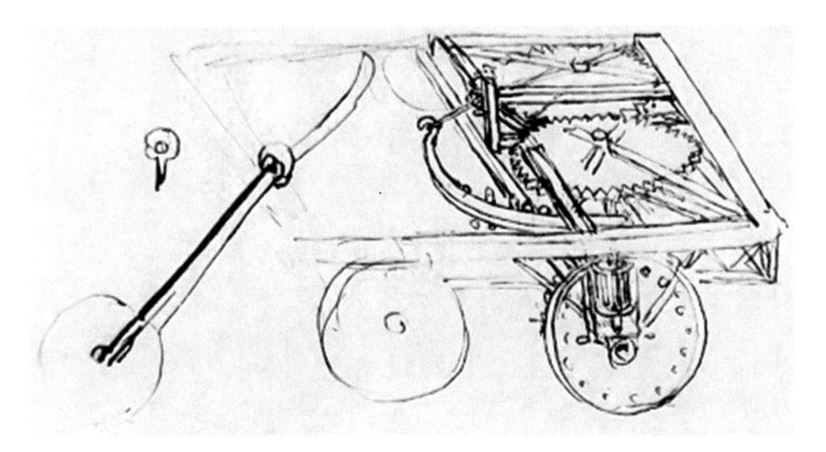

Fig. 7.15 Project of an automaton cart, side view, detail from the upper part of the folio. Leonardo da Vinci. Codex Atlanticus, folio 812r. BAM

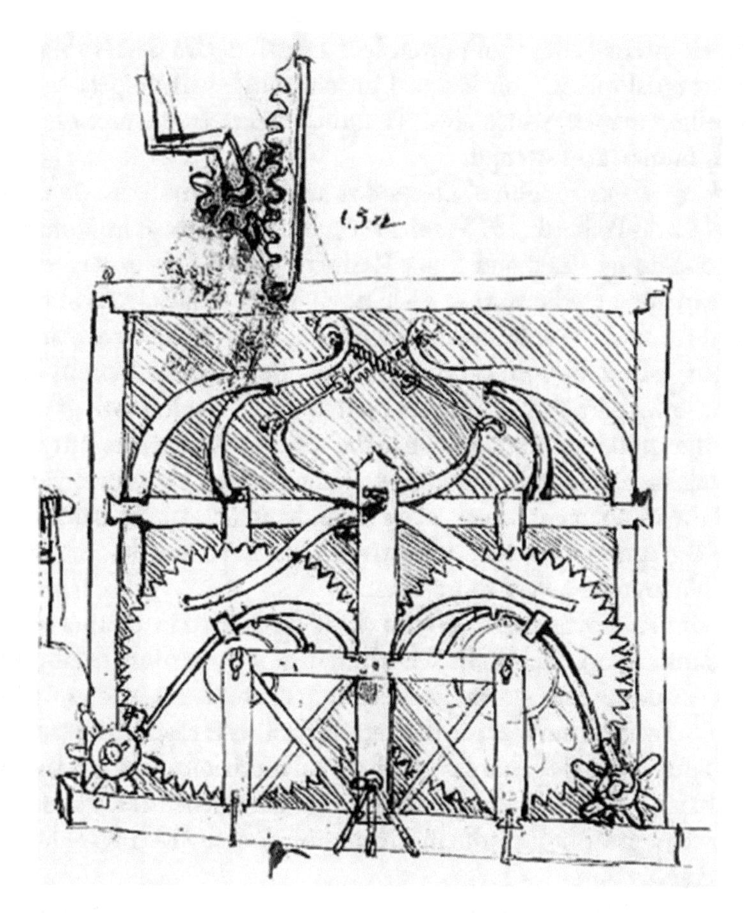

Fig. 7.16 Project of an automaton cart, top view, detail from the central part of the folio. Leonardo da Vinci. Codex Atlanticus, folio 812r. BAM

way to protect themselves from the copying of others; in other cases, the rough sketch simply served as a memo, quickly recording an idea or an interesting project with the intention of returning to it later. In addition, due to the wide dispersion and presumable loss of some of Leonardo's notes, it is very likely that additional details for some projects were lost to history. Indeed, only in limited cases is it possible to realize a proper combination between the various items scattered among the different notebooks.

In the case of the automaton car, in particular, some details of the mechanics can probably be traced in other codices. Leonardo used his knowledge of mechanics to design the automaton, and, as the many designs found in this chapter ably show, in part, the work of his predecessors must be acknowledged, at least for the general idea. Leonardo's car, however, is characterized by a great innovation not found in the designs of his predecessors and contemporaries: an internal source of power. It is a small device that is able to move without external forces. This is a real breakthrough with respect to previous self-propelling models, which always required an external source of power, be it wind or human arm strength.

A working reconstruction of Leonardo's automaton car based on the observations of Carlo Pedretti (1975 and 1996), one of the most important scholars of Leonardo da Vinci and Mark Rosheim (2001),[4] an expert in robotics was made in 2004 by the Istituto e Museo di Storia della Scienza in Florence.[5] It was realized that the automaton's engine was not formed by a system of leaf springs but, rather, by spiral spring engines. The two small knobs on either side of the lower part of the figure are used to load the spiral springs contained in two drums under the cart, and once the spring is loaded, the cart is capable of traveling independently some tens of meters. An important part of the device is the escapement which plays a fundamental role in controlling the release of the spring force that otherwise would be so fast that the car would only be able to move a very short distance.

A very similar device, this time for a clock, is reported in another part of the *Codex Atlanticus*, on folio 863r. We will return to it in more detail in Chap. 13. The core of the mechanism for Leonardo's automaton is the spiral spring in the drum, which provides the driving force. This is exactly the same device that Leonardo applied for his design of several mechanical clocks. Thus we see that Leonardo's skill as an engineer and his comprehension of mechanical devices owes a great deal to the mechanics of clocks, which was already well developed at his time.

It is interesting also to observe that the modern reconstructions of Leonardo's machines, such as his "car," can be quite different from one another. Because of the vague descriptions and depictions of Leonardo's projects, a certain amount of choice and imagination in reconstruction are necessary to realize a working model. This has led to numerous interpretations of Leonardo's machine. For example, some tend to privilege the three-dimensional projection (Fig. 7.15) over the top view drawing (Fig. 7.16). In some recent reconstructions, the small front wheel, intended to be used to guide the vehicle, has disappeared.[6]

[4] Rosheim ME (2001) L'automa programmabile di Leonardo. Codex *Atlanticus*, f. 812 r, ex 296 v-a. XL Lettura Vinciana, 15 aprile 2000, Città di Vinci, Biblioteca Leonardiana, Firenze, Giunti

[5] A detailed analysis of Leonardo's car can be found in the website, together with an exhaustive bibliography on the topic: http://brunelleschi.imss.fi.it/automobile

[6] A computer reconstruction of Leonardo's "car" is described in: Taddei M, Zanon E, Laurenza D (2013) Le macchine di Leonardo. Segreti ed invenzioni nei Codici da Vinci. Giunti.

But if Leonardo thought about a driving system, and if we look carefully at the small free space left in the front of the "engine room" housing the mechanical system, then it may be that it wasn't just for theatrical performances but a real small car … Perhaps … and maybe leaving space to the imagination and to the abstraction of the project gives us back that fantastic dimension which is part of the endless charm exercised by Leonardo.

8

Perpetuum Mobile

The dream of a machine able to remain indefinitely in motion without any external energy source—i.e., the possibility of perpetual motion—is one of those die-hard ideas that still haunts the imagination of many amateur inventors. The history of the search for perpetual motion is full of frustrating attempts, failures, and reattempts in spite of the laws of physics. Max Planck (1858–1947) put a tombstone on the possibility of building such a type of dream machines: "It is impossible to get perpetual motion by mechanical, thermal, chemical methods or any other method, that is, it is impossible to build an engine that continually works and produces work from nothing or from kinetic energy."[1] Even in the quantum world, perpetual motion machines remain impossible to achieve simply because they violate the laws of thermodynamics.

During the Renaissance, however, the physical impossibility of perpetual motion was not so obvious, and the problem, at least from a mechanical point of view, could not fail to attract the attention of engineers like Taccola and then, through him, Leonardo. Regarding the problem of perpetual motion, Leonardo took a clear position and stated, without any appeal, that those dedicated to such research were only foolish gold miners. The topic, however, despite his protestations, was very attractive even for him, and he extensively studied it despite his strong declaration of principles. As we will see, Leonardo's notebooks, in fact, are filled with schemes and notes about systems that could exploit perpetual motion.

[1] Planck M (1903) Treatise on Thermodynamics. London, Longmans & Green

P. Innocenzi, *The Innovators Behind Leonardo*, https://doi.org/10.1007/978-3-319-90449-8_8

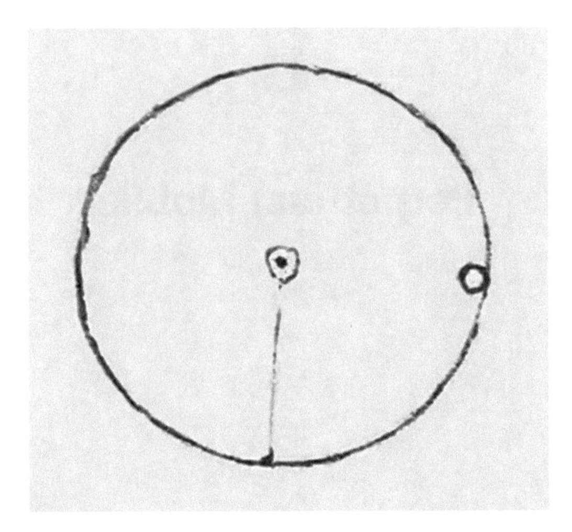

Fig. 8.1 A simple wheel with an attached weight to illustrate the impossibility of perpetual motion machines (detail). Leonardo da Vinci. Codex Forster II, folio 92v. Victoria and Albert Museum. (Milan, about 1495–1497)

Leonardo started with a simple scheme (folio 184v, *Codex Forster II*), a circle with an attached weight (Fig. 8.1). As alluded to above, the study of this basic system inspired him with some important considerations which he recorded in the notebook: "Any weight will be attached to the wheel, which is the cause of the wheel motion, undoubtedly the center of this weight will stop under the center of its pole; and no instrument manufactured with human genius …can remedy this effect. Speculators of continuous motion, how many unnecessary designs you have created in this quest! Get your partners with gold prospectors."[2]

Basically, what Leonardo means is that it is impossible to fight against the force of gravity: If a weight is used to set a wheel in motion, it will find its balance point and inevitably will stop. However, at the same time that he took the opportunity to mock all the people involved in this impossible search, in contradiction to his own statement, he himself continued to explore different solutions. In many pages of the *Codex Forster II*, for instance, notwithstanding his skepticism, he drew many perpetual motion machines with detailed comments.

[2]Original text: "*Qualunque peso sarà appiccato alla rota, il qual peso sia causa del moto d'essa rota, sanza alcun dubbio il centro di tal peso si fermerà sotto il centro del suo polo; e nessun istrumento che per umano ingegno fabbricar si possa col. suo polo si volti, potrà a tale effetto riparare. O speculatori dello continuo moto, quanti vani disegni in simile cerca ave' creati! Accompagnatevi colli cercator dell'oro*"

Early Designs

The search for perpetual motion began well before Leonardo. Villard de Honnecourt[3,4] already in the thirteenth century was among the first to propose a system capable of achieving perpetual motion.

Villard de Honnecourt

Villard de Honnecourt was one of the European forerunners of those polymaths who would revolutionize technology starting from Brunelleschi. The figure of Villard, in whom many even see a precursor of Leonardo, is difficult to define because of the lack of biographical information.[5] He lived from the beginning of the 1200s until at least the middle of the century and was probably from the village of Honnecourt-sur-l'Escault in the French region of Picardie, from which he took his name. The language he used, a dialect of the region, is another clue to support his origin. Although he presented himself as an architect and a soldier, there are also doubts about his true activities. He owes his fame to the notebook full of architectural drawings and machines known as the *Livre de portraiture* (Sketchbook)[6,7,8,9] The extant part of the manuscript is composed of 33 pages containing approximately 250 drawings.[10] These illustrations are probably the result of his travels in Europe. de Honnecourt noted carefully during the trips his observations, which were collected in the notebook to form the sketchbook we know. This is basically the same method the Sienese engineers and Leonardo had used many centuries later.

His notes were related to themes of architecture, and human and animal figures as well as various types of machines, including a permanently overbalanced hammer for perpetual motion.

Villard de Honnecourt's machine (Fig. 8.2) features a spoked wheel and is accompanied by the following description: "Many days the masters have disputed on how to spin a wheel alone. Here's how to do using odd mallets or mercury."

---

[3] Barnes CFJ (2009) *The Portfolio of Villard de Honnecourt*. Ashgate Publishing Limited, UK

[4] Bechmann R (2000). Entry on Villard de Honnecourt. In: Enciclopedia dell'Arte Medievale. Treccani.

[5] The exact name is not known for sure because he appears in the documents variously as *Wilars Dehonecort* or *Vilars de Honecourt*.

[6] The *Livre de portraiture* is composed of 33 parchment pages written on both sides, recto and verso, with notes in the old dialect of Picardie.

[7] Lassus JBA (1858). *Album de Villard de Honnecourt. Architecte du XIIIe siècle*. Publication of the manuscript in facsimile with annotations, Paris

[8] de Honnecourt V (2006) The Medieval Sketchbook of Villard de Honnecourt. Introduction and captions by Theodore Bowie. Dover Publications. Mineola, New York.

[9] A digitalized copy is available at the Gallica (BNF) online library: http://gallica.bnf.fr/ark:/12148/btv1b10509412z.r=villard%20de%20honnecourt

[10] Biblioteque Nationale de France, Manuscript Fr. 19093

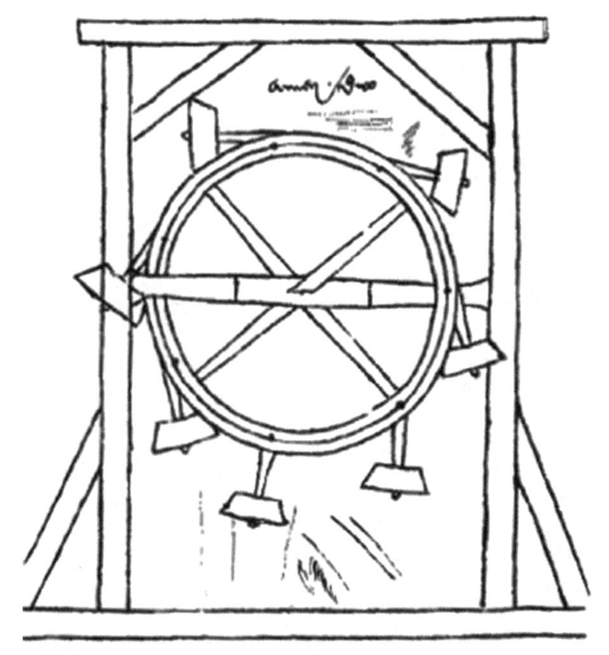

Fig. 8.2 Perpetual motion machines from Villard de Honnecourt, about 1230. Folio 5f (Planche VIII), Fr.19093, BNF. Under the drawing is written: "Maint ior se sunt maistre dispute de faire torner une ruee par li seule. Ves ent ci «con en puet faire par mailles non pers ou par vif argent»." (The transcription in modern French is: *Maint jour, se sont maîtres disputés pour faire tourner une roue par elle seule. Voici comment on peut le faire par maillets non pairs ou par vif-argent*)

As with the illustrations in the *Livre de portraiture* in general, the drawing of the perpetual motion machine lacks the depth and three-dimensionality that would allow an immediate understanding of how the system works. The machine consists of two wooden posts supporting a central axis that bears a wheel that is equipped with four additional spokes. The wheel is free to rotate around the central axis. The crucial point, according to Villard, is the odd number of mallets—three at the top and four at the bottom. This is to ensure that the center of gravity is always outside the center of the wheel, allowing for a continuous rotation. This reasoning works well up to a point, but eventually fails because the hammers are mobile in both directions, and when the wheel starts turning, those in descent aid the motion, but the others fall backward, which counteracts the movement until it stops altogether.

What was the likely origin or inspiration for Villard's design? He was not, in fact, the first to have had a similar insight: The Indian mathematician Bhaskara (Bijjada Bida, 1114–1185) described a wheel with cylinders filled with mercury around its rim as a device for perpetual motion. Bhaskara's idea was that the movement of the mercury within the cylinders, which is activated by the rotation of the wheel, would make the system always heavier on one side, thus resulting in perpetual motion. From India, this idea spread in the Arabian peninsula and then probably in Europe. Villard's mention of mercury for possible inclusion in the machine points to Bhaskara's design as his likely original model. His second design, with the mallets, may have been inspired by the many churches in Villard's time which featured a wheel with small hammers instead of bells. The observation that these wheels were still spinning well past the time when they had been put in motion may have inspired Villard. Each explanation has its charm....

The next intriguing figure in this story is a medieval French scholar, Petrus Peregrinus of Maricourt (c. thirteenth century), also known as Peter the Pilgrim from Maricourt. As was the case with Villard, not so much is known about his life. What is known is that he was the author of an important manuscript, the *Epistola Petri Peregrini de Maricourt ad Sygerum de Foucaucourt, militem, de magnete* ("Letter of Peter Peregrinus of Maricourt to Sygerus of Foucaucourt, soldier, on the magnet")[11,12] generally known by the abbreviation *Epistola de Magnete* ("Letter on the magnet").[13] It is a treatise describing the properties of magnets in the form of a letter that Petrus Peregrinus addressed to his friend Sygerus. The document, written in Latin in 1269,[14,15] is divided in two parts of ten and three chapters each. The first part contains a description of the properties of magnetism that recognizes the existence of two inseparable poles; the second part is dedicated to applications including a compass and a perpetual motion system based on a magnetic needle (Fig. 8.3).

[11] De Maricourt P (2012) The letter of Petrus Peregrinus on the Magnet, a D 1269. Hardpress

[12] An English translation of the full text of the letter is available online: https://archive.org/stream/letterofpetruspe00pieriala/letterofpetruspe00pieriala_djvu.txt

[13] Sparavigna, AC (2016) Petrus Peregrinus of Maricourt and the Medieval Magnetism, Mechanics. Materials Science & Engineering Journal, 2: 1–8.

[14] Thirty-nine manuscript copies of the *Epistola de Magnete* survive; in only one of these copies is written as closing note: *Actum in castris in obsidione Luceriæ anno domini 1269° 8° die augusti* ("Done in the military camp for the siege of Lucera [a town in the South of Italy], in the year of god 1269, eight day of August"). Based on this note, this would suggest that Peregrinus had served in the army of Charles of Anjou, King of Sicily, who in 1269 put Lucera in Apulia under siege. Peregrinus's nickname may derive from his participation in a crusade or maybe to a pilgrimage to the Holy Land.

[15] The first printed edition was published in 1558 in Augsburg by Achille Gasser.

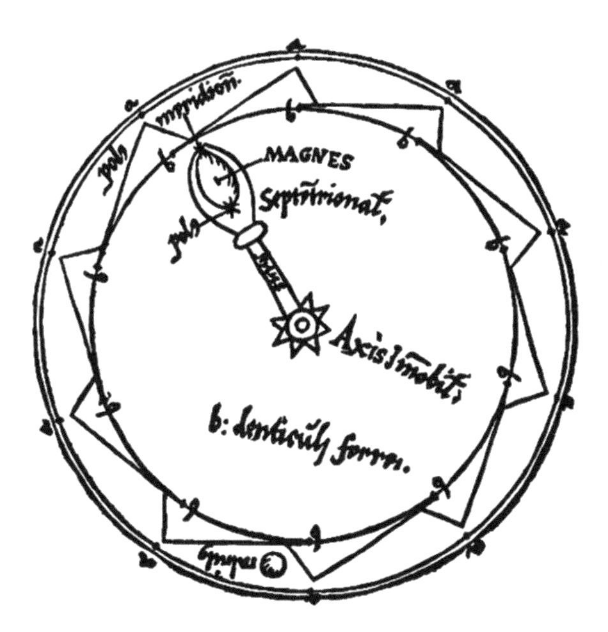

Fig. 8.3 Project for a magnetic perpetual motion system. Petrus Peregrinus. Epistola de Magnete

Petrus Peregrinus was the first to describe the phenomena of magnetism in such detail, and his passion for the subject led him to formulate the idea of achieving perpetual motion by exploiting the properties of the magnetic poles. The last chapter of the letter is therefore dedicated to "The art of making a wheel of perpetual motion." Peregrinus wrote[13]: "In this chapter I will make known to you the construction of a wheel which in a remarkable manner moves continuously. I have seen many persons vainly busy themselves and even becoming exhausted with much labor in their endeavors to invent such a wheel. But these invariably failed to notice that by means of the virtue or power of the lodestone all difficulty can be overcome." He designed a machine with a silver hand fastened in the center of a wheel. A magnet, with one pole pointing outward and the other toward the center of the circle, was positioned at the end of the hand. The magnet is used to push some iron nails placed in the rim of the wheel. The nails in the drawing are placed along the inner circle system, while the small counterpoise, made of brass or silver to avoid corrosion, observed at the bottom of the wheel is used to improve the performance.

Even putting aside this magnetic perpetual motion machine which was ultimately doomed to fail, Peter Peregrinus's *Epistola* remains an extraordinary treatise. Peregrinus was the first to define the position of the poles in a lodestone and to prove that different poles attract, while similar poles repel. He also understood that every fragment of a loadstone keeps its magnetic properties and realized a series of magnetic compasses such that "you will be able to direct your steps to cities and islands and to any place whatever in the world." In an intriguing coincidence, one of the first people to make a critical translation of Peregrinus's *Epistola* was Guglielmo Libri,[16] the book thief that we met in Chap. 3.

By the mid-1200s, then, already three possible models for perpetual motion had been proposed, one mechanical, another fluid, and a third magnetic. This futile search would occupy busy generations of inventors up to the present day, when we still find some who insist on trying to violate the basic principles of thermodynamics.

Perpetual Motion Machines of the Renaissance

The basic idea of perpetual motion arrived in Europe, as we have seen, very likely from India via Arab mathematicians. For a while, after Peregrinus, there is a gap, at least in the extant sources, in designs for such machines. However, the challenge could not avoid to tickle the Renaissance engineer-artists, first and foremost Taccola and then Leonardo da Vinci.

Taccola's device for perpetual motion (Fig. 8.4) appears in the center of a page of *De Ingeneis*. At the bottom of the same page, several siege machines are illustrated, probably drawn after the central sketch. The perpetual motion machine consists of a wheel with a central drum with 12 small rods protruding from its edge. Each of these rods is divided into two parts of equal length connected by a hinge. In its design, the system is similar to that of Villard de Honnecourt. During rotation, the rods are free to bend, and when they reach the maximum vertical point, they fall down pushed by gravity and turn the wheel. The fact that the rods are hinged prevents backward motion that could slow the wheel.

This model, like the previous ones, doesn't actually work, but it is the likely source of inspiration for Leonardo's studies. Leonardo, as we have noted, did not believe in the possibility of perpetual motion, but in his notebooks, several variants of the perpetual motion wheel of Taccola can be found. Indeed, he explored quite systematically the possible mechanical responses of a rotating system with unbalanced weights. On folio 148r of *Codex Madrid I*, Leonardo reprocessed the model of Taccola, studying it in much detail

[16] Libri G (1838) Histoire des sciences mathematiques en Italie, 2: 487–505. Paris

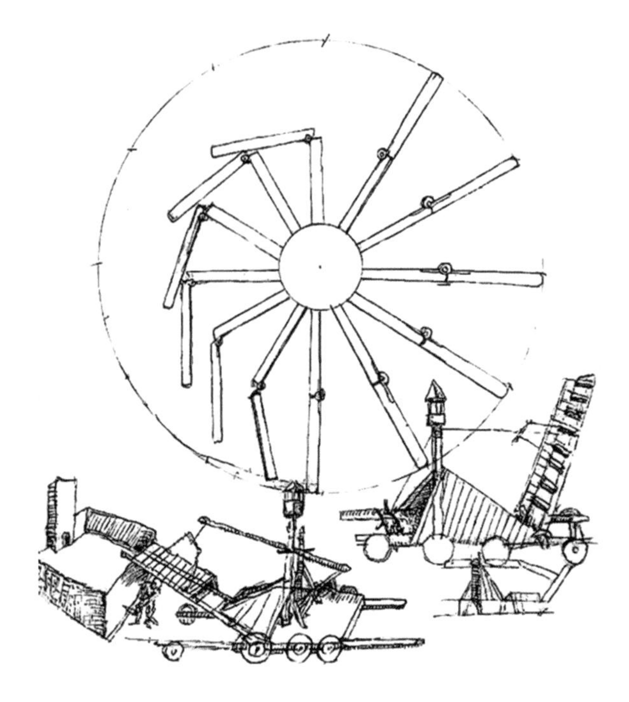

Fig. 8.4 Perpetual motion machine. Mariano di Jacopo (Taccola). De Ingeneis, folio 36v. BSBM

(Fig. 8.5). The system is a little more complex but the operating principle is quite similar.

In Leonardo's system, 12 weights are also present, but they are locked with an inner drum, so that the length of the small rod brings the weight right to the edge of the outer toothed wheel. Backward motion of the wheel is prevented by the vertical block located at the point marked "g" at the bottom left of the toothed wheel. The system of weights and counterweights is designed to ensure that, once the wheel is set in motion by the fall of one of the weights, it should be able to continue its run in the clockwise direction.

Although it does not guarantee the impossible to achieve perpetual motion, this design allows some control of the rotating mechanical system. Indeed, the *Codex Forster I* contains many iterations of similar wheels with weights, each accompanied with quite detailed comments. These drawings combine to form a rather systematic study of rotational motion.

So, as we see, although Leonardo was convinced of the impossibility of perpetual motion, he was nonetheless undoubtedly much fascinated by the mechanics of these wheels with movable weights, and, in time, he managed to elaborate several different models.

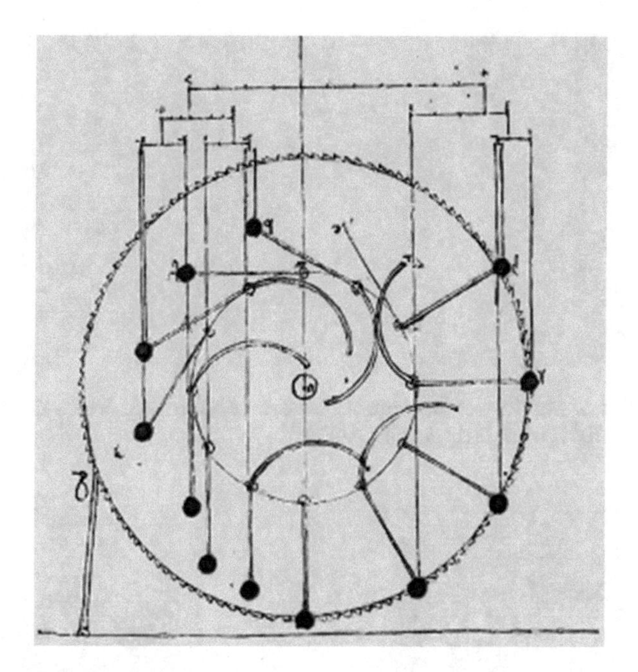

Fig. 8.5 Model of a machine for perpetual motion. Leonardo da Vinci. Codex Madrid I, folio 148r (detail). BNE

Leonardo's studies of wheels with counterweights and balances of weights in the *Codex Foster I* differed significantly from those of his predecessors and contemporaries. The approach is systematic and rigorous, based on models and mathematical calculations, at least within the limits allowed by his level of knowledge.

A wheel for perpetual motion, nearly identical to that found on folio 148r of the *Codex Madrid I* but in a simpler variant with no counterweights, appears in the *Codex Forster II* on folios 90v, 90r, and 104v (Fig. 8.6) and finally on folio 133v. In the same Codex, on folio 91v (Fig. 8.7), another very charming version can be found. The system consists of 12 half-moon-shaped adjacent channels which allow the free movement of 12 small balls as a function of the wheel's rotation. In its form, this model recalls the children's game where balls in a small box are delicately made to all fall into their little holes. In this case, the mobile balls take the place of the weights anchored to a fixed point. Once again, however, Leonardo reprimanded that despite the fact that everything might seem to work, "you will find the impossibility of motion above believed."[17]

[17] Original text: "*troverai la impossibilità del moto di sopra creduto.*"

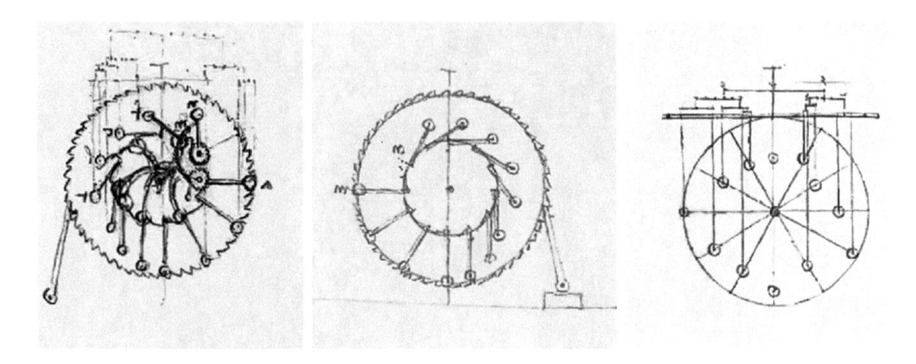

Fig. 8.6 Series of studies for perpetual motion. Leonardo da Vinci, Codex Forster II, folios 90r, 90v, 104v, from left (details). VAM

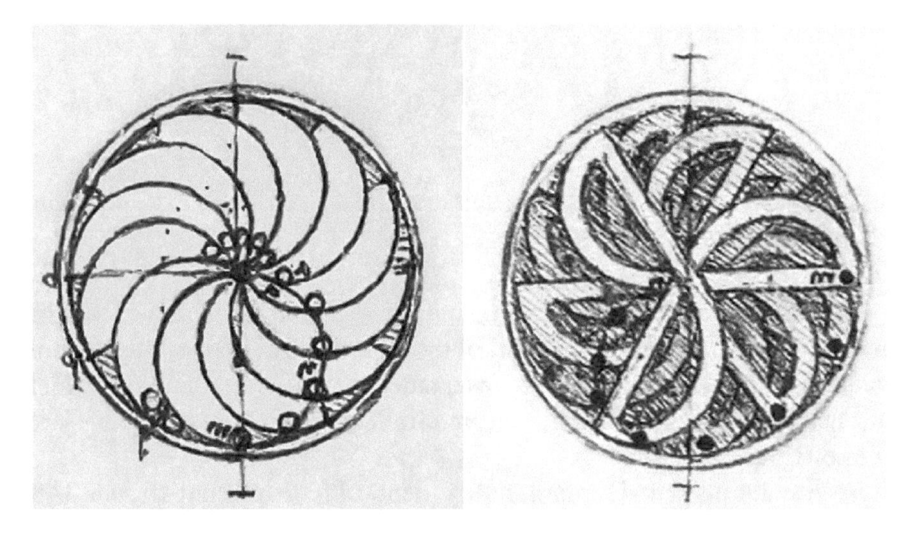

Fig. 8.7 Studies for perpetual motion using a system of wheels with half-moon channels and small balls. Leonardo da Vinci, Codex Forster II, folios 91r and 91v (details), about 1495–1497. VAM

At one point during the rotation, an imbalance will be created whereby more balls will find themselves on one side than the other, as shown in folio 91r.

The opportunities to play with the movement of balls during the rotation of a ribbed wheel are virtually endless, and Leonardo explored a more complex variation with eight balls (folio 91r, *Codex Forster II*). Several other studies of perpetual motion wheels with channels and balls appear also on folio 1062r of the *Codex Atlanticus*.

Among Leonardo's many versions of unbalanced wheels for perpetual motion with "*pallotte* (small balls)," there is a detailed study on folio 44v of the *Codex Arundel* (Fig. 8.8). It is probably an analytical scheme of the system

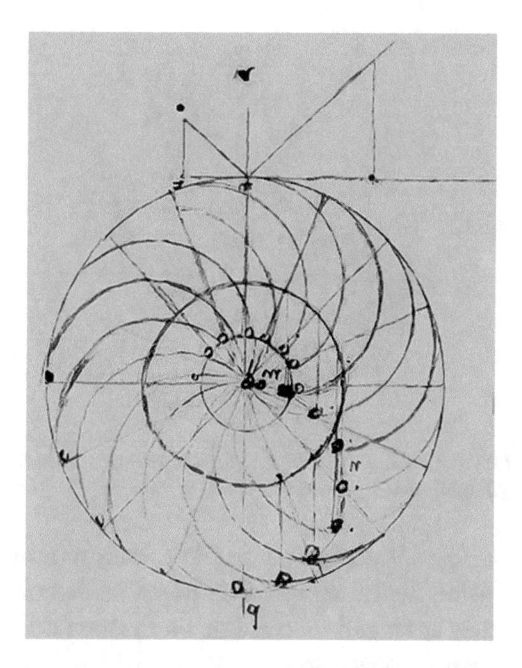

Fig. 8.8 Analysis of perpetual motion in a wheel with small balls. Leonardo da Vinci. Codex Arundel, folio 44v (detail). BL (about 1500)

with channels and spheres which Leonardo used to study the geometry of motion with its precise trajectories.

The mechanical model is not the only one that Leonardo considered. He also considered perpetual motion in fluid devices. On folio 44r of *Codex Forster I* (Fig. 8.9) we find a small system in which water could flow indefinitely. Both the sketch and the explanation leave some ambiguity: "Notice that when water is in the center of the flat screw S p, it enters into the pyramidal screw n c, and from c to p is doing as an equidistant lever half screw, and from p is going up to the center m e, and so from m to c, redoing a similar path."[18]

It is very interesting to note how Leonardo was making use in his mechanical projects of letters to indicate specific points. This is the first time that such an approach for technical drawing had been employed. In general, Leonardo used letters when, in his opinion, comprehension of the device solely from the sketch would be too complicated. The inclusion of such letters and accompanying descriptions supports the hypothesis that, behind Leonardo's work, there was the final intention to publish thematic volumes on the different subjects.

[18] Original text: "*Sappi che quando l'acqua è nel centro della vite piana S p, ell'entra nella vite piramidale n c, e da c a p fa lieva equidistante a semivite, e da p saglie al centro m e, e così da m in c, rifacendo il simile.*"

Fig. 8.9 Study for a hydraulic system of perpetual motion. Leonardo da Vinci. Codex Forster I, folio 44r (detail). VAM

The drawing in Fig. 8.9 depicts a possible variant of a hydraulic machine for raising water using helical screws. The water, as described by Leonardo, should rise from the center and then fall down through a system of screws to produce a closed loop for water flow.

On folio 24v of the *Codex Arundel*, Leonardo studied yet another creative device that made use of the pressure of a fluid on an immersed body to fuel perpetual motion, with several sketches and notes about the device filling the page (Fig. 8.10). The system is formed by wheels with movable arms that end in jars filled with air. As described by Leonardo, once the wheel starts moving, it should take advantage of the momentum created: "a, b, c, d are returned into the wheel pushed from the air which is locked up into them."[19]

In general, observing Leonardo's various models, we get the general impression that they were theoretical studies without any true experimental basis. We also have no records of Leonardo ever attempting to realize any of the perpetual motion devices he designed. His conviction of the impossibility of perpetual motion was probably the result of a combination of his own studies and direct observation of failed attempts. Despite the studies being dispersed across many codices and their somewhat fragmentary nature, they appear systematic and analytic in nature. The different projects and models must be seen as tools for thought experiments. Regardless of whether Leonardo's conclusion as to the impossibility of perpetual motion was the result of experimental work or perhaps just a hunch, it was certainly the right one.

[19] Original text: "*a, b, c, d son rientrati nella rota sospinti dall'aria che in loro è rinchiusa*"

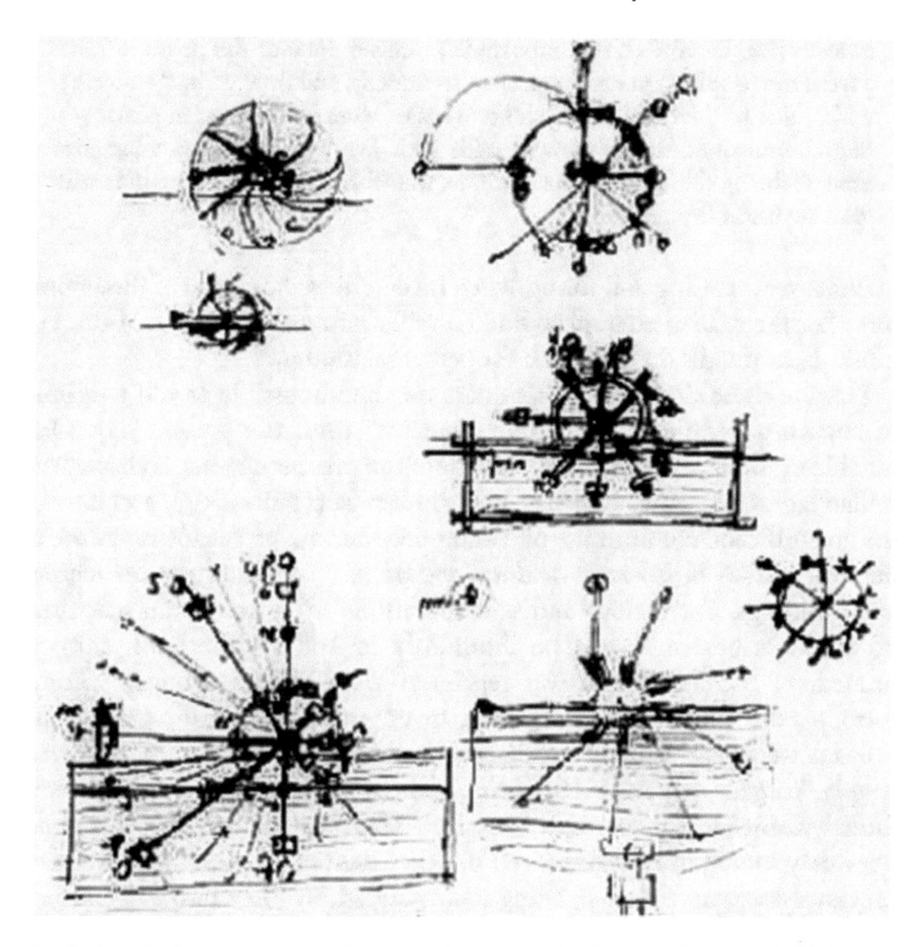

Fig. 8.10 Models for perpetual motion. Leonardo da Vinci. Codex Arundel folio 24v, (detail). BL

On the first page of the *Codex Madrid I*, we can find a long comment from Leonardo on the search for perpetual motion, perhaps intended as the opening of a dedicated work to be published on the subject:

> I found the search for the continuous motion, which for some is called perpetual wheel, one of the greatest impossible beliefs of men. This has kept busy for many centuries, with extensive research and experimentation and great expense, almost all men who delight of hydraulic machines and wars and other clever wits. And in the end as the alchemists for a small part they lose everything. Now I intend to make this gift to this sect of investigators, i.e. to give them peace in this research, for how long this will work. And besides this, what they will promise to others will have the desired end, and they will not always

> need to flee, because of the impossible promises done to the principles that govern the peoples. I remember having seen many, and from various countries, which, due to their childish credulity went to Venice with great hopes of getting commissions, and fabricated mills with firm water. Failing, after great expense, being able to move the machine, were forced to move themselves with great swiftness from the air.[20]

The note is cutting and funny, when he describes the escape of the impostors after their failed attempt to build a mill that moves using still water. An ironic de profundis on the search for perpetual motion.

Despite these drawbacks, this quest for the impossible would continue to engage generations of amateur inventors until the present day. One notable example with which we will close the present chapter is that of the Italian late Renaissance engineer Vittorio Zonca (1568–1602), architect of the magnificent community of Padua and author of the *Novo teatro di machine et edificii. Per varie at sicure operationi* ("New theater of machines and buildings. For various and safe operations"),[21] a treatise on machines of various kinds published posthumously in 1607.[22] The book enjoyed immediate popularity and was reprinted several times. Zonca's volume introduced a remarkable innovation in graphics: For the first time, the projects were represented in scale and using axonometry. Zonca, however, simply "forgets" to reveal that many of the machines that he depicts were directly copied by Francesco di Giorgio.[23] Zonca's machines, in turn, would be widely looted to be reproduced in other treaties, eventually, as we have previously seen in Chap. 1, being incorporated, in 1726 into the Chinese Encyclopedia.

---

[20] Original text: "*Io ho trovato intra l'altre superchie e inpossibile credulità degli omini la cierca del moto continuo, la quale per alcuno è decta rota perpetua. Questa ha ttenuto moltissimi seculi, co llunga cierca e sspe[ri]mentatione e grande spesa, ocupati quasi tutte quelli omini che si dilettano di machinamenti d'acqua e di guerre e altri sottili ingiegni. E ssenpre nel fine intervenne a lloro come alli archimisti che per una piccola parte si perdea il tutto. Ora intendo fare, questa limosina a questa setta d'investigatori, cioè di dare loro tanto di quiete in tale cierca, quanto durerà questa mia picola opera. E oltre a di questo, ciò che di sé a altri imprometteranno arò il disiderato loro fine, e none aranno senpre a stare in fughe, per le cose inpossibile promesse ai rpincipi regitori di popoli. Io mi ricordo avere veduti molti, e di vari paesi, essere per loro puerile credulità, essersi condotti alla cità di Vinegia con grande speranza di provisioni, e fare mulina in acqua morta. Che non potendo dopo molta spesa movere tal machina, eran costretti a movere con gran fu[ri]a sé medesimi di tal aer.*"

[21] Zonca V (1607) Novo Teatro di Machine et Edificii 1607. Edited by Poni C. (1985) Edizioni Il Polifilo, Milano

[22] The 1656 edition is available online thanks to Cornell University: http://ebooks.library.cornell.edu/k/kmoddl/toc_zonca1.html

[23] Reti L (1963) Francesco di Giorgio Martini's Treatise on Engineering and Its Plagiarists. Technology and Culture, 4:287–298.

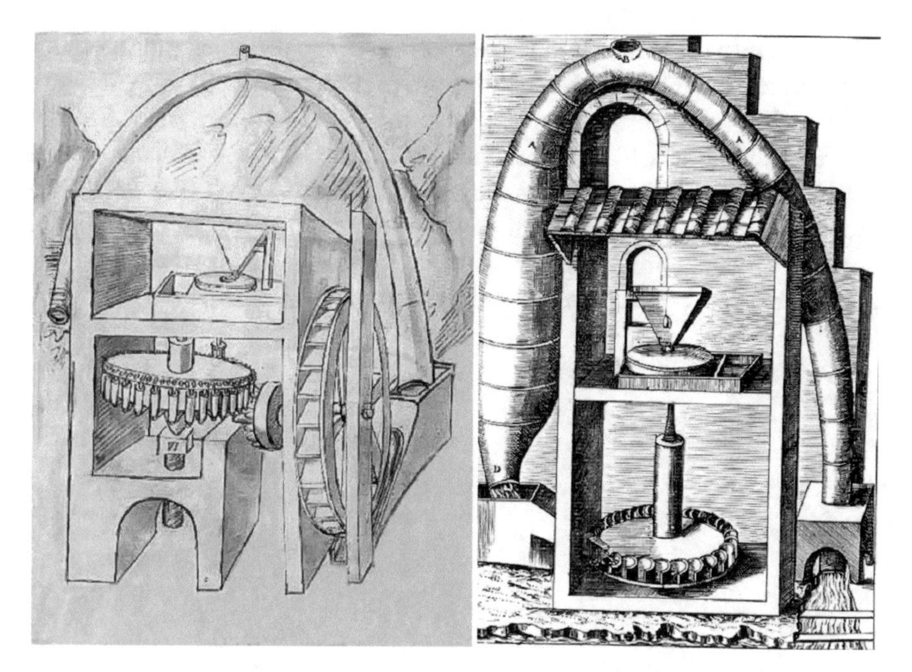

Fig. 8.11 Project of mill with firm water (*left image*). Francesco di Giorgio. Opusculum de Architectura (Ms. 197.b.21), folio 74r. BML. Mill with firm water with perpetual flow (*right image*). Vittorio Zonca. De Machine et Edificii, table 115

Among Zonca's machines, there is one system which is described as serving "to raise water with a perpetual motion"[24] (Fig. 8.11, right). It is a mill that works with firm water and is built with a large asymmetric U-tube extended on one side. Once filled with water through the top hole, the greater thrust exerted by the left side should start up the system with the central turbine, ensuring the mill's operation. Essentially, the mechanism represents a kind of impossible siphon—indeed, even in this case, the system does not work because the water level between the two pipelines will tend to balance. Evidently, given its impossibility, the system was never built and tested even though Zonca seems to be very sure of how it works: "To raise water with perpetual motion proves in this table an invention never again used, by which you can make a continuous motion, which by itself will raise water, after sending water only for the first time, and in that table we're showing that will be able to rise so much water, that will turn a mill"[25] (*Novo Teatro Di Machine et Edificii*, table 115).

[24] Original text: "*A levar acqua con un moto perpetuo*"

[25] Original text: *Per levar acqua con moto perpetuo si dimostra nella presente tavola una inventione non mai più usata, con la quale si potrà fare un moto continuo, il quale da se stesso laverà l'acqua, come farà inviato solamente la prima volta, & in quella tavola noi dimettiamo, che ne leverà tanta, che girerà un mulino*

What Zonca does not say in his description of the perpetual motion mill is that the idea is not solely his sack of flour, as it were. As with so many other plates in his book, this project was inspired—that is to say, copied—from Francesco di Giorgio. Indeed, on folio 36v of the *Codex Ashburnham*, there is a little drawing showing a system almost identical to Zonca's (Fig. 8.11, left). The latter revised his perpetual motion mill without acknowledging di Giorgio's original authorship. As we have noted many times, recognition of intellectual property is one of the eternal battles which is still going on today.

9

Homo ad Circulum

Among the most famous images of Leonardo, next to the paintings of the Mona Lisa and the Last Supper, we can certainly place the *Vitruvian Man*, a figure that perfectly represents the Renaissance ideal of man as the measure of all things and the center of creation.[1] Few other images have so captured the popular imagination to become a global icon and source of continuous inspiration for generations of artists.

The drawing in pencil and ink was done on a small paper (34 × 24 cm) in Milan probably around the year 1490[2] although the exact dating is somewhat uncertain: Carlo Pedretti, for example, assigns the drawing to the beginning of the fifteenth century. The *Vitruvian Man* is part of the collection of Leonardo's scattered papers that has been kept in the cabinet of drawings and prints of the Gallerie dell'Accademia in Venice since 1822. The drawings were bought from Austria from the heirs of Giuseppe Bossi, a collector artist from Milan, and then reached Venice, at that time part of the Austro-Hungarian Empire. After the return of Venice to Italy, Leonardo's drawings remained in the art collection of the Gallerie where they are still scrupulously conserved.

The name of the famous sketch derives from Vitruvius (Marcus Vitruvius Pollio, 80–15 BC), a Roman architect who lived at the time of the Emperor Augustus. In the third book of his treatise *De Architectura*, he described how to simultaneously inscribe a human figure in a circle and a square. It is from here that the story of the *Vitruvian Man* begins.

---

[1] Torrisi A (ed.) (2009) Leonardo. L'uomo vitruviano tra arte e scienza. Marsilio.

[2] Pedretti C (1978) *Leonardo architetto*. Mondadori Electa. Milano.

P. Innocenzi, *The Innovators Behind Leonardo*, https://doi.org/10.1007/978-3-319-90449-8_9

A Forgotten Manuscript: Vitruvius's *De Architectura*

Not much is known about Vitruvius. Both his place of birth and his activities are uncertain. From the limited information which it is possible to glean and deduce from his work, it seems that he was the architect of the Basilica in Fano (*Colonia Iulia Fanestris*) and probably official superintendent for war machines under Julius Caesar. He owes his great fame to his treatise *De Architectura*, composed of ten books and written between 29 and 23 BC during the rule of the Roman Emperor Augustus.

The treatise, through various vicissitudes, was able to arrive in nearly complete form in Renaissance Europe. Its rediscovery was extremely important because the great heritage of Greek and Roman classical architecture was thereby rescued from the clutches of obscurity. While the book may not have been particularly remarkable or famous in its own time, today its importance is owing to the fact that it is the only evidence on Roman architecture that has been transmitted to us virtually untouched.

Several copies of the treatise survived the fall of the Roman Empire and the Middle Ages,[3] during which they attracted only a little attention, at least until the beginning of 1400 when the importance of the heritage the work represented was finally recognized. The editio princeps was published in Rome by Sulpizio da Veroli between 1487 and 1492, and in 1511 the first printed version illustrated by Giovanni Giocondo was published in Venice, followed in 1521 by the first edition in Italian with translation and illustrations by Cesare Cesariano.

Among the first to realize the importance of *De Architectura* were the Sienese architect engineers Mariano di Jacopo (Taccola) and, especially, Francesco di Giorgio, who, as mentioned in Chap. 1, was the author of a fragmentary translation in Italian probably used by Leonardo himself. They would not be the only ones to carefully study the *De Architectura* and to use it as a working tool. Since then, Vitruvius's book has served to represent, in fact, the landmark for generations of architects.

The ten books of *De Architectura* treat different topics, and the third book, in particular, is dedicated to sacred architecture. According to Vitruvius, the architectural composition rules are based on symmetry and proportion: "the composition of the temples is based on symmetry ... this arises from proportion ...." Vitruvius believed that the dimensions of the human body could be used as a natural reference to determine appropriate and attractive ratios for

[3] Several copies of *De Architectura* were realized in the Scriptorium of Charlemagne during the ninth century.

the dimensions of buildings and that the different architectural elements would exhibit in this way a "divine" proportion.

Hence the famous human figure at the center of a perfectly defined geometric universe formed by a circle and a square: "Thus, the center of the body is naturally the navel; in fact if a man is put lying on his back with his hands and feet opened and if the center of the compass is put in the navel, describing a circle, the fingers and toes of his two hands and feet be tangentially touched. But that's not all: beyond the outline of the circle, in the body it will also be found the figure of a square. In fact, if it is measured from the sole of the feet up to the top of the head, and then this measure is compared to the outstretched hands, a length equal to the height will be found, as it is the case for a square drawn with a set square."[4]

Vitruvius noticed that a man's height corresponds to the length of his outstretched hands, and this allows one to fit the human figure within both a circle and square. The human body becomes, thereby, the benchmark of the perfect proportions to be used as an aesthetic standard in architecture. It is from this starting point that the pursuit of aesthetic perfection of the Renaissance architects and their research on the *Vitruvian Man* begins.

The surviving copies of *De Architectura* do not contain the illustrations that were probably present in the original manuscript. According to the notes in the book, at least a dozen of drawings should have been present. This lack of the originals, however, left a lot of space for free interpretations and sparked the imagination of the illustrators of the early versions, such as Fra Giocondo from Verona and, especially, Cesare Cesariano from Milan. The works described in *De Architectura*, including the *Vitruvian Man*, thus returned to life and became a reference model for most of the Renaissance architects.

As we know so well, Leonardo himself took the challenging task posed by Vitruvius and realized his own version of the *Vitruvian Man*. As we will see in the course of this chapter, he was not the first to undertake this exercise; before him, other attempts were realized, by, among others, Taccola and especially Francesco di Giorgio, who would study in detail the correspondence between the proportions of the human body and architectural ones. None of these studies would reach the aesthetic beauty and mysterious fascination of Leonardo's *Vitruvian Man*, but, undoubtedly and as we will see, his drawing was largely inspired by these earlier works.

---

[4] "Item corporis centrum medium naturaliter est. umbilicus. Namque si homo conlocatus fuerit supinus manibus et pedibus pansis circinique conlocatum centrum in umbilico eius, circumagendo rotundationem utrarumque manuum et pedum digiti linea tangentur. Non minus quemadmodum schema rotundationis in copore efficitur, item quadrata designatio in eo invenietur. Nam si a pedibus imis ad summum caput mensum erit eaque mensura relata fuerit ad manus pansas, invenietur eadem latitudo uti altitudo, quemadmodum areae, quae ad normam sunt quadratae."

The Human Figure of Villard de Honnecourt

As we have mentioned, the idea behind the Renaissance version of the *Vitruvian Man* is the pursuit of the best proportions in architecture using the dimensions of the human body to represent the perfection of creation. This was paired with an appreciation for geometry as a measure of things and a tool for research, something which Villard de Honnecourt, whom we have already met in the previous chapter as the inventor of a mechanical system for perpetual motion, clearly understood.

Villard used simple geometric shapes such as triangles, circles, and squares to determine the proportions for the figures of animals and men, including the mysterious figure of four intertwined stonecutters to form a swastika-like shape (Fig. 9.1). As noted earlier, it is not known whether Villard de Honnecourt was truly an architect or simply an observant traveler. Regardless of the specifics of his training and occupation, the fascinating stylized illustrations shown in the *Livre de portraiture* are notable for their use geometry as a foundation for drawings and schemes. This method anticipates the efforts of

Fig. 9.1 Methods for drawing human figures and animals. Villard de Honnecourt. Livre de portraiture, folio 19v (Planche XXXVII) (*left*) and folio18v (Planche XXXV) (*right*) (1230 circa). BNF

Leonardo and Renaissance architects to revolutionize architecture by underpinning it with geometry.

Villard used various geometric shapes as a guide for the architectural composition for his human, animal, and other figures. It is important to stress, however, that they served only as a trace for the drawing, to make it faster and simpler, rather than forming the basis for a systematic search for universal laws for the proper proportions of figures. This approach would completely change when we come to the Renaissance artists, including Leonardo, who were instead deeply fascinated by the search for perfection and harmony that could be expressed through mathematical and geometrical laws.

Taccola's Vitruvian Man

As with many things, Taccola was probably the first in the Middle Ages or the Renaissance to make a real attempt to represent the *Vitruvian Man* on paper (Fig. 9.2). Taccola, who had a strong grasp of Latin and composed his treatises in that language, carefully studied Vitruvius's *De Architectura* and achieved a good level of understanding of that complex and difficult to read text. Taccola's drawing of a man inscribed within a square and a circle certainly lacks the charm of Leonardo's drawing or the beauty of Francesco di Giorgio's version but is nonetheless notable for being the first of the three.

The tools used to trace the lines of the drawing, a compass for the circle and a square ruler for the straight edges and angles, are represented carefully at the top of the paper. The center of the naked man falls in the pubic area rather than in the navel, and the arms lie along the body without touching the circle or the square. From the left-hand bottom side, small curves were drawn with a compass centered at a small circle traced under the man's feet. On the right-hand side, in contrast, we find instead a series of corresponding line segments that were probably used to study the proportions.

As we can see, Taccola was unable to resolve the problem of inscribing the man both in the circle and in the square. Although the human figure is well within the circle, it does not fit into the square, with the head and the feet falling clearly outside. The whole sketch, given the poor quality of the drawing, does not appear particularly nice and was probably just a sketched study, a simple attempt without any particular deep ambitions to accurately represent what was described by Vitruvius. However, Francesco di Giorgio and Leonardo soon after him must have seen this drawing, which represents the starting point for their versions of the *Vitruvian Man*.

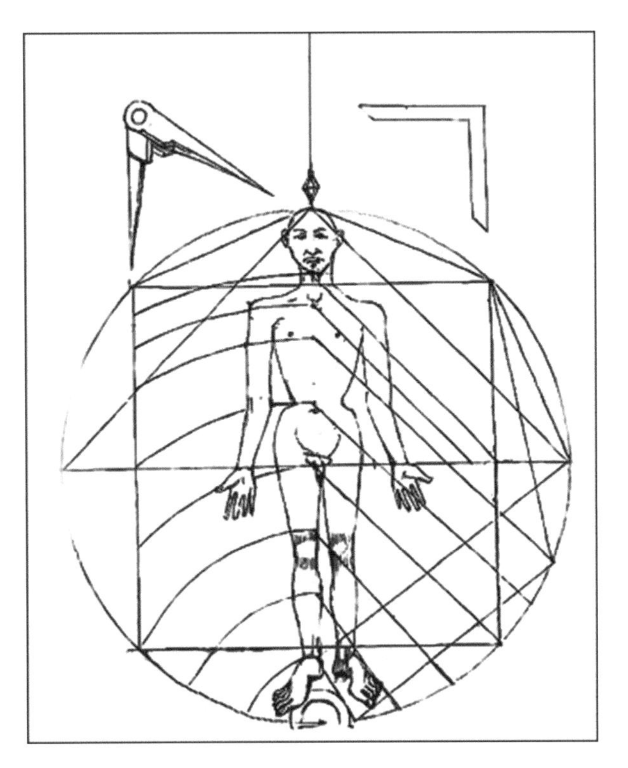

Fig. 9.2 The Vitruvian Man in the interpretation of Taccola, (detail). De ingeneis, books I-II, f.36v, Cod. Lat. Monacensis 197 II (BSBM), (1419–1450 ca)

Francesco di Giorgio: The Man is the Measure of all Things

Taccola's *Vitruvian Man* led the way to the work of Francesco di Giorgio, who performed a careful and systematic study, which, in turn, probably formed the basis for the later ones of Leonardo. As with so many of Leonardo's interests and studies, the key role of the Siena engineers as Leonardo's precursors and inspiration in exploring the idea of the *Vitruvian Man* is frequently overlooked even in scholarly circles.

Francesco di Giorgio widely studied the possibility of using the human body to prorate the dimensions of buildings and architectonical structures. One notable example is the cruciform church in which a man with outstretched arms is used as a template (Fig. 9.3, *left*). He also used the human body as the model for the proportions for the façade of a building (Fig. 9.3, *right*). Man's attractive proportions serve as an expression of the creation and

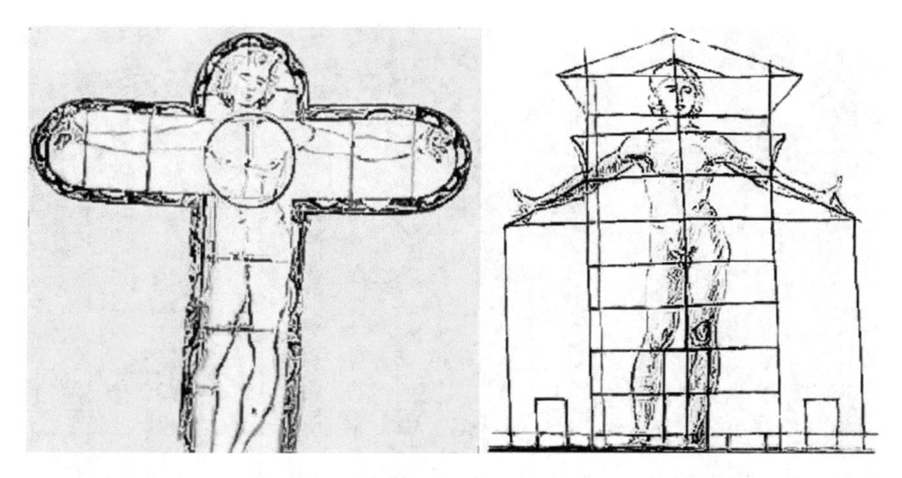

Fig. 9.3 Francesco di Giorgio. Studies of dimensions of the human body to derive models for architecture. Codex Ashburnham 361, folio 10v (detail), and folio 21r (detail). BMLF. (1481–1485)

the perfection of God, and this perfection is further reflected in the perfection of the geometrical forms. The plan of the church is built on symmetries, in the center of which is located the man whose open arms form the imprint of the cross.

di Giorgio also used the human figure to design individual architectural elements such as columns or facades (Fig. 9.4). This is clearly also part of the Vitruvian lesson and Vitruvius's recommendations on how temples should be built. Several such examples can be found in di Giorgio's manuscripts.

di Giorgio made detailed study of the correspondence between the size of the human face and architectural elements such as columns and capitals (Fig. 9.5). The standing male becomes a template for Doric columns and the female for Ionic ones, while the human face is geometrically dissected to provide the individual elements of the capital (or top section) of the column. This distinct use of men and women as models for different architectural elements is quite original. Leonardo's approach to the masculine and feminine worlds, on the other hand, is instead more complex and contradictory. In general, the drawings in Leonardo's notebooks reflect a primarily rude and masculine universe, which is in contrast to his paintings where women are frequently the focus.

Francesco di Giorgio engaged, therefore, with the *Vitruvian Man* and realized his own version, which has great aesthetic appeal: a naked man with his head slightly tilted and his eyes closed (Fig. 9.6). The man is at the center of a

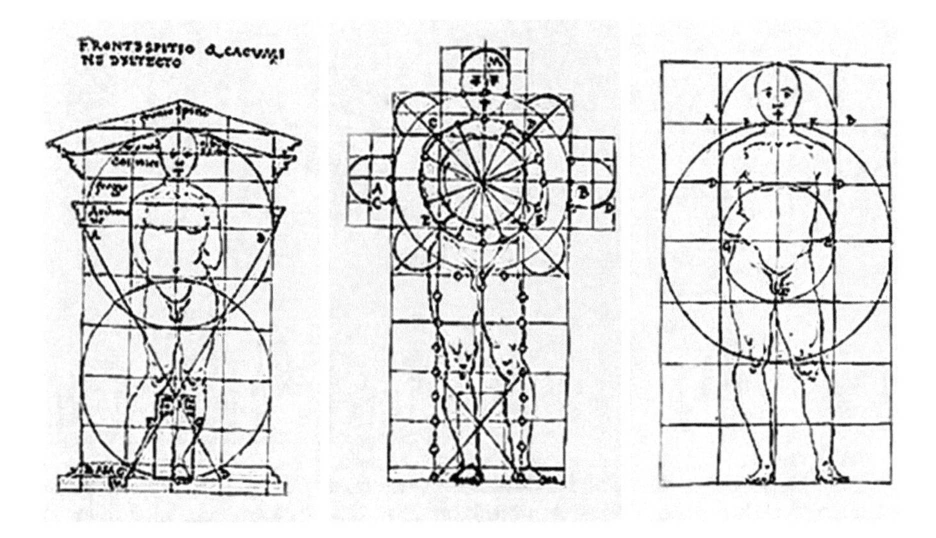

Fig. 9.4 Studies of architectural proportions based on the human body. Francesco di Giorgio. Trattato di Architettura, folio 36, details (*left*), folio 42, (details), (*center* and *right* images). BNCF

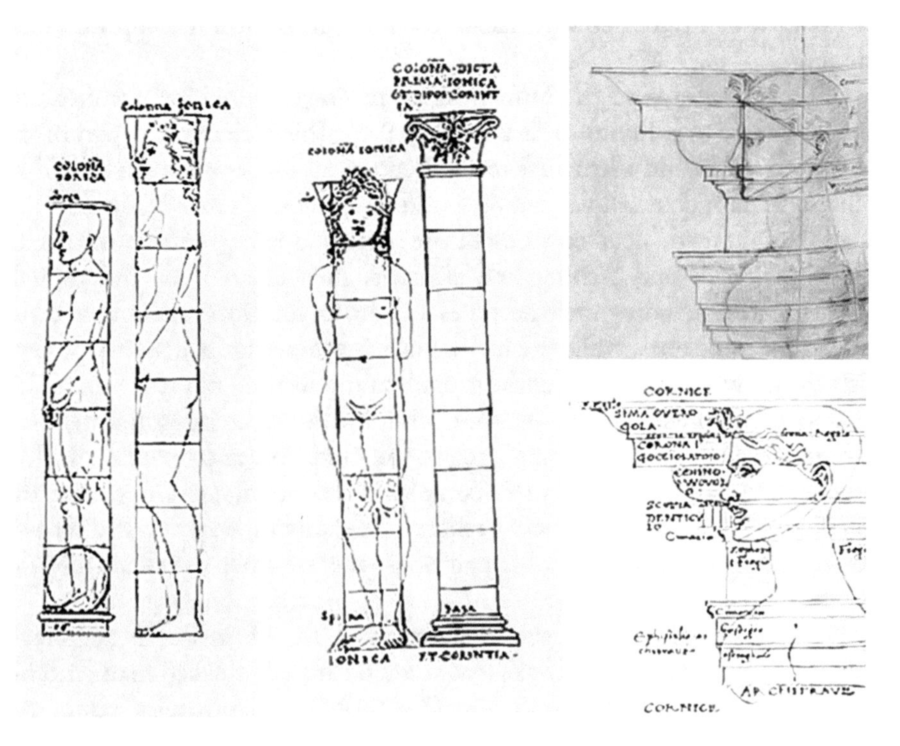

Fig. 9.5 Studies of architectural proportion based on the human body. Francesco di Giorgio. Trattato di Architettura, details from folio 25 (*left*), folio 26 (*center*), and folio 35 (*bottom right*). BNCF. Codex Ashburnham 361, folio 20v (*top right*). BMLF

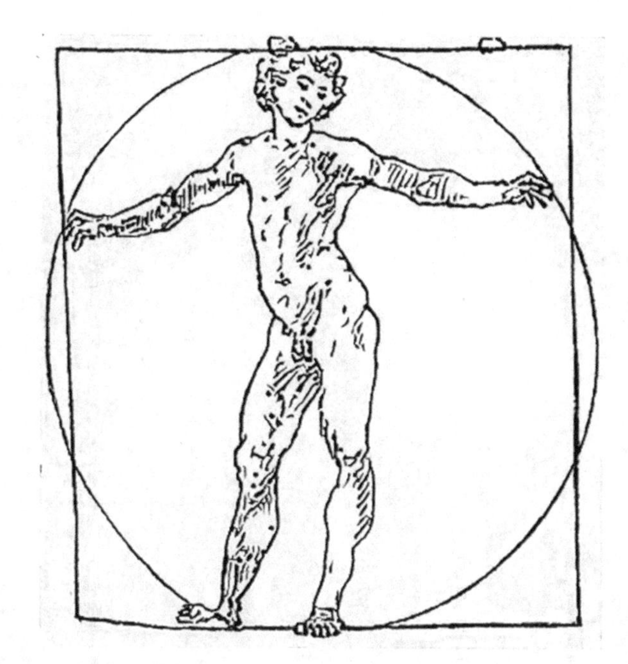

Fig. 9.6 Francesco di Giorgio's version of the Vitruvian Man. Codex Ashburnham, folio 5r (detail). BMLF. (1481–1485)

circle which is partially inscribed into a square, tangent to its top and bottom, which touch the man's head and feet. The man gazes to his bottom left, rather than watching us with the magnetic, epic but also violent force that only Leonardo is able to transmit, but reminds us of a world, friendly and a little sore, where the circle and the square are playful elements and at the same time a trap that encloses the space and the possibilities.

The Other Man

In 2010, a story attracted media attention from around the world: the discovery by an Italian researcher, the architect Claudio Sgarbi, during his studies into the surviving copies of *De Architectura*,[5] that in an anonymous manuscript copy of *De Architectura,* there was an unknown and astonishing version of the *Vitruvian Man* (Fig. 9.7). Who was the author of the drawing? The manuscript, which postdates Taccola's figure discussed above, is preserved in the Biblioteca Ariostea in Ferrara and was written between the late fifteenth

[5] Sgarbi C (2012). All'origine dell'uomo ideale di Leonardo. DISEGNARECON. https://doi.org/10.6092/issn.1828-5961/3166.

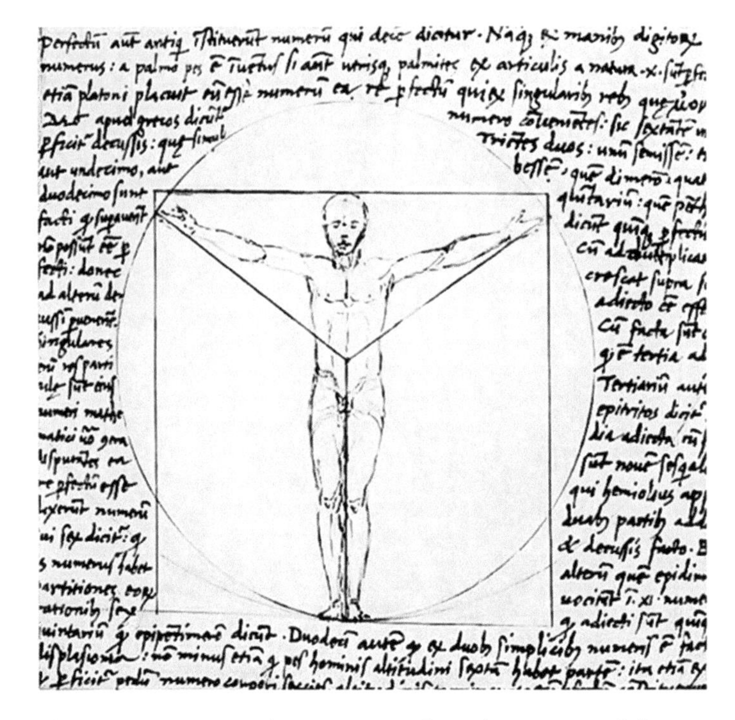

Fig. 9.7 The Vitruvian Man of Giacomo Andrea da Ferrara. Biblioteca Ariostea, Ferrara. (Cart. Sec. XVI, Fol. Figurato, Classe II, N. 176, fol. 78v)

and early sixteenth centuries. It contains 127 drawings and would be the first illustrated version of Vitruvius's *De Architectura*. Sgarbi examined the manuscript very carefully and attributed the anonymous copy to a Renaissance Italian architect, contemporary, and friend of Leonardo da Vinci, Giacomo Andrea da Ferrara.[6] Giacomo Andrea's *Vitruvian Man* would have been "copied" by Leonardo for his version,[7] because of the several analogies with his representation.[8]

Giacomo Andrea's *Vitruvian Man* is not particularly aesthetically pleasing and closely resembles that of Taccola, from which may have been inspired; however, the solution adopted for inscribing a man within a circle and a square is identical to that of Leonardo, as we shall see in detail later.

---

[6] *Did Leonardo da Vinci copy his famous Vitruvian Man?.* Science (2012). http://www.msnbc.msn.com/id/46204318#.TyhDoIHMNc8.

[7] Sgarbi C (1993) A Newly Discovered Corpus of Vitruvian Images. RES: Anthropology and Aesthetics 23:31–51. Published by: The President and Fellows of Harvard College.

[8] Schofield R (2015). Notes on Leonardo and Vitruvius. In: Moffat C, Tagliagamba S (eds.) Illuminating Leonardo, Brill.

The suspicion of a direct correlation is therefore fully legitimate: Among other things, Leonardo and Giacomo Andrea knew each other very well and, indeed, had a brotherly friendship according to Leonardo's other friend Luca Pacioli, who described Giacomo Andrea as being "suo quanto fratello (as his brother)" with Leonardo. Their friendship dated back to around 1490, during Leonardo's time in Milan, where Giacomo Andrea was in the personal service of Ludovico Sforza to whom he was very loyal. This loyalty had a dramatic cost for him: When Ludovico was imprisoned by the French after the fall of Milan, Giacomo Andrea, probably because of his support for Ludovico's return, was hanged on 12 May 1500, and his body quartered and exposed in public as a warning. The figure of Giacomo Andrea then fell into oblivion until Claudio Sgarbi's discovery of the anonymous manuscript in the library of Ferrara.

Leonardo and Luca Pacioli, on the other hand, kept a safer distance from the Lord of Milan, and as soon as they saw the situation turning for the worse, they thought well to load all their belongings onto carts and head for as safe shores as they could expect to find in an Italy shaken by continual internal wars and foreign invasions. In Leonardo's *Manuscript L*, there is a brief note about the fall of the Duke of Milan: "the Duke lost the state, the belongings and liberty, and no work was finished for him."[9] The regret of Leonardo was also due to the fact that the French invasion prevented him from completing one of his biggest projects, the equestrian statue dedicated to Ludovico Sforza's father to which he had devoted great efforts for many years.

Returning now to Giacomo Andrea's *Vitruvian Man* (Fig. 9.7), we find that, while in geometrical structure it matches that of Leonardo, aesthetically the rough figure is more akin to that of Taccola. A curious detail is the modest thong appearing around the body to cover the male penis, the only drawing where this expedient is used; in other versions realized by different authors, the *Vitruvian Man* sometimes even shows off a prominent erection without any shame.

Giacomo Andrea's figure, like that of Francesco di Giorgio, has closed eyes and a minute and hardly sketched body. As becomes apparent, a comparison between the different ways that the Vitruvian Man was depicted tells us a lot about the character of the artists who drew him. Francesco di Giorgio's version presents a mannered and elegant individual, while that of Giacomo Andrea is modest and essential, without the disruptive and magnetic force of Leonardo's *Vitruvian Man*, a triumph of human power and will.

---

[9] Original text: "Il Duca perso la Stato, la roba e la libertà, e nessuna sua opera si finì per lui."

Because of the great number of Renaissance sketches of men inscribed into circles and squares, the attempt to carefully delineate who copied whom is relatively meaningless. The Vitruvian enigma is better viewed as a collective challenge open to the free imagination of the artists. In their attempts is reflected not only the ability to tackle the problem from the geometric standpoint but also and above all a vision of the role of man in an expanding world which leaves behind the boundaries of the medieval universe. This new Renaissance man is the master of a will to power that would have a long history, perhaps especially tragic.

The Vitruvian Man of a Friar Architect

The first illustrated edition of *De Architectura* was realized in 1511 by Giovanni Giocondo from Verona, known as Fra Giocondo[10] (1433–1515), and this printed version contains two different drawings related to the *Vitruvian Man*. The edition was published in Venice and contained 136 woodcuts. Fra Giocondo, a civilian and military architect, on whose life we have only fragmentary information, played a very important role in the dissemination of Vitruvian architecture, especially in Venice. He studied Vitruvius's treatise in depth and acquired a considerable reputation as a connoisseur of *De Architectura*, whose principles he extensively applied in his architectural works. His activities as an architect and engineer took him across Italy and France, and he worked in Rome, Venice, and Naples where he met Francesco di Giorgio in June 1492.[11] At that time, Francesco di Giorgio was in Naples to oversee the fortifications of the city. It is likely that Fra Giocondo aided Francesco di Giorgio in his Italian translation of *De Architectura* and discussed the interpretation of the text with him. Their meeting happened subsequent to Francesco di Giorgio's meeting Leonardo in Pavia in 1490, when the two were together for a consultation on the proposed construction of a cathedral.

It is interesting to see how all the protagonists of these various stories are interconnected to form a network of relationships that helped the spread of the ideas and culture of the Renaissance from the rediscovery of classical knowledge. It is also impressive that they were always so available to move around Italy and Europe, where the geography of power was constantly chang-

[10] Pagliara PN (2001) Entry on: Giovanni Giocondo da Verona. In: Dizionario Biografico degli Italiani. Treccani. Volume 56.

[11] Hughes J, Buogiovanni C (2015) Remembering Parthenope: the reception to Classical Naples from antiquity to the present. Pag. 199. OUP Oxford.

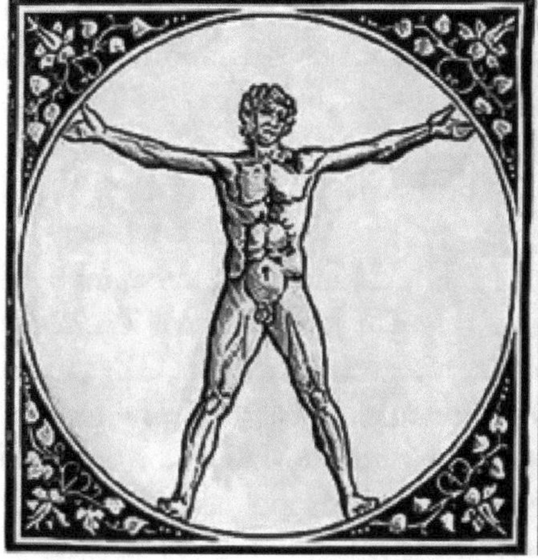
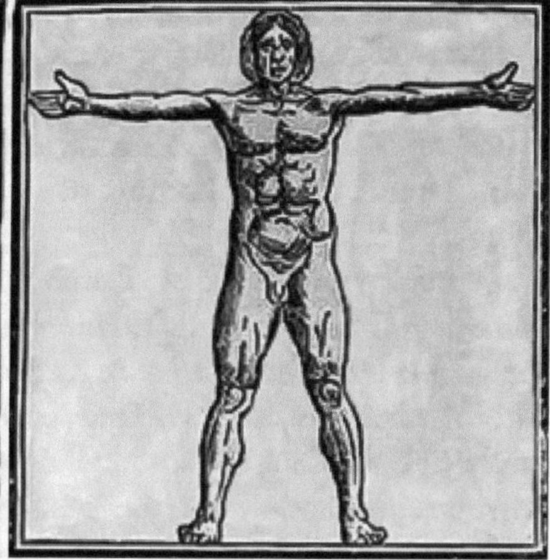

Fig. 9.8 Vitruvian Man, Fra Giocondo. (Illustrations from *De Architectura* of Vitruvius)

ing, to pursue the best opportunities. Leonardo da Vinci himself did not hesitate to do the same even if in many cases his moves, such as his abandonment of Milan with Luca Pacioli, were driven mainly by need rather than preference.

In Fra Giocondo's version of the *Vitruvian Man*, the naked man with open arms is at the center of a circle perfectly inscribed within a square (Fig. 9.8). This is a different interpretation compared to the previous representations that we have seen. Moving away from the indications of Vitruvius, the man only touches the circle despite its location in the center of the two geometric figures. Fra Giocondo's second drawing depicts the *homo ad quadratum* who, with open arms and spread legs, remains perfectly inscribed within a square.

Cesare Cesariano, the Other Man

The first release of a complete Italian translation of *De Architectura* arrived in 1521, about 10 years after Fra Giocondo's Latin version. This volume, which was also richly illustrated, was edited by a cultivated person, Cesare Cesariano (1475–1543).[12] Cesariano[13] was an architect and painter, a pupil of Donato Bramante (1444–1514), and active as an artist in Reggio Emilia, Parma,

[12] Gatti Perer ML, Rovetta A (1996). Cesare Cesariano ed. il classicismo di primo cinquecento. Vita e Pensiero, Università Cattolica. Milan.

[13] Samek Ludovici S (1980) Entry on: Cesare Cesariano. In: Dizionario Biografico degli Italiani. Treccani. Volume 24.

Rome, and then in Milan, until being appointed by the Spanish government as chief architect of the city, working at the Fabbrica del Duomo and the Castello Sforzesco. His reputation today, however, is mainly due to his Italian version of *De Architectura*, which included both commentary and illustrations. Behind this publication there is another small story that is worth telling.

The book was printed in Como on 15 July 1521 in the typography of Gottardo da Ponte with the title: *L'opera preclara de Lucio Vitruvio Pollione de Architectura traducta de latino in vulgare, istoriata e commentata* ("The famous work of Lucio Vitruvio Pollione de Architectura translated from Latin to Italian, illustrated and interpreted"). The name of the author of the work, the individual responsible for the translation, interpretation and illustrations, only appears on the first page and almost covertly: *Cesare Cesariano, cittadino mediolanense, professore di Architectura* etc. ("Cesare Cesariano citizen of Milan, Professor of Architecture, etc."). The final result of Cesariano's efforts is a real masterpiece, which historian Carlo Pedretti has declared to be the most beautiful book of the sixteenth century.[14]

The publication of the work in 1312 copies was made possible thanks to the financial and organizational contribution of a Milanese noble Luigi Pirovano, who forced Cesariano to accept contributions from two other scholars, Benedetto Giovio[15] from Como and Bono Mauro from Bergamo. The collaboration, however, proved immediately difficult, in part because the main role played by the two new participants was mainly that of inspectors on behalf of the funder Pirovano. Upon the completion of Book VIII, Cesariano abruptly ended the relationship with these two scholars and moved away, taking with him his own versions of the two final books, IX and X.

These two volumes were recently rediscovered in Madrid, and the close similarities between them and books I–VIII has allowed us to understand the true, essentially supervisory, and supernumerary nature of Benedetto Giovio and Bono Mauro's contributions to the project. Cesariano not only translated the Latin text into Italian but also interpreted in detail Vitruvius's complex descriptions of architecture—not by any means a trivial task—enriching it with beautiful illustrations. It thus becomes clear that Cesariano did most of the work alone and must have felt cheated and disheartened, despite the several letters of apology and explanation that Benedetto Giovio wrote to him.

---

[14] Pedretti C (2007). Leonardo & iO. Chapter: "Uomo vitruviano, anche donna". Pag. 210. Mondadori.

[15] Foà S (2001) Entry on: Benedetto Giovio. Dizionario Biografico degli Italiani. Treccani. Volume 56.

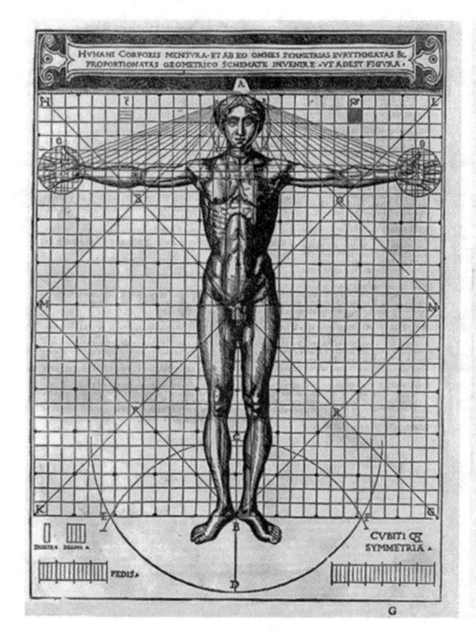

Fig. 9.9 Cesare Cesariano's Vitruvian Man, *De Architectura* of Vitruvius, Book III. 1521, Como

These letters must not have had much of an effect in appeasing Cesariano's wrath given that the manuscripts found in Madrid[16] contain a note dismissing Giovio as "notaruzo" (a "small notary"). Cesariano subsequently brought a successful lawsuit asserting that the work was the product of his own efforts.

Let us now return to the two depictions of man's geometrical proportions to be found in Cesariano's translation of *De Architectura* (Fig. 9.9). According to its title "Humani corporis mesura et ab eo omnes symmetrias eurythmiatas & proportionatas geometrico schemata invenire ut adest figura" or "Measure of the human body and from this all the well proportionate symmetries and geometrical schemes can be obtained from this figure", the first illustration, which shows a naked man standing with his outstretched arms forming a cross, serves as a reference for obtaining symmetrical and proportionate measures of the human body. This is followed by a second illustration of a naked man with an erect member at the center of a square inscribed in a circle. The diagonals of the square perfectly meet at the man's navel.

However, in order to make the figure fit within the two shapes, Cesariano had to resort to the artifice of giving the man a pair of monstrously long and deformed feet. The result is that, while the man is symmetric, his limbs and certain other body parts are unnaturally oversized. Aside from the foot, his

[16]Discovered in Madrid in 1986 and published in 1996. Madrid, Real Academia de la Historia, ms. 9.2790.

head is too small compared to his torso, and, taken as a whole, the figure looks rather awkward, if not downright disturbing. He certainly cannot be taken as a model of classic beauty.

Thanks to the good circulation of the printed copies, Cesariano's *Vitruvian Man* reached as far as Germany, with German scholar, Walter Hermann Ruff (c. 1500–1548) publishing in 1547 the first German version of Vitruvius's text with commentary, the *Vitruvius Teutsch*, which was based on Cesariano's work. In Ruff's text, the two figures created by the Milanese artist for the Vitruvian man were reproduced without any modifications.

Leonardo, Homo ad Quadratum and ad Circulum

We finally arrive at the famous *Vitruvian Man* by Leonardo (Fig. 9.10). If we compare this drawing with the previous versions, the man fits perfectly, standing with legs and arms outstretched.

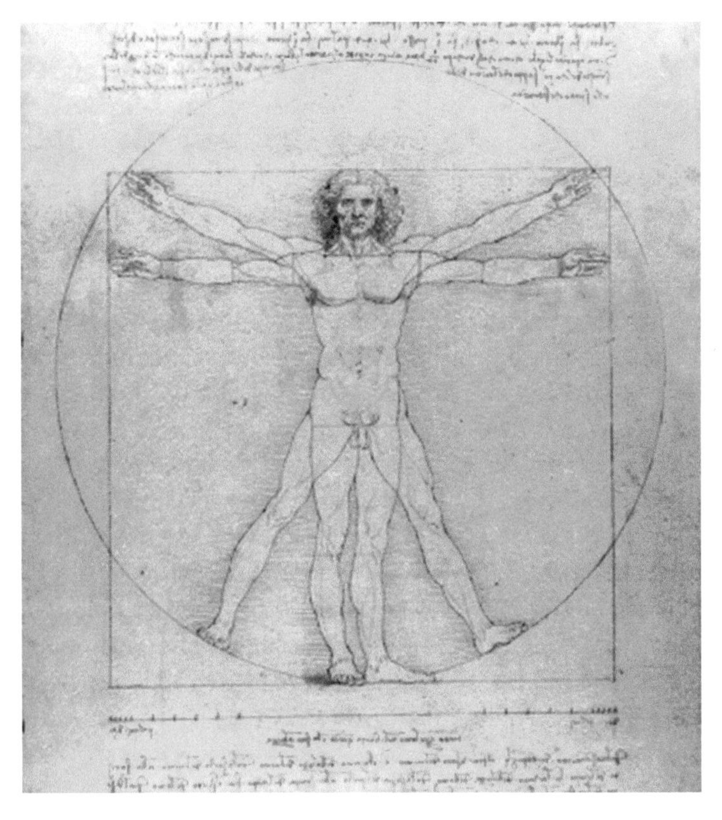

Fig. 9.10 Vitruvian Man of Leonardo da Vinci. Gabinetto dei Disegni e delle Stampe delle Gallerie dell'Accademia in Venice

The figure is at the center of the page and Leonardo included comments both above and below: "Vitruvius, architect, puts in his work of architecture, those measures of man that are spread by nature in this way, 4 dita (fingers) make a palmo (palm), and 4 palmi make 1 piede (foot), 6 palmi make a cubito (elbow), 4 cubiti make a man, 4 cubiti make a passo (pace), 24 palmi are 1 man, and these measures are in its structures. If you open both legs so that your height decreases from the head of 1/14 and open and lift your arms so that your stretched fingers touch the line at the top of the head, you should know that the center of the open ends of the limbs will be the navel. The space that lies between the legs is an equilateral triangle."[17]

So we see that Leonardo gave a precise description of the measures and used the figure to represent them; the units employed by Leonardo are *dita*, *palmi*, *cubiti*, *passi*, and *uomo* in the following equivalences:

$$4\,\text{dita} = 1\,\text{palmo};\quad 4\,\text{palmi} = 1\,\text{piede};\quad 6\,\text{palmi} = 1\,\text{cubito};$$
$$4\,\text{cubiti} = 24\,\text{palmi} = 1\,\text{passo} = 1\,\text{uomo}$$

The *cubit* (cubitum in Latin, namely, elbow) used by Leonardo is a unit of measure that had been in use since antiquity. It was calculated with reference to the length of the forearm from the tip of the middle finger to the elbow. For the Romans, this corresponded to 44.4375 cm. In general, the length of a cubit fluctuated at around half a meter depending on the country and era. If we take as our reference the Roman cubit, then it turns out that the ideal height for a well-proportioned man is 178 cm. At the bottom of the page, Leonardo goes on to provide a detailed, even if somewhat pedantic, description of the measures of various body parts.[18]

---

[17] Original text: "Vetruvio, architecto, mecte nella sua op(er)a d'architectura, chelle misure dell'omo sono dalla natura disstribuite inquesto modo cioè che 4 diti fa 1 palmo, et 4 palmi fa 1 pie, 6 palmi fa un chubito, 4 cubiti fa 1 homo, he 4 chubiti fa 1 passo, he 24 palmi fa 1 homo ecqueste misure son ne' sua edifiti. Settu ap(r)i ta(n)to le ga(m)be chettu chali da chapo 1/14 di tua altez(z)a e ap(r)i e alza tanto le b(r)acia che cholle lunge dita tu tochi la linia della somita del chapo, sappi che 'l cie(n)tro delle stremita delle ap(er)te me(m)bra fia il bellicho. Ello spatio chessi truova infralle ga(m)be fia tria(n)golo equilatero."

[18] Original text: "Tanto ap(r)e l'omo nele b(r)accia, qua(n)to ella sua alteza. Dal nasscimento de chapegli al fine di sotto del mento è il decimo dell'altez(z)a del(l)'uomo. Dal di socto del mento alla som(m)ità del chapo he l'octavo dell'altez(z)a dell'omo. Dal di sop(r)a del pecto alla som(m)ità del chapo fia il sexto dell'omo. Dal di sop(r)a del pecto al nasscime(n)to de chapegli fia la sectima parte di tucto l'omo. Dalle tette al di sop(r)a del chapo fia la quarta parte dell'omo. La mag(g)iore larg(h)ez(z)a delle spalli chontiene insè [la oct] la quarta parte dell'omo. Dal gomito alla punta della mano fia la quarta parte dell'omo, da esso gomito al termine della isspalla fia la octava parte d'esso omo; tucta la mano fia la decima parte dell'omo. Il menb(r)o birile nasscie nel mez(z)o dell'omo. Il piè fia la sectima parte dell'omo. Dal di socto del piè al di socto del ginochio fia la quarta parte dell'omo. Dal di socto del ginochio al nasscime(n)to del memb(r)o fia la quarta parte dell'omo. Le parti chessi truovano infra il me(n)to e 'l naso e 'l nasscime(n) to de chapegli e quel de cigli ciasscuno spatio p(er)se essimile alloreche è 'l terzo del volto."

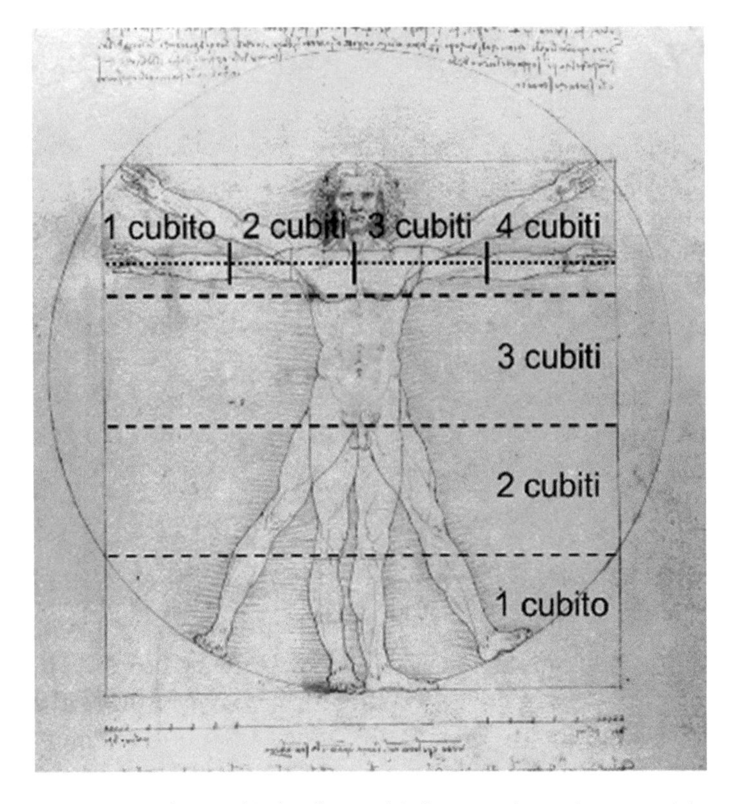

Fig. 9.11 Leonardo's four cubit by four cubit homo ad quadratum, with measures marked

In the drawing, the references with respect to the different parts of the body to be used for the calculation of proportions are well marked, with a scale expressed in *dita*, *palmi*, and *cubiti*. Taking as reference the scale from left to right, there are four small signs corresponding to the four *dita* forming a *palmo*, then another five more widely spaced signs indicating the 6 *palmi* which are a *cubito*, and then the four *cubiti* which form the man (the latter marked in Fig. 9.11).

At this point, it is easy to see that the standing man with his arms opened to form a cross is perfectly inscribed in the square and that he measures so exactly four *cubiti* by four *cubiti*: "The extent of man's open arms corresponds with his height."[19] Exactly as Vitruvius himself enunciated: "Indeed, if it is measured from the mounting surface of the feet up to the top of the head, and then you will compare this measure with those of the outstretched hands, one will find a length equal to the height, as is the case of a square sketched using a set square."

[19] Original text: *Tanto ap(r)e l'omo nele b(r)accia, qua(n)to ella sua alteza.*

But is it actually true that a man's height corresponds to the length of his open arms? Even though it may seem nonintuitive, on average, the answer is yes.

There is one final detail which we should note with regard to Leonardo's *homo ad quadratum*: The center of this human figure is not the navel but the pubis.

Homo ad Circulum

It remains now to inscribe the man in the circle. Leonardo solved the problem by raising the man's arms up to the top of the head ("hold up both arms that with the long fingers you [can] touch the top of the head")[20] and spreading the legs so that his height decreases by 1/14 ("open so much the legs to reduce by 1/14 your height [calculated] from the head").[21] If the side of the square is split into 14 parts, this decrease can easily be seen. At this point the man is perfectly inscribed in a circle centered at his navel.

If we stand back and carefully look again at the figures drawn by Cesariano, we realize that, in these as well, when the standing man is positioned in the square, the diagonals are centered at the pubis, whereas in the *ad circulum* figure, they meet at his navel. There is another similarity between Leonardo's and Cesariano's drawings, that of the units of the drawing scale, with the *digiti* (or *dita* in Leonardo's form for the plural of *dito* (finger)[22]), *palmi*, and *cubiti*. This is not particularly surprising: Before the introduction of the modern, scientific metric system, scales of measures were mostly anthropomorphic, taking the average human dimensions as reference. However, the similarities end here, for, as we have noted, Cesariano needed to use two different figures, and, in the second one, he was forced to introduce a distorted, figurative artifice to get the man to fit into both the square and the circle.

Considering that the design of Leonardo's drawings can be dated to around 1490 and that the Italian version of Cesariano's *De Architectura* was printed in 1521—2 years after Leonardo's death in 1519—it is difficult to imagine that there could have been some mutual influence. Instead, it is more likely that they arrived independently at a similar idea, although, as we have seen, Cesariano completely failed to solve the geometrical problem.

[20] Original text: "alza tanto le braccia che colle lunghe dita tu tocchi la linea della sommità del capo."

[21] Original text: "apri tanto le gambe da far calare dal capo la tua altezza di 1/14."

[22] *In Italian the word finger is irregular; the singular is masculine "dito" the plural becomes feminine "dita" when is used to indicate all the fingers in general. The plural form for a specific finger, such as the ring fingers, is instead the masculine "diti".*

However, if we compare Leonardo's *Vitruvian Man* with the sketch of Giacomo Andrea (Fig. 9.7), it appears rather clear that these two figurative solutions have some significant contact points. Although Giacomo Andrea's *Vitruvian Man* appears as merely an ungainly draft in comparison to Leonardo's detailed and eye-catching illustration, the simple human figure again occupies the center of the scene, almost a depiction of Christ on the cross. Indeed, the simplicity of the picture exercises a strange fascination on the viewer.

The geometric solution adopted for the representation of *Vitruvian Man* in the two drawings is similar. It is likely that Leonardo discussed the question with his friend, perhaps even seeing his drawing, and that it served as his inspiration. As usual, however, Leonardo went far beyond. He scientifically analyzed the human figure and, at the same time, was able to add to his illustration the evocative force that made of it the global icon that is recognized today worldwide.

Yet Another Man

There is an interesting manuscript from the late sixteenth century, the *Codex Huygens*[23,24] which contains an impressive collection of studies about human body movement. The manuscript is attributed to Carlo Urbino da Crema (born between 1510 and 1520—sometime after 1585), a polymath scholar from the North of Italy. The extant manuscript is composed of five sections with a total of 128 folios. Some parts of the work are supposed to have been lost, and what remains should, therefore, not be supposed to be the full version. The strong modern-day interest in this manuscript is due not only to the quality of the content—especially the illustrations which have been traced by an expert hand—but also to the hypothesis that some of the drawings may reproduce lost originals from Leonardo.

Unsurprisingly, perhaps, then, we find that one of illustrations in the manuscript (folio 7) shows yet another version of the *Vitruvian Man* (Fig. 9.12). However, as we will see, Carlo Urbino presents a different solution to that of Leonardo.

This version of the *Vitruvian Man* exhibits a series of geometrical figures—triangles as well as circles and squares—that are used to measure and deter-

[23] The manuscript is conserved at The Morgan Library & Museum of New York, Codex M. A. 1139. The name of the Codex derives from a previous owner, the secretary of the King of England William III, Constantin Huygens. The manuscript is available online: www.themorgan.org/collection/Codex-Huygens.

[24] Panofsky E (1940), *The Codex Huygens and Leonardo da Vinci's art theory, the Pierpont Morgan Library codex M.A. 1139*, The Warburg Institute, London.

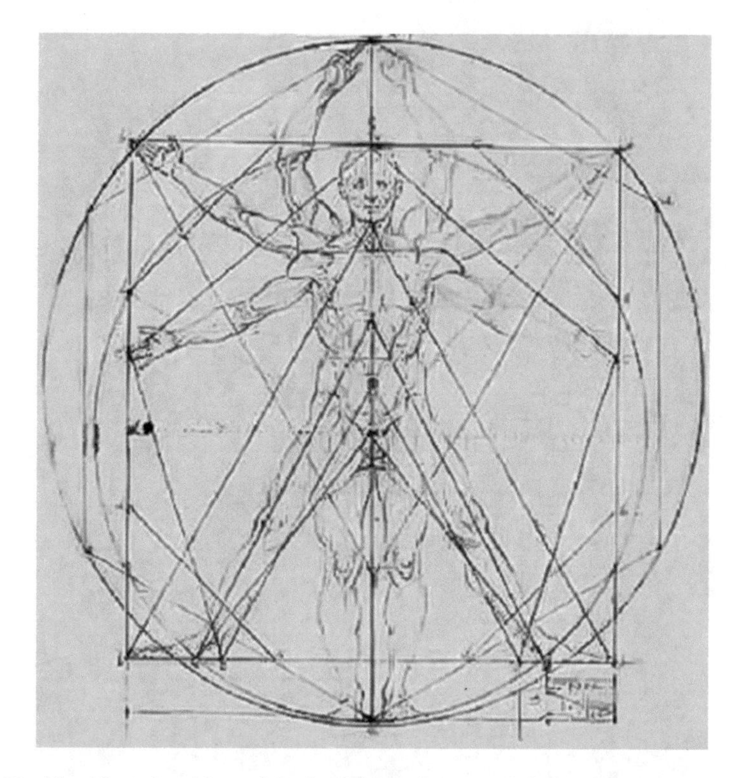

Fig. 9.12 The Vitruvian Man of Carlo Urbino da Crema. Codex Huygens M. A. 1139, folio 7. The Morgan Library & Museum of New York

mine the appropriate proportions for the human body. As we can see, Carlo Urbino's solutions were not as brilliant as Leonardo's, and he also needed to use some tricks to fit the man within both circle and square, such as deforming the feet, as Cesariano had also done. This drawing is part of a rather impressive collection of studies of human body movements which explore a variety of different possibilities but always restricting these movements to within the limits of an artificial geometrical enclosure (Fig. 9.13).

A Geometrical Comparison

The range of graphical versions of the *Vitruvian Man* realized by different people in a short time span are the clear evidence of the enormous interest that the Renaissance architects had in *De Architectura*. It is also notable that most of these solutions seem to have been independently explored, even if

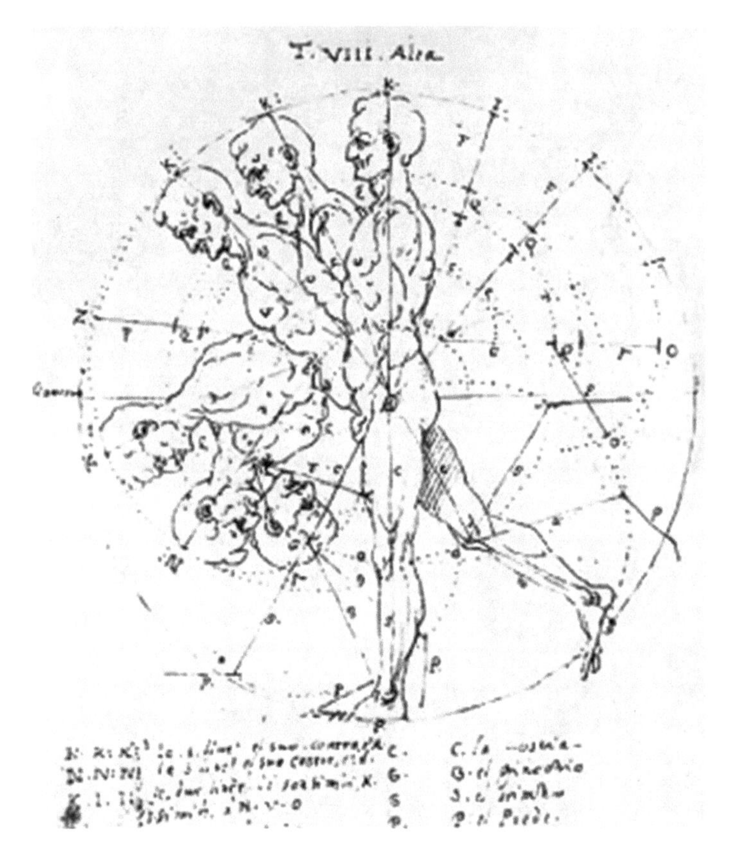

Fig. 9.13 Studies of human body in movement. Carlo Urbino da Crema. Codex Huygens M. A. 1139, folio 29, detail. The Morgan Library & Museum of New York

clearly some of the authors—in particular, Leonardo—could take advantage of previous works.

Let us now reconsider in geometric terms the various solutions proposed for combining the circle and the square (Fig. 9.14). While some versions, such of those of Taccola and Cesariano, overlap from a geometrical point of view, five distinct solutions are to be found.

In the approach shared by Leonardo and Giacomo Andrea, the square and the circle are independent, rather than one being inscribed within the other. In fact, if we carefully reread Vitruvius' text, it clearly states a very precise correspondence: A man with open arms is enclosed in a square as well as in a circle centered at his navel. But must the circle and the square have the same center?

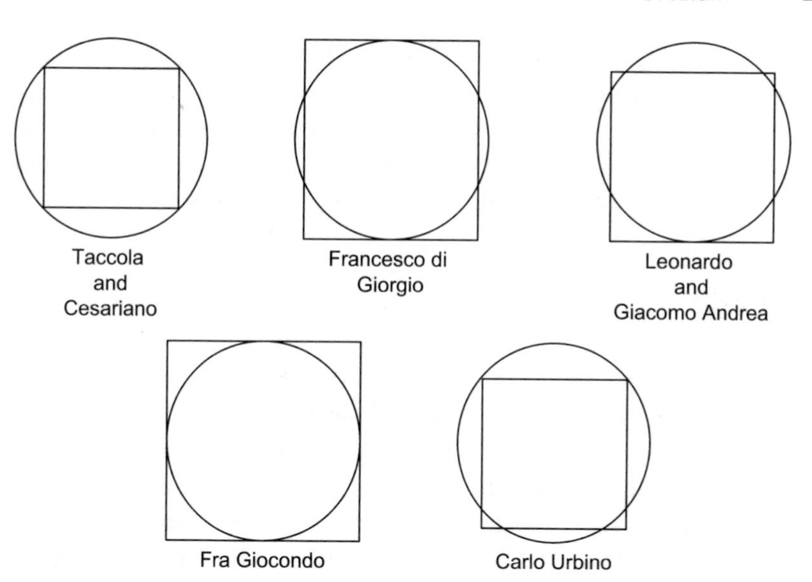

Fig. 9.14 The different geometrical solutions adopted for the Vitruvian Man

Vitruvius does not specify, and in Leonardo's rendering, the circle is centered at the man's navel but the square at his pubis. Indeed, the two shapes are independent. Leonardo's trick allows the overlap of the two men in the same drawing, creating a visual illusion that simulates movement and helps create the somewhat arcane charm of the drawing. For others, like Cesariano, the solution is that the center of both figures should be the navel. As we have seen, most of the authors need to resort to some personal tricks to make the man fit—for example, the deformation of the man's body in Cesariano's drawing and Francesco di Giorgio elongation of the square into a rectangle.

The Legacy of Leonardo's Vitruvian Man

Although Leonardo's *Vitruvian Man* is now a global icon, it remained forgotten for a long time. It was only just after the turn of the twentieth century that the drawing began to be part of the collective imagination. The real discovery, however, starts with the publication in 1949 of the first edition of the *Architectural Principles in the Age of Humanism* by R. Wittcower, a fundamental work on the aesthetics of Renaissance architecture.

A lot has been written about the significance of Leonardo's *Vitruvian Man*, but without going into Leonardo's subconscious, perhaps the best interpretation is given by the notes that accompany the drawing: For him, it is just a challenge, a study on the dimension of the human body to derive universal laws of beauty to be applied to painting, sculpture, and architecture.

A Mystic Vision

As a coda to the present chapter, it is worth noting that circles and squares are elements of a mystical and symbolic iconography that has deep roots, and the great evocative strength of the *Vitruvian Man* is also due to this old tradition. It is not by chance that one of the first iconographies to represent man as being at the center of the universe is due to another extraordinary personality: Saint Hildegard of Bingen (Bermersheim vor der Höhe, Bingen am Rhein, 1098–1179), whose name refers to the monastery she founded on the Rhine. When she was very young, Hildegard began to have visions that she transcribed. These visions pushed her to monastic vows and found the monasteries of Rupertsberg and, later, Eibingen. For her cultural contributions and diversity of interests, she is one of the most extraordinary figures of the European Middle Ages, having written treatises on music, natural sciences, and medicine and even inventing a new alphabet, the *Lingua Ignota*.

The visions of Hildergard, which were probably of neurologic migraine-induced origin, were collected in a so-called prophetic trilogy: the *Scivias* (completed in 1151), the *Liber Vitae Meritorum* (begun in 1158), and, finally, her most visionary work, the *Liber Divinorum Operum* (finished in 1174).

It is this final manuscript that demands Hildergard's presence in this chapter. The *Liber Divinorum Operum*, which was very finely painted in miniature, describes ten visions with accompanying illustrations. One of these illustrations depicts a naked human figure in the center of a sphere that represents the universe (Fig. 9.15). The illustration represents the complex relationship between man, seen as a microcosm, and the macrocosm of the universe. In the colorful miniature, the naked man with open arms is also surrounded by symbols which represent the different spheres of the universe.

The human figure occupies a central position for Hildergade: "With arms and hands outstretched to the sides of the torso, the height of the human figure coincides with his width, just like the height of the firmament is equal

Fig. 9.15 Hildegard of Bingen (1098–1179), *Liber Divinorum Operum*, thirteenth century. Lucca, Biblioteca Statale, Ms. 1942, c. 9r. (This work is in the public domain in its country of origin and other countries and areas where the copyright term is the author's life plus 70 years or less. https://en.wikipedia.org/wiki/Hildegard_of_Bingen#/media/File:Hildegard_von_Bingen_Liber_Divinorum_Operum.jpg)

to its width." The man, as evidenced by his size, is an element of the universe itself and a small reproduction, a microcosm, of the divine space. So we see that, well before Leonardo, man had already been placed at the center of a geometrical universe symbolically represented by a rectangle and a circle.

10

A Friend, an Enigmatic Portrait, and Two Duchesses

Leonardo was an affable and gentle person, and during the course of his whole life, he was surrounded by the affection of his many pupils and friends. In Milan, as we have seen in the previous chapter, he enjoyed something akin to brotherhood with Giacomo Andrea, and he also had a very special friendship with the Tuscan mathematician Luca Pacioli[1] who would open to him the doors of the universe of geometry and mathematics.

The lives of Leonardo and Pacioli were deeply interwoven for a relatively long time, and they shared not only some happy years in Milan but also the escape from the city and the subsequent wanderings around Italy. Leonardo's work and influence are to be found in Pacioli's work, in many cases very prominently, as, for example, for the illustrations he made for Pacioli's *Divina Proportione*. At other times the connection is more difficult to decipher but nonetheless apparent.

The two shared a fruitful relationship that was rooted in mutual and sincere admiration. The friar taught Leonardo mathematics and geometry and had a profound influence on the formation of his scientific personality. Many annotations in Leonardo's notebooks contain observations and mathematical games taken from Pacioli's work.

A dedication paragraph in Pacioli's *De viribus quantitatis* (*On the power of numbers*)[2] expresses all the admiration and affection that he had for Leonardo, wistfully recalling the time they spent together in Milan: "... in the world it is impossible doing better the Platonic solids and mathematical bodies using the

[1] The name of Luca Pacioli is also found in the form Luca Paciolo o Paciuolo e Luca di Borgo.

[2] Codex 250, Biblioteca Universitaria di Bologna. A digitalized version is available online: http://www.uriland.it/matematica/DeViribus/Pagine/index.html.

P. Innocenzi, *The Innovators Behind Leonardo*, https://doi.org/10.1007/978-3-319-90449-8_10

drawing in perspective ... that were made by that ineffable left hand, talented in all disciplines of mathematics of the prince among mortals, our Florentine Leonardo da Vinci, in that happy time we were together in the city of Milan with the same salaries."[3] This admiration was definitely reciprocated by Leonardo who saw in Luca Pacioli the Maestro who could initiate him to the secrets of arithmetic and geometry: "learn the multiplication of root from Maestro Luca." Indeed, Leonardo had used Pacioli's mathematical texts even before meeting him in person. As soon as Pacioli's treatise of mathematics, the *Summa*, was published, he purchased a copy whose cost he noted carefully: "119 [coins] for arithmetic [book] of maestro Luca."[4]

A Copycat Mathematician

Like Leonardo, Luca Pacioli (c. 1445–1517)[5,6] was born in Tuscany, in Borgo San Sepolcro, a village near Arezzo. He occupies a special place in the history of mathematics because it is to him that we owe the first extensive treatise of financial mathematics with a description of double-entry bookkeeping.

Orphaned at the age of 14, he was raised by a wealthy family from Sansepolcro and later moved to Venice (probably around 1465), where he learned mathematics by attending the Rialto School. At the same time, he worked as a live-in tutor for the three children of a merchant. At the age of 23, he became a reader in mathematics and, from that moment on, began his brilliant career that would lead him to achieve great fame among his contemporaries.[7] After Venice, Pacioli continued to travel around Italy, moving to Florence, Perugia, Urbino, Zara, Naples, Assisi, and Rome. In 1471, he returned to Borgo Sansepolcro to become a Franciscan friar, but after a while, he resumed his wanderings around Italy. In 1493, his superiors demanded that he return to Assisi; however, they were able to keep him in the monastery for just one year.

[3] Original text: "supraeme et legiadrissime figure de tutti li platonici et mathematici corpi regulare et dependenti che in prospectivo disegno non è possibile al mondo farli meglio.... facte et formate per quella ineffabile senistra mano a tutte discipline mathematici acomodatissima del principe oggi fra mortali pro prima fiorentino Lionardo nostro da Venci, in quel foelici tempo che insiemi a medesimi stipendij nella mirabilissima citta di Milano ci trovammo."

[4] *Codex Atlanticus* folio 288r.

[5] Parisi D (2012) entry on Luca Pacioli. In Enciclopedia Treccani.

[6] Napolitani PD (2013) entry on Luca Pacioli. The Italian contribution to the history of thought. Enciclopedia Treccani, Sciences.

[7] A (2003) Luca Pacioli e la matematizzazione del sapere nel Rinascimento. Cacucci editore.

In 1494, Pacioli traveled to Venice for the publication of his compendium on mathematics in 615 pages, the *Summa de arithmetica, geometria, proportioni et proportionalita*[8] (Summary of arithmetic, geometry, proportions and proportionality)[9,10]. The description, mentioned above, of the double-entry bookkeeping system appears in one of the chapters (*Tractatus de computis et scripturis*) of the *Summa.* This large treatise is a kind of encyclopedia of mathematics, where Pacioli collected the knowledge of the time drawing on a variety of sources. The work was written in the Tuscan vernacular with some small terms and short sentences in Latin and Venetian. The work was dedicated to Guidobaldo da Montefeltro, Duke of Urbino, one of the main patrons of Renaissance science, whom Pacioli had met in Rome in 1489.

The publication of the *Summa* in Venice in 1494 would contribute significantly to the spread of Pacioli's fame. During his continuous, if not frenetic, movements around the leading Italian cultural centers, he taught mathematics and knew powerful patrons. Together with the publication of the *Summa*, this certainly helped to increase his reputation as a scholar and mathematician. It is thanks to this fame that Pacioli arrived in 1496 at the court of Ludovico il Moro in Milan, where he was called by the Duke as a public reader in mathematics at the Palatine School. Pacioli would remain in Milan, though not continuously, until 1499, when, together with Leonardo, he moved to Mantua, because of the fall of Milan into the hands of the French.

Pacioli's Milanese period was very fertile, thanks also to his friendship with Leonardo, who, as we have said, would work with him to illustrate his famous treatise on the golden ratio, the *Divina Proportione*,[11] written between 1496 and 1497 and printed in Venice in 1509 (see Chap. 11).

During this same period in Milan, between 1496 and 1499, Pacioli also wrote his most controversial book, the *Libellus corporum regularium* (small book of the regular bodies), which is dedicated to the study of the regular polyhedra. The controversy stems from the fact that something very similar had already been written by Piero della Francesca (c. 1416–1492), the famous

---

[8] The work was published by the Venetian printer Paganino Paganini, who would establish a long and fruitful collaboration with Luca Pacioli. Paganino Paganini (or De Paganinis) of Brescian origin moved to Venice in 1483 and printed for Pacioli both the *Summa de arithmetica* (20 November 1494) and that other famous work, the *De divina proportione* (1509). The printed version is illustrated with woodcuts of the drawings of Leonardo da Vinci's polyhedra and the geometrically constructed alphabet with uppercase letters. In 1509 Paganini also printed Euclid's text edited by Luca Pacioli.

[9] The first edition (*Editio princeps*) of the *Summa de arithmetica, geometria, proportioni et proportionalita* was published in Venice in 1494 and the second in 1523 in Toscolano. This was the first printed treatise on arithmetic and algebra.

[10] A digitalized copy of the *Summa* is available online: http://jeremycripps.com/docs/Summa.pdf.

[11] Chadwick D, Odifreddi P, Panta A (2009) Antologia della Divina Proporzione di Luca Pacioli, Piero della Francesca e Leonardo da Vinci. Aboca Museum Edizioni.

Florentine painter who dedicated a small treatise to the five regular bodies, the *Libellus de quinque corporibus regularibus* (small book about the five regular bodies)[12,13]. Luca Pacioli's book was therefore considered by his contemporaries to be a work of plagiarism, and the first to accuse him was Giorgio Vasari (1511–1574), who, in his section on Piero della Francesca's life, wrote: "Because Maestro Luca dal Borgo, Franciscan friar, who wrote about the geometry of regular bodies, was one of his disciples: and with old age of Piero, who had written several books, Maestro Luca printed all them pretending to be his own work... as already said above"[14,15,16]. It was later shown that these accusations were essentially correct: Luca Pacioli drew extensively from other manuscripts to compile his work. Even in the case of double-entry bookkeeping, to which most of his posthumous fame is due, he drew on previous sources but seldom mentioned them properly.[17] In this regard, it is worth mentioning the comment by Leonardo scholar K. D. Keele on Pacioli's ambiguous character: "Luca Pacioli was a man of great ambition, questionable integrity, and a remarkable ability to use some of his predecessors."[18] This comment is, however, a little ungenerous with respect to the friar who played such a fundamental role in diffusing and transmitting the knowledge of mathematics during the Renaissance.

An Enigmatic Portrait

Luca Pacioli's likeness is captured in an oil portrait on a panel, *Il Ritratto di fra' Luca Pacioli con un allievo (Portrait of friar Luca Pacioli with a student)*, executed around 1500 (Fig. 10.1). The identity of the artist is uncertain, with the

[12] Codex Urb. Lat. 632. Biblioteca Apostolica Vaticana.

[13] Cervantes A (2009) Luca Pacioli tra Piero della Francesca e Leonardo. Aboca Museum Edizioni.

[14] Original text: "Perché il Maestro Luca dal Borgo, frate di San Francesco, che scrisse sulla geometria dei corpi regolari, fu un suo discepolo: e con la vecchiaia di Pietro, che aveva scritto molti libri, il Maestro Luca li fece stampare tutti usurpandoli come sue opere...come si è già detto sopra."

[15] Giorgio Vasari. *Le vite de' più eccellenti architetti, pittori, et scultori italiani, da Cimabue insino a' tempi nostri: descritte in lingua Toscana da Giorgio Vasari Pittore Aretino. Con una sua utile et necessaria introduzzione a le arti loro.* Florence, 1550.

[16] Mancini G (1915) L'opera "De corporibus regularibus" di Piero Franceschi detto Della Francesca, usurpata da fra Luca Pacioli. "Memorie della Regia Accademia dei Lincei." Classe di scienze morali, storiche e filologiche, vol. XIV: 441–580.

[17] Picutti E (1989). Sui plagi matematici di frate Luca Pacioli. Le Scienze. February 1989: 72–79.

[18] Keele KD (1983) Leonardo da Vinci's elements of the science of man. Academic Press.

Fig. 10.1 Portrait of Luca Pacioli with a student. 1500. Commonly attributed to Jacopo de' Barbari. Pinacoteca of Capodimonte, Naples, N. Q. 58. (This work is in the public domain in its country of origin and other countries and areas where the copyright term is the author's life plus 70 years or less. https://it.wikipedia.org/wiki/Luca_Pacioli#/media/File:Pacioli.jpg)

most likely candidate being the painter Jacopo de' Barbari (1440–1515)[19,20]. This painting remained forgotten for quite a long time. The first mention of it is in a 1631 inventory of the Ducal Palace in Urbino as part of the assets of the Della Rovere family. The picture would reappear centuries later in Naples as part of the legacy of the Medici family and is now displayed in the Farnese Gallery of the Museo di Capodimonte in Naples, even though it is not really part of the Farnese collection, having been purchased in the early twentieth century by the Italian state exercising its right of first refusal on a sale destined abroad.

Thus, we see that this picture had a curious fate, passing from obscurity to being at the center of an almost morbid attention which has led scholars and lay public enthusiasts alike to study every detail and to formulate many hypotheses, some very imaginative. One reason for this curiosity, as we will see, is the mysterious shadow of Leonardo, the great friend of Luca Pacioli, emerging from some suggestive clues.

[19] Gilbert C (1964) Entry on: Iacopo de' Barbari. In: Dizionario Biografico degli Italiani. Treccani, Volume 6.

[20] Cervantes A (2011) Il doppio ritratto del poliedrico Luca Pacioli. De Computis, Spanish Journal of Accounting History. 16: 107.

The Tuscan mathematician appears in the portrait dressed in a gray friar's robe and impressively occupying the central space, standing behind a table covered with a green cloth. With one hand he traces geometric figures on a slate, and with the other, he indicates a paragraph in an open book in front of him. On the edge of the slate is clearly written EVCLIDES, and the book has been identified as Euclid's *Elementa in artem geometriae et Campani commentationes* (Elements in the art of geometry with comments of Campano). The reproduction of the open page is so accurate that it is even possible to identify which printed edition it represents: that edited by Erhard Ratdolt in Venice in 1482. The geometric figure that Luca Pacioli is drawing with his right hand on the slate is taken from the proposition from Euclid's book XIII to which he is pointing with his left.

On the right, on top of the table, there is a large volume, a copy of Pacioli's own *Summa*, bound in red with the Latin inscription: LI. RI. LUC. BUR., which, when spelled out, becomes *Liber Reverend Luca Burgensis* (Book of Reverend Luca from Borgo). All of the necessary tools for a mathematician can be found on the table: a sponge, a pen with ink, a pencil case, a piece of chalk, a protractor, and a compass. Two geometrical solids appear, at the two ends of a diagonal from the top left of the painting to the bottom right: At the bottom part of the picture, a Platonic solid, the dodecahedron, is placed over the *Summa*, while at the opposite end on the left side of the picture is hung a convex solid formed from 18 squares and 8 triangles, the rhombicuboctahedron.

Pacioli is not alone in the picture; behind him appears the figure of an elegantly dressed young man who must be one of his students. His identity has not yet been determined, although one possibility is Guidobaldo da Montefeltro, Duke of Urbino from 1482–1508, to whom in 1494 Pacioli dedicated the *Summa*. Other possibilities are that the figure is representative rather than specific or, even, that he might be a self-portrait of the artist of the painting. In any case, the figure of the elegantly dressed young man wearing blue gloves provides a strong contrast with the mathematician in his simple and severe friar's dress, whose eyes seem to fix on a vague point in front of him. The mysterious pupil, whose face stands out thanks to the light that illuminates him, directly captures the eyes of the viewer.

The Guidobaldo da Montefeltro thesis is strongly undercut by a comparison with an undisputed likeness of the Duke (Fig. 10.2), a portrait attributed to Raffaello Sanzio (Raphael, 1483–1520). Even to the nonexpert, the difference between the two portraits catches the eye. The diaphanous and lean figure depicted by Raphael does not bear a great resemblance to Jacopo de' Barbari's portrait.

Fig. 10.2 Raffaello Sanzio. Guidobaldo da Montefeltro. 1507. Uffizi Gallery, Florence. Oil on panel. 52 × 69 cm. (This work is in the public domain in its country of origin and other countries and areas where the copyright term is the author's life plus 70 years or less. https://upload.wikimedia.org/wikipedia/commons/4/48/Guidobaldo_montefeltro.jpg)

As we have said, there are many hypotheses about the identity of the young student, and it bears pausing to explore some of the others. First, let us return for a moment to the probable author of the painting, Jacopo de' Barbari, an important artist even if he is not among the most famous of the Renaissance. He plays, however, at least in this story, an important role.

Jacopo de' Barbari's date of birth is not known for sure, and there is no other information about his life before he became active as an artist in Venice. Few of his works remain, but one in particular, the *View of Venice*,[21] a woodcut of great artistic quality showing a detailed cartography of the lagoon city, has at least allowed his fame to pass into posterity (Fig. 10.3).

The fate of Jacopo de' Barbari intersects briefly in Venice with the name of a great protagonist of the Renaissance, Albrecht Dürer. The German artist made two trips to Italy, the first in the period between the autumn of 1494 and May 1495, spending that winter in Venice. During this period, Albrecht Dürer took lessons on human proportions from Jacopo de' Barbari, who showed him, without, however, disclosing all the secrets of the technique, how to draw a man and a woman using the correct dimensions. At the end of

[21] Virtual Archive veneziano-ESPRIT Project 20,638, www.tridente.it/venetie/fhome.htm.

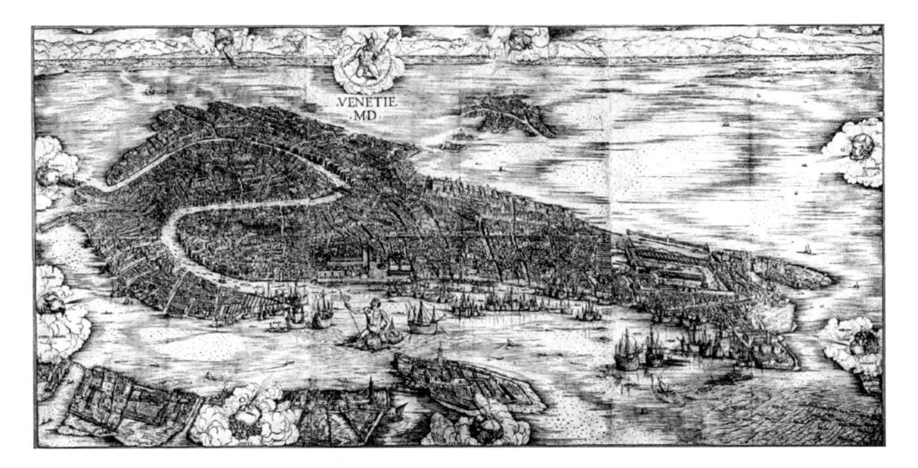

Fig. 10.3 Jacopo de' Barbari. View of Venice. 1500. Woodcut. Minneapolis Institute of Art. (This work is in the public domain in its country of origin and other countries and areas where the copyright term is the author's life plus 70 years or less. The image has been modified to reduce the noise. https://commons.wikimedia.org/wiki/File:Jacopo_de%27_Barbari_-_View_of_Venice_-_Google_Art_Project.jpg)

1494, Luca Pacioli was also present in Venice, for the printing of the *Summa*, and it is plausible that the German artist, so eager to study geometry and proportions, may have taken some lessons from the Italian mathematician.

So, at the same time, Luca Pacioli, Jacopo de' Barbari, and Albrecht Dürer were present together in Venice. It is possible that the Venetian painter wanted to immortalize the mathematician giving a lecture to the German artist. The painting is dated 1495, so the dating is certainly compatible with this theory. Another intriguing aspect is the striking resemblance of the young man depicted in the picture to two known self-portraits by Albrecht Dürer, where he appears with long red curly hair.[22]

This resemblance is definitely much higher than that to Raffaello's portrait of Guidobaldo, where the Duke of Urbino has a sad and dull appearance, perhaps a foreshadowing of the death that would seize him shortly thereafter.[23] Of course, this resemblance could just be a coincidence, and is certainly not a proof, but there is another strange correlation between Luca Pacioli and Albrecht Dürer which has been discovered recently: Luca Pacioli had, despite his darkly serious appearance in his portrait, a special passion for games and, in fact, wrote two books on this subject. One of them, dedicated to chess,

[22] Albrecht Dürer: 1) Self Portrait (1493), Louvre, Paris; 2) Self Portrait at 28 (1500), Alte Pinakothek, Munich.

[23] Guidobaldo da Montefeltro, Duke of Urbino (Gubbio, 1472–Fossombrone, 1508).

disappeared for 500 years; the other one, *De viribus quantitatis*[24,25] (1496–1508), remained forgotten in the library of the University of Bologna. It is one of his most intriguing works and contains a collection of scientific and mathematical recreations.[26]

One of these problems is the construction of a magic square using the numbers from 3 to 9.[27] A magic square is a square table in which the sum of the numbers in each row, column, and diagonal gives the same value, the so-called magic constant. Magic squares were well known in antiquity, but in Europe no examples were reported before 1500. In one of the most famous engravings of Albrecht Dürer, *Melencolia I* (see Chap. 11), however, a 4 × 4 magic square is clearly shown. The two numbers in the middle of the last row, 15 and 14, give the year in which the work was made, while 1 and 4 represent the signature of the author, 1 = *A* and 4 = *D* for Albrecht Dürer. Because Luca Pacioli's *De viribus quantitatis* had remained for a long time in oblivion, Dürer's magic square has long been considered to be the first to have appeared in North Europe. However, when we consider the relationship between the artist and the mathematician, it is also plausible that the original source of Dürer's magic square was, in fact, the Tuscan friar. It is worth noting here that, within the context of the limited sources available in Italy at the time, Pacioli also had very good knowledge of the Arabic texts of mathematics.

Another interesting detail in Dürer's engraving is a strange solid, a truncated rhombohedron, also known as Dürer's solid (see Chap. 11). As we will see in the next chapter, 15 years before the realization of this engraving, Leonardo had drawn several examples of such solids for Luca Pacioli's *Divina Proportione*. Thus, we see that the German artist could have been inspired by the Pacioli-Leonardo work in this as well, which undoubtedly increases the mystery and fascination of the portrait.

Now let us return to the rhombicuboctahedron in Luca Pacioli's portrait, because the similarities of this figure with those drawn by Leonardo for the *Divina Proportione* are really remarkable. The figure is suspended by a thread in the same manner used by Leonardo to represent the different solids in Luca Pacioli's book (see Chap. 11 for a longer discussion of these). In the *Divina Proportione*, the rhombicuboctahedron is made of glass and appears in the painting partly filled with water. This part of the picture is a work of great qual-

[24] Luca Pacioli (2006) De viribus quantitatis. Transcription from the codex 250 of the University Library of Bologna by maria Garlaschi Peirani. Giunti.

[25] Amedeo Agostini (1924) De viribus quantitatis by Luca Pacioli. Journal of mathematics, vol. IV:165–192.

[26] Bano D, Tonato S (2011) I giochi matematici di Fra' Luca Pacioli. Edizioni Dedalo, Bari.

[27] Problem LXXII.

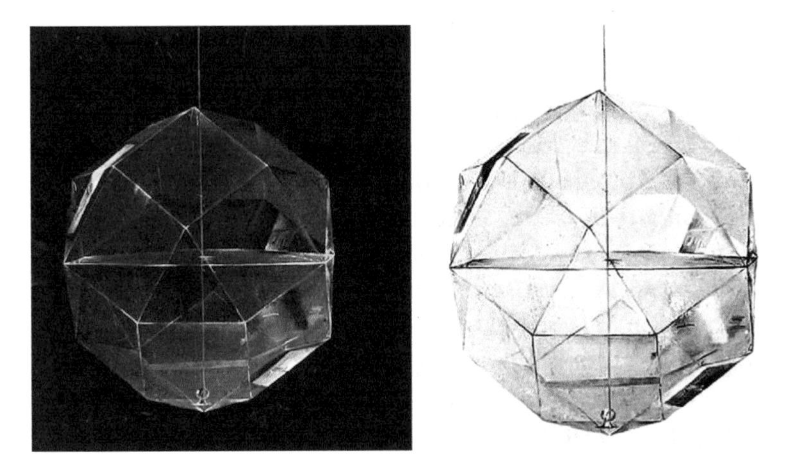

Fig. 10.4 Enlargement of the rhombicuboctahedron in Luca Pacioli's portrait (*left*) and the inverted image which shows the geometrical structure (*right*)

ity and required a high skill to be realized (Fig. 10.4). It is hard to say if Leonardo's personal hand is behind this masterpiece or if it was rendered by someone from his workshop. In any case, it is abundantly apparent that the author of the picture must have been well aware of Leonardo's drawings for Pacioli.

There is also another detail to complicate the story: In a reflection on its surface, the rhombicuboctahedron shows what has been identified as the Ducal Palace of Urbino, thus providing a good element to couple the young figure in the painting with the Duke Guidobaldo da Montefeltro. The drawing quality of the rhombicuboctahedron is however undeniable, and Mackinnon, a modern-day scholar, writes of it: "I think there was only one man in Europe able to paint this subject, and [he] is not the author of the painting."[28] So, we see that the evidence points us in multiple and conflicting directions.

There is another interesting detail in the picture, and it is the little scroll near the book on the table (Fig. 10.5). If we look at it in detail, it is possible to read the following text in capital letters: IACO. BAR. VIGENNIS. P. 1495.

As can easily be seen, an annoying fly sits right next to the date. Might this insect have a symbolic and recondite meaning? The simplest interpretation is that the cartouche is the signature of the artist, namely, the painter Jacopo (IACO) de' Barbari (BAR), with the date of execution of the work (1495). The abbreviation P. stands for Pictus (painted), while Vigennis would indicate not so much the age as the place of origin, for example, by Vigano or di Vige.[29]

[28] Mackinnon N. The portrait of Fra Luca Pacioli. The Mathematical Gazette, 77,130–219. The Mathematical Association.

[29] Chastel A (1986) *A fly in the pigment*. FMR English Edition, April/May: 61–79.

Fig. 10.5 Detail of Luca Pacioli's portrait. The cartouche can be read as: IACO. BAR. VIGENNIS. P. 1495. A fly is also clearly visible on the right

There is also another strange detail, however: Jacopo de' Barbari was, in fact, also known as the Master of the Caduceus, for his mania for signing his paintings by placing a symbol of a stick and two snakes intertwined beside his name. In Luca Pacioli's portrait, however, this symbol does not appear. In keeping with the taste for mystery that always surrounds Leonardo, a rather fantastic interpretation has been proposed: that the cartouche must contain the key to resolving a historical puzzle, the suspicious death of Galeazzo Sforza (1444–1476), the rightful heir to the Dukedom of Milan, which was followed by Ludovico il Moro's rise to power in Milan, symbolized by the fly. The acronym of the cartouche, duly interpreted, would indicate precisely in Ludovico Sforza the author of the crime.[30]

Leaving aside these mysteries and possible encryptions, the fly probably refers to a famous anecdote reported by Vasari about Giotto di Bondone (known as Giotto, c. 1266–1337), who wanted to demonstrate his extraordinary skill as a painter by adding a fly on the nose of a figure painted by his master Cimabue (c. 1240–1302). This fly was so lifelike that Cimabue tried in vain to drive it away with his hands. The fly in Luca Pacioli's portrait might therefore simply be included as a demonstration of the skill of the painter, and then, following the simplest interpretation, the author of the cartouche is probably Jacopo de' Barbari.

This is, however, not the last word about this picture.

This is likely not the only painting depicting Luca Pacioli. In Piero della Francesca's *Madonna and child with Saints and angels and Federico da Montefeltro kneels in front of her*,[31] the face of St. Peter Martyr (the second from the right) has been seen as a Luca Pacioli's portrait[32] realized by his mas-

[30] Glori C (2013). Il misterioso doppio ritratto di Luca Pacioli e dell'allievo e la duplice decifrazione del cartiglio di Capodimonte. www.foglidarte.it/images/stories/013e/glori-pacioli.pdf.

[31] The *Pala di Brera*, which is located in the Pinacoteca di Brera in Milan, is a tempera on panel (248 × 170 cm) attributed to Piero della Francesca and executed between 1472 to 1474.

[32] The attribution has been done by the art historian Ivano Ricci. Ricci I (1940) Fra Luca Pacioli l'uomo e lo scienziato (con documenti inediti). Sansepolcro. Stabilimento tipografico Boncompagni.

Fig. 10.6 Piero della Francesca. Madonna and child with Saints and angels (Pala di Brera). 1472. Pinacoteca di Brera, Milan. (This work is in the public domain in its country of origin and other countries and areas where the copyright term is the author's life plus 70 years or less. https://upload.wikimedia.org/wikipedia/commons/9/9e/Piero_della_Francesca_046.jpg)

ter and friend when the friar was 27 years old (Fig. 10.6). The comparison between the two portraits is merciless, the fixity of gaze, and the static pose of the friar in Jacopo de' Barbari's portrait is in stark contrast to the face painted by Piero della Francesca. Thus, an almost mystical aura encompasses the dual personality of Pacioli, one side turned to the mystery of God and the other immersed in the mathematical world.

There is one last little strange detail in Jacopo de' Barbari's portrait: If we look closely at the bottom left of the slate on which Luca Pacioli is drawing the geometric figure, it is possible to observe an addition problem: 478 + 935 + 621 = 2034. What do the terms of the addition mean? According to a recent hypothesis, these could perhaps be the calculations necessary to define the relationship between the diameter of a sphere and the side of a rhombicuboctahedron inscribed inside it.[33]

Regardless of the final conclusions that we draw as to the portrait's attribution and meaning, the many and intriguing quotes and references to mathematics and geometry found throughout bear a special witness to the relationship between art and science peculiar to Renaissance culture. At the same time, these citations are an expression of the Renaissance taste for symbolism and puzzles, which were also a significant part of the spirit of the time.

Precipitating Events

As we have seen, after the fall of Ludovico il Moro[34] in 1499, Leonardo da Vinci and Luca Pacioli left Milan and headed first to nearby Mantua to the Court of Isabella d'Este (1474–1539)[35,36]. Their departure, however, was neither rushed nor panicked: In fact, for a while they were waiting and watching the events to understand what was the best choice for the future, to stay or to leave.

Milan at the time of Ludovico was a magnificent Renaissance court, a center of attraction for artists and literati, and a place where Leonardo and Pacioli could enjoy the pleasure of a rich intellectual life under the powerful protection of the Duke. Around their happy world, however, history was preparing tumultuous events. In 1498, Louis XII (1462–1515) ascended the throne of France and claimed the Duchy of Milan because of old rights based on the marriage of an ancestor with the daughter of the first Duke of Milan, Gian Galeazzo Visconti (1351–1402). The situation of Ludovico il Moro became then rather difficult, in part because his constant changes of alliances, combined with great ambition and political opportunism, had alienated him from the support of an important ally, the Republic of Venice, which instead sided with France. Even in Milan itself, he was not particularly popular, and, in fact,

[33] Cesaroni FM, Ciambotti M, Gamba E, Montebelli V (2010) Le tre facce del poliedrico Luca Pacioli. Quaderni del centro internazionale di studi Urbino e la prospettiva.

[34] Ludovico Sforza so called Il Moro, Duke of Bari from 1479, the Duke of Milan from 1480 to 1499. (Vigevano, July 27 1452–Loches (France) July 27, 1508).

[35] Isabella d'Este (Mantua 1474, Ferrara 1539) was the eldest daughter of the Duke of Ferrara, Ercole d'Este I, and Eleanor of Aragon, daughter of the King of Naples. After the death of her husband, Francesco II Gonzaga, in 1519, with whom she had nine children, she became regent of Mantua on behalf of her firstborn son Federico II Gonzaga. It was thanks to the deft political and diplomatic actions of his mother Isabella that Francesco II from 1530 gained the elevation to Duke of Mantua by Emperor Charles V.

[36] Raffaele Tamalio. Voice over Isabella d'Este, Marchioness of Mantua. Dizionario Biografico degli Italiani. Volume 62 (2004).

a home-grown rebellion facilitated the entry into the city of the French army, which occupied the Duchy without having to fight. Ludovico il Moro fled to Innsbruck seeking the protection of the Emperor Maximilian and from there tried to regroup an army to regain Milan. This attempt took place in 1500 but failed almost immediately, resulting in the capture of Ludovico il Moro, who would be held in France until his death in 1508.

Luca Pacioli and Leonardo were never particularly attracted by politics or developed any special affection for any of the various Lords in the mosaic of small states in which Italy was divided at the time. They regularly dedicated books and works celebrating the qualities of their protectors to these powerful Lords with great fanfare, but did not hesitate one moment to change place and loyalties as soon as things began to go wrong or they found someone who offered them better terms.

Thus, upon the arrival of the French in Milan in September 1499, they studied the situation for a while. The King of France, in particular, was very impressed with Leonardo's Last Supper. The King even had in mind to detach the fresco and take it away, but, fortunately, he did not act upon this desire. Leonardo received an attractive offer for a position in the new administration of the city, as a military engineer. However, despite the blandishments of Louis XII, he decided to leave Milan. The atmosphere was not the same anymore, times had changed, and the wonderful small court of the Duke, one of the centers of the Italian Renaissance, no longer existed. Without much more hesitation, Leonardo set his business in order, delivered his savings to Florence, and, before the end of the year, had left the city with his friend Luca Pacioli. This ended a period of 18 years, from 1482 to 1499, which Leonardo spent in Milan, where he achieved extraordinary results. In this period, he realized some of his most beautiful paintings and had the opportunity to concentrate on his beloved scientific and technical studies.

The departure from Milan would mark the beginning of a more uncertain period right at the threshold of Leonardo's old age. However, it was not a final goodbye, and in 1508 Leonardo would be back in Milan for another 5 years until 1513, reflecting the great charm that the city exercised upon him in spite of the political turmoil.

Portraits and Lost Books

Leonardo did not remain long in Mantua, moving after a few months to Venice and, subsequently, to Florence in 1501. The choice of these places was not done by chance, since Leonardo had to find protection and wealthy patrons, not easy at that time of instability and uncertainty across the small states that made up the Italian peninsula. The stay in Mantua was of particularly short duration—Venice and Florence indeed appeared much richer in opportunities than the small marquisate.

Isabella d'Este, the wife of Francesco II Gonzaga (1466–1519) Marquis of Mantua, was a woman of great charm and culture and must have found the appropriate motivation to attract the two famous intellectuals who were fleeing from Milan. She knew them rather well, at least by reputation, since her sister Beatrice d'Este (1475–1497) had become the young wife of Ludovico il Moro. Isabella d'Este's sister in Milan played an important part in making of the city one of the most splendid courts of Italy, and she liked to be surrounded by famous artists and literati. Unfortunately, Beatrice died in childbirth in 1497 at the age of 22, leaving two sons, but at least without having to watch the sad ending of the domain of her husband Ludovico il Moro.

Isabella d'Este, during her visit to Milan in 1498, shortly after the death of her sister, had been able to observe the beautiful portrait that Leonardo da Vinci had realized for Cecilia Gallerani, Ludovico il Moro's mistress before his marriage. The Duke, inter alia, collected lovers and children and had two legitimate and at least six more illegitimate sons. In addition to the portrait of Cecilia Gallerani,[37] Leonardo also made one for him of Lucrezia Crivelli (*La Belle Ferronière*)[38] another of his lovers (Fig. 10.7).

The portrait of Cecilia Gallerani was realized between 1488 and 1490, and it is known as the *Dama con l'ermellino* (*Lady with an Ermine*).[39] The portrait of the young lady is of extraordinary beauty and is considered one of Leonardo's masterpieces. It depicts a young woman in a pose that flouts the canonical conventions of fifteenth-century portraitures by placing the torso and the head, illuminated by an external light from above, in two different directions. The enigmatic smile is the unmistakable touch of Leonardo, who managed to fix forever the indefinite and ambiguous situation. Beatrice d'Este borrowed this portrait for a short time and was fascinated by it.[40] This experience likely inspired her desire to, at all costs, get her own portrait by Leonardo.

It is unknown whether Leonardo and Luca Pacioli went to Mantua because of a specific invitation from Isabella d'Este. In any case, the refined Marchioness took advantage of the opportunity to have at her court the two friends, and, as we will see, their stay, even if rather short, would prove to be somewhat fruitful.

---

[37] Pizzagalli, D (1999) La dama con l'ermellino. BUR, Milano.

[38] La *Belle Ferronnière*, or *Portrait of a lady* is an oil on panel (63 × 45 cm) by Leonardo da Vinci, dated between 1490 and 1495. The painting is now at the Louvre Museum in Paris.

[39] The painting by Leonardo da Vinci is now located in Krakow in Poland. It is an oil on panel of size 54.8 cm × 40.3 cm, realized between 1488 and 1490.

[40] Kemp M (2004) Lezioni dell'occhio: Leonardo da Vinci discepolo dell'esperienza. Vita e Pensiero, Milano.

Fig. 10.7 Leonardo da Vinci. The Lady with an Ermine, Oil on wood panel (1489). Czartoryski Museum, Kraków (Poland) (*left*). Leonardo da Vinci. La belle ferronnière. Oil on wood panel (1490–1496). Louvre, Paris (France) (*right*). (The image of the Lady of Ermine is in the public domain in its country of origin and other countries and areas where the copyright term is the author's life plus 70 years or less. https://upload.wikimedia.org/wikipedia/commons/e/e1/The_Lady_with_an_Ermine.jpg. The image of La Belle Ferronnière is in the public domain in its country of origin and other countries and areas where the copyright term is the author's life plus 70 years or less. The image has been modified to reduce the noise. https://upload.wikimedia.org/wikipedia/commons/c/c3/Leonardo_da_Vinci_%28attrib%29-_la_Belle_Ferroniere.jpg)

Isabella persisted in pressing Leonardo for the portrait, and eventually she seemed to have convinced him to work on it.[41] Part of this work is preserved in a preliminary drawing executed in charcoal in 1500, now preserved at the Louvre (Fig. 10.8). It is likely that Leonardo left the drawing behind in Mantua upon his departure, only to be given away by her husband without any warning, much to Isabella d'Este's displeasure. The artists of Leonardo's workshop, who were following him, would, however, make a copy, which Leonardo brought with him at least as far as Florence, perhaps to finish the full portrait later. This version of Isabella d'Este's portrait is kept at the Ashmolean Museum in Oxford.

These are not the only remaining copies: at least six can be counted—in addition to the two we have just mentioned, there are copies at the British Museum and at the Uffizi in Florence and two copies at the Staatliche

[41] Ames-Lewis F (2012) Isabella and Leonardo: The Artistic Relationship between Isabella d'Este and Leonardo da Vinci, 1500–1506. Yale University Press.

Fig. 10.8 Leonardo da Vinci. The portrait of Isabella d'Este, 63 cm × 46 cm, charcoal, blood dye and yellow crayon on paper. 1500. Louvre Museum. (This work is in the public domain in its country of origin and other countries and areas where the copyright term is the author's life plus 70 years or less. The image has been modified with respect to the original to reduce the noise, https://upload.wikimedia.org/wikipedia/commons/3/38/Da_Vinci_Isabella_d%27Este.jpg)

Graphische Sammlung in Munich. Of the six, there is only certainty that the one at the Louvre was done by Leonardo himself.

Once he had left Mantua, without completing the coveted portrait for Isabella d'Este, Leonardo never worked on it again. Isabella penned a dense correspondence with Leonardo over the course of many years, urging him to complete the portrait, but she never got further than receiving general reassurances. In 2013, however, the famous Leonardo scholar Carlo Pedretti announced to general amazement and disbelief the discovery of the famous portrait of Isabella d'Este realized by Leonardo: The painting was part of a private collection kept in the vault of a Swiss bank. Carbon 13 dating of the painting confirmed at least a time frame compatible with Leonardo's activity. Following the script that is repeated virtually identically in the case all such discoveries and rediscoveries, there immediately started a controversy between "experts," and the question still remains open.[42]

[42] Kemp M (2013) Leonardo and the supposed portrait of Isabella d'Este, http://martinkempsthisandthat.blogspot.jp/2013/10/leonardo-and-isabella-deste.html.

Another Isabella

The fate of Leonardo had previously woven in Milan with that of another Isabella of whom he was probably personally very fond, the Duchess Isabella of Aragon (1470–1524).[43] The Duchess was the daughter of Alfonso II of Aragon (1448–1495), eldest son of the King of Naples Ferrante and his wife Ippolita Maria Sforza, sister of Ludovico Sforza and Galeazzo Maria. Her fate was written as a child when she was betrothed to his cousin Gian Galeazzo Sforza, whom she finally married in the Cathedral of Milan on 2 February 1489. But things did not go as perhaps Isabella might have hoped. Her husband was a weak person, and when he inherited the dukedom, his uncle Ludovico il Moro actually took possession of power, relegating the pair to the castle of Pavia. Isabella, however, did not give up so easily to her sad fate and wrote to her father-in-law describing the situation and asking for his help. In response, the angry King of Naples declared war on Ludovico Sforza in 1494. It is at this point that the unexpected death of her husband occurred; he was probably poisoned by order of the same Ludovico who then became Duke of Milan, usurping the title that rightly should have belonged to Isabella's son. This event would mark Isabella, who was forced to leave Milan for the Duchy of Bari in the South of Italy and never ceased, for the rest of her life, to try to get back the legitimate inheritance for herself and her son. As we saw above, the instigator of the killing went unpunished despite the rumors accusing Ludovico il Moro.

The arrival of the Duchess in Milan and her marriage to the heir to the Duchy were greeted by a large crowd and much festivity. Leonardo was commissioned by Ludovico il Moro to celebrate the event by staging a great feast, the *Festa del Paradiso* (Feast of Heaven) which took place on 13 January 1490, in fact nearly a year after the real wedding date. For this feast, Leonardo used a surprising representation of Heaven, designing extraordinary theatrical machines that reproduced the motion of the planets, with lighting effects and music which deeply amazed all the attendees.[44]

Leonardo would later visit Pavia, and there remain in his manuscripts traces of a bathroom designed for the Duchess Isabella (Fig. 10.9), one of which is annotated with the statement "in the bathroom of the Duchess Isabella." The bathroom was likely designed to control the flow of hot and cold water.

A portrait of the Duchess Isabella remains preserved at the Veneranda Biblioteca Ambrosiana in Milan (Fig. 10.10). It is attributed to one of the

[43] Isabella of Aragon was born in Castel Capuano on 2 October 1470 and died in Naples on 11 February 1524.

[44] There is a description of the Feast of Heaven by Jacopo Trotti, ambassador of the Este family in Milan.

Fig. 10.9 The bath of the Duchess Isabella. Leonardo da Vinci. Codex Atlanticus, folio 289r. BAM

Fig. 10.10 Santa Barbara (Isabella d'Aragona?). 1498–1502. Giovanni Antonio Boltraffio, Ambrosian Library in Milan. Drawing on paper, 54.5 × 40.5 cm. (This work is in the public domain in its country of origin and other countries and areas where the copyright term is the author's life plus 70 years or less. The image has been modified with respect to the original to reduce the noise. https://upload.wikimedia.org/wikipedia/commons/b/bb/Female_leonardo.jpg)

students from Leonardo's workshop, Giovanni Antonio Boltraffio (1467–1516).[45] The portrait depicts Santa Barbara, but her face corresponds to that of the young Duchess.[46] The drawing has great charm, showing a young and beautiful woman who exudes an undefined sadness, perhaps an omen of her unhappy fate. The portrait closely resembles Leonardo's *Dama con Ermellino* or *La Belle Ferronière*. As we can see, the students learned well the lessons of their Maestro.

The two Isabellas somehow would share a common destiny: They are both listed as possible candidates for the subject of Leonardo's Mona Lisa. In the case of Isabella of Aragon, however, the fantasy has gone well beyond any reasonable imagination, with some theorists even assigning to her the role of Leonardo's secret wife from whom he would have had five children.[47]

The Lost Schifanoia

While Leonardo in Mantua was struggling with the insistence of Isabella d'Este and her mania for a personal portrait, Luca Pacioli's existence was no less hectic. The Marchioness in fact was not only an avid art collector but also dabbled in various forms of intellectual entertainment, including the game of chess. As we know, Luca Pacioli was a big fan of games, particularly mathematical ones, to which, as noted above, he dedicated a whole treatise.

In Mantua, to dispel the boredom of his chess-loving host, Luca Pacioli wrote a book in Latin on the game, the *De Ludis in genere,* cum *illicitorum reprobatione, specialmente di quelli de scachi in tutti i modi (About games in general, with exclusion of the illicit (games) and especially about the chess (game) and all the ways (to play))*, dedicated to the Marquis and Marchioness of Mantua. The book is also known as *Schifanoia* (Boredome dredger). Unfortunately, the memory of this manuscript was lost for a long time, as indeed also happened to Isabella's portrait by Leonardo, with just an autograph quotation of the work on chess, dedicated to "Segnor Marchese e Marchegiana di Mantoa, Francesco Gonzaga e Isabela Extense," included in the introduction of *De viribus quantitatis*.

As we can see, the two friends met a similar fate at the court of Isabella, and their work dedicated to their kind and brilliant guest disappeared quickly

[45] Giovanni Antonio Boltraffio or Beltraffio (1467–1516) joined Leonardo's workshop on his arrival in Milan together with Marco d'Oggiono and Gian Giacomo Caprotti (Salai).

[46] Besides the portrait of Boltraffio, there exists a remarkable marble bust by Francesco Laurana, sculptor active in the Neapolitan Court, which reproduces with probability Isabella of Aragon.

[47] The mystery of the Mona Lisa, the answer in the DNA? 20/04/2012-culture/www.nationalgeographic.it/popoli/news/monna_lisa_isabella_leonardo_figli-977873.

from memory. Luca Pacioli's treatise, however, though for a long time considered to be permanently lost, almost by a miracle was rediscovered in 2006 in Gorizia (a small city in the Northeast of Italy) by a bibliophile, Duilio Contin. Contin was studying the 22,000 volumes which were the bequest of the Count William Coronini Cronberg to the Foundation of the same name, when he discovered the lost text.

In the manuscript, 114 illustrated chess schemes of some complexity are reproduced. They were probably realized over quite a long period of time, and, in fact, the manuscript appears as a collection of scattered papers. The treatise documents, among other things, a key time in Europe when new chess rules were developed: It is between 1400 and 1500 that the Queen was introduced as the most powerful piece on the board. The new rules are referred to as "*a la rabiosa technique* (mad chess technique)" and spread quickly in Europe to supplant the old modes of play. These new rules are the ones that are still used today.

There is a special feature that strikes in this manuscript: the illustrations to represent the game schemes. In most chess treatises of the time, in fact, the pieces were represented by letters or numbers, but in Luca Pacioli's manuscript, symbols in red or black to differentiate between the two sides were used instead (Fig. 10.11).

The attribution of the manuscript to the Tuscan mathematician, at least in this case, has obtained universal approval from all the critics, but the subsequent assertions that Leonardo might be responsible for drawing the chess symbols, unsurprisingly, exploded in the usual controversy[48,49] In fact, the symbols appear to have been drawn with an uncertain and ungainly hand (though probably left-handed) and seem hardly attributable to Leonardo. However, despite all of these legitimate doubts as to the attribution, numerous versions of "Leonardo's Chess" were soon produced and marketed, yet again confirming the irresistible attraction which even the most tenuous reference to his name still exercises today.

Putting aside the question of whose hand drew the schemes, it would be even more interesting to know whether Leonardo had any role in designing the problems. The problems in *De Ludo Scachorum* are quite difficult, especially in light of the fact that new game rules had been introduced only a few years before. Designing these problems in such a short time must have required considerable intelligence and knowledge of the game, which was very popular in the courts.

---

[48] The comment of one of the most famous scholars of Leonardo, Martin Kemp, Emeritus Professor of art history at Oxford University, was terse and final: "The silly season on Leo never closes."

[49] Rocco F (2013) Leonardo e Luca Pacioli, l'evidenza. Fondazione Palazzo Coronini Cromberg.

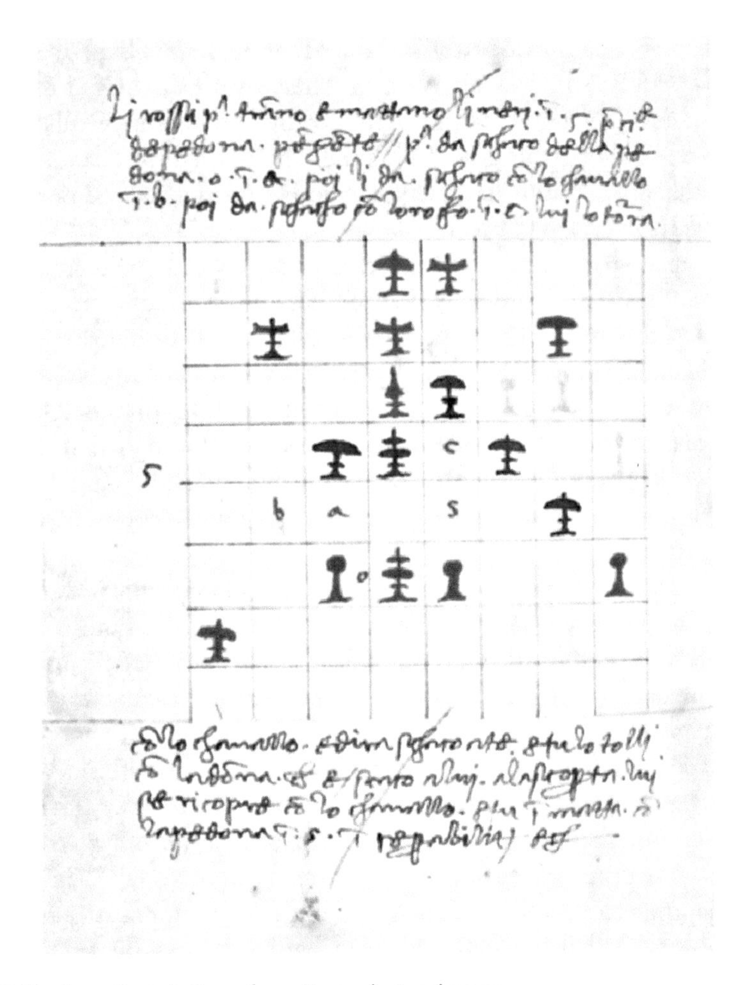

Fig. 10.11 Luca Pacioli. Page from *De Ludo Scachorum*

While some clues point toward a possible contribution of Leonardo to Pacioli's chess studies, how sure are we that Leonardo played the game? There is a single reference about it in his writings, a rebus on folio 12692r at the Royal Library, Windsor Castle[50] (Fig. 10.12). The rebus is: "*I a (drawing of a tower) ro*" or, in modern Italian, "io arroccherò," "I will castle." This is enough to demonstrate the strong likelihood of Leonardo's knowledge of the game and the new rules including that of castling which was introduced to limit the excessive power of the Queen in the play "a la rabiosa." Since this is all we have to go on, it seems to be too great a leap to deduce that Leonardo was an expert in chess or that he regularly played it. Among other things, this folio is entirely occupied by dozens of Leonardo's rebuses, showing once more his passion for games.

[50] Sanvito A (2009) *Leonardo da Vinci e gli scacchi*. Scacchi e scienze applicate. vol. 27.

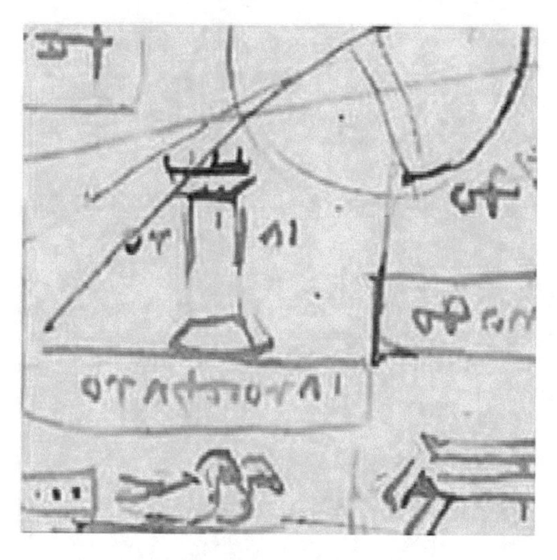

Fig. 10.12 The chess rebus of Leonardo. Leonardo da Vinci, folio RL 12692r, Royal Library of the Windsor Castle. (c. 1487–90)

The lack of other notes about chess allows us to suppose that, even if Leonardo knew the game, he was not sufficiently fascinated to practice it with any regularity, perhaps just due to lack of time. He was much more attracted by the study of natural phenomena and by mathematics. As we will see in the next chapter, the two friends would find in this discipline a shared passion, and this would bring Leonardo to collaborate directly with the Tuscan mathematician to prepare one of his most famous treatises, the *Divina Proportione*.

11

The Divine Proportion

The close relationship between art and science is a distinctive feature of the Renaissance, and, as we have observed again and again, Leonardo represents one of the highest points of this synthesis. The scientific knowledge that he obtained via his studies became a fundamental instrument also for his work as an artist and vice versa. As discussed in the previous chapter, it was thanks to his dear friend Luca Pacioli that Leonardo was able to fully appreciate the beauty of geometry and arithmetic. Luca Pacioli's support was also critical, more generally, in enabling him to achieve a satisfactory knowledge of mathematics.[1]

Nonetheless, as with so many of his pursuits, Leonardo was not the only Renaissance artist with a passion for mathematics, and several others studied the possible applications of geometry to art. Piero della Francesca, in particular, was one of the first artists to dedicate serious and fundamental attention to the study of geometry and mathematics. Later on, Leonardo and Albrecht Dürer would devote detailed studies to perspective and human proportions and their transposition into painting and architecture. As we saw in the earlier chapter on the Vitruvian Man, this endeavor aimed at employing geometry to capture the perfection of the forms of creation.

In order to fully appreciate the significance of this work, we must recall that, at the time, the perfection of God was implicitly associated with mathematics. The beauty of the dimensions of the human body, created in God's image, was seen as an emanation of divine perfection. Thus, the mathematical

[1] Folicaldi F (2005) *Il numero e le sue forme: storie di poliedri da Platone a Poinsot passando per Luca Pacioli.* Nardini editore.

P. Innocenzi, *The Innovators Behind Leonardo*, https://doi.org/10.1007/978-3-319-90449-8_11

study and reproduction of these perfect proportions provided access to a divine language which made possible the representation of the harmony of creation.

Moreover, the Renaissance artist-engineers found that mathematics lends itself to many possible transcendental interpretations, where it is possible to find the hand of God. This was particularly true for the mathematical constants such as the Greek π, which assume a well-defined but somehow mysterious value. Another one of these constants which has tormented and captured the imaginations of generations of artists and architects, including with Leonardo himself, is the golden ratio, also known as the golden mean or the divine proportion, to which Luca Pacioli would dedicate one of his most famous treatises: the *Divina Proportione* (On the Divine Proportion).

The Golden Ratio[2]

What is so special about this constant for it to have attracted so much attention since ancient times? Let us start first with its definition. If we take any two numbers a and b, with a greater than b, then the ratio between their sum ($a + b$), and the larger number (a), forms a golden ratio if it is equal to the ratio of a to b.

Translating this into an equation, the language of mathematics, the golden ratio φ is defined as:

$$\frac{a}{b} = \frac{a+b}{a} = \varphi, \quad \text{where } a > b \tag{11.1}$$

The value of the golden ratio is therefore a constant, and its numerical value is $\varphi = 1.6180339887...$. This constant, however, is an irrational number and thus cannot be written as a fraction a/b with a and b both integers (i.e., whole numbers) and b not equal to zero. As with π, the decimal expansion of φ never ends and does not repeat to form a periodic sequence.

Starting from the golden ratio, it is possible to construct line segments and rectangles in golden proportion. In the case of a segment, we need to divide it into two parts, a and b (with $a > b$), such that the ratio of $a + b$ to a is the same as the ratio of a to b (Fig. 11.1).

[2] Livio M (2002) *The Golden Ratio: The Story of Phi, The World's Most Astonishing Number.* New York, Broadway Books.

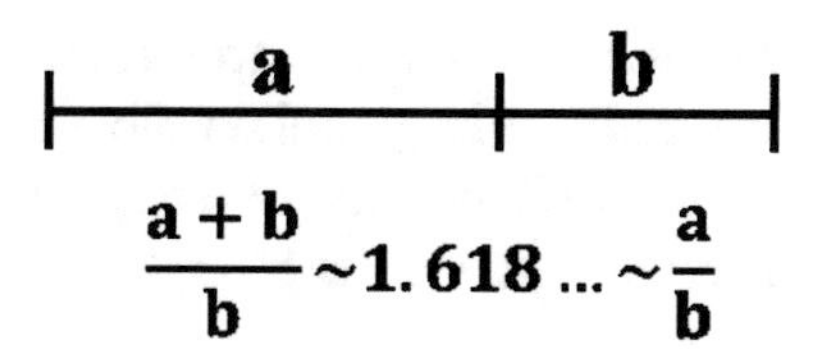

$$\frac{a+b}{b} \sim 1.618\ldots \sim \frac{a}{b}$$

Fig. 11.1 Line segments in the golden ratio

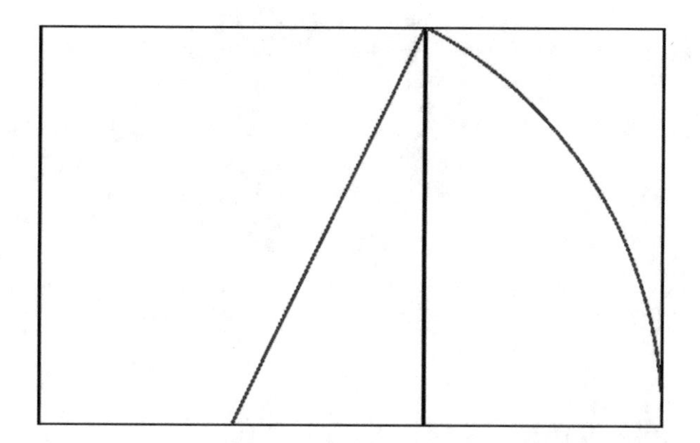

Fig. 11.2 Euclid's method for constructing a golden rectangle with a ruler and compass

The construction of a golden rectangle is relatively simple, and the first method, using only a ruler and compass, was shown by Euclid in Proposition 2.11 of his *Elements* (Fig. 11.2).[3] The first step is the construction of a simple square. Then, from the midpoint of one side of the square, a line must be traced to the opposite corner. This line is the radius of an arc which determines the length of the additional line segment to the right of the square. The sum of these two figures gives us a golden rectangle. What happens if we remove the square? The rectangle that remains is exactly another golden rectangle with the same golden ratio, and the procedure can be repeated indefinitely to produce ever smaller and smaller golden rectangles.

It was Pacioli who spurred renewed interest in the ratio with his coining of the term the "divine proportion," writing: "As God cannot be properly defined nor understood with words, so this proportion can never be defined with any understandable number, nor expressed through a rational amount, but always will be hidden and secret and called by mathematicians irrational."[4] Pacioli's

[3] "Come dividere un segmento in modo che il rettangolo che ha per lati l'intero segmento e la parte minore sia equivalente al quadrato che ha per lato la parte maggiore".

[4] Original text: "Commo Idio propriamente non se po diffinire ne per parolle a noi intendere, così questa nostra proportione non se po mai per numero intendibile asegnare, né per quantità alcuna rationale exprimere, ma sempre fia occulta e secreta e da li mathematici chiamata irrationale."

use of the adjective divine was by no means accidental: It crucially linked the irrationality of the golden ratio with the unknowable mystery of God.

This magical relationship would become an obsession for many generations of not only mathematicians but also painters and architects, as well as untrained, independent enthusiasts.

How Golden Is it? Assessing the Appeal and Legacy of the Divine Proportion

The golden rectangle has long been considered the aesthetic ideal, in the absence, however, of any scientific evidence to support this predilection. Recently, scientific studies have been undertaken to see if people looking at a series of different rectangles really had any preference for those realized using the golden ratio. No particular fondness for the "divine" figures was observed.[5]

Despite this, the golden ratio has captured the hearts and minds of many architects well past the Renaissance. Le Corbusier (1887–1965), for example, devised a special system, the *Modulor*, which used the golden ratio and the Fibonacci sequence as a template to create architectural structures that would reproduce the harmony of the divine proportion.

In the Renaissance, the golden ratio was in great vogue, and one of its strongest proponents was Luca Pacioli's master himself, Piero della Francesca. An example of the use of the golden ratio in the work of the great Tuscan master can be found in one of his most famous paintings, *La Flagellazione di Cristo* (The Flagellation of Christ),[6] where the proportion of the two halves of the painting is determined using the golden ratio (Fig. 11.3).

And what about Leonardo? There are many interpretations that suggest the use of the golden rectangle in his most famous paintings, including the *Mona Lisa* and the *Last Supper*, as well as in the drawing of the *Vitruvian Man* which we encountered in Chap. 9. However, putting aside the lack of any documentary evidence indicating Leonardo's reliance on the golden ratio in these works, the sheer arbitrariness in the choice of where to draw the golden rectangle in each work makes this exercise somewhat doubtful and unconvincing.

This also applies to the buildings most frequent cited as examples of the application of the golden ratio. Among these, the most famous is the Parthenon in Athens. In this case, only a strange and unnatural selection of points allows proportions close to the golden ratio to be found.[7]

---

[5] Zocchi A (2005) *La sezione aurea, Gli esperimenti psicologici per verificare la bellezza del rapporto aureo.* Available online: www.cicap.org/new/articolo.php?id=101948. Accessed 8 August 2016.

[6] The *Flagellation of Christ* (probably 1455–1460) is an oil and tempera work realized by Piero della Francesca between 1455 and 1460 and is now in the Galleria Nazionale delle Marche in Urbino. The interpretation of the painting is still controversial because of the complexity of the iconography. Piero della Francesca used linear perspective with an unprecedented level of mastery; for the art historian Kenneth Clark, the Flagellation is "the Greatest Small Painting in the World" (Clark K (1951) Piero della Francesca. Phaidon-Press London).

[7] Hoge H (1997) The Golden Section Hypothesis—Its Last Funeral. Empirical Studies of the Arts 15 (2), pag 233–255.

Fig. 11.3 Piero della Francesca. The Flagellation of Christ (c. 1444–1469). Tempera on panel (58.4 × 81.5 cm). Galleria Nazionale delle Marche, Urbino. Italy. (This work is in the public domain in its country of origin and other countries and areas where the copyright term is the author's life plus 70 years or less. https://upload.wikimedia.org/wikipedia/commons/8/85/Piero%2C_flagellazione_11.jpg)

The golden ratio can also be found in many geometrical forms besides the rectangle, in particular in pentagons. In a regular pentagon (i.e., all five sides of the same length and all five angles of the same size), for instance, the ratio of the diagonal to the side is the golden ratio. As an aside, it is intriguing to note that the pentagon held a special fascination for the mystical Pythagoreans who had a special fondness for the number five, to which they attributed an important symbolism: The number five is the sum of a male element (the number 2) and a female (the number 3) and therefore represents the number of love and marriage.

This link between the golden ratio and the number five has certainly helped to increase its mystique and charm. From a regular pentagon, one can also derive another fascinating polygon associated with golden ratio: the pentagram, a polygon composed of five segments that intersect to form a five-pointed star.

At this point, we should introduce a rather bizarre contemporary of Leonardo and Luca Pacioli, the German occultist Heinrich Cornelius Agrippa (1486–1535). Agrippa's main work is the book *De occulta philosophia libri tres*

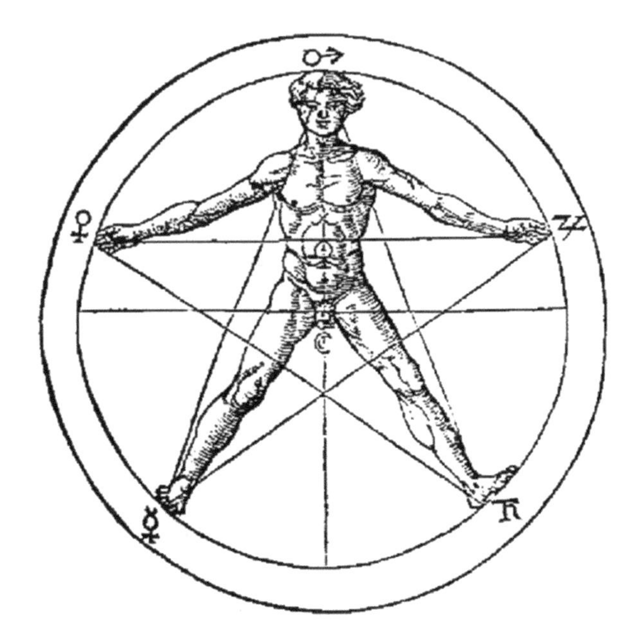

Fig. 11.4 Image of a man inscribed into a pentagram. "Libri tres de occulta philosophia" of Heinrich Cornelius Agrippa. (This work is in the public domain in its country of origin and other countries and areas where the copyright term is the author's life plus 70 years or less. The image has been modified with respect to the original to reduce the noise. https://upload.wikimedia.org/wikipedia/commons/d/dc/Pentagram_and_human_body_%28Agrippa%29.jpg)

(*Three Books of Occult Philosophy*),[8] a kind of encyclopedia of occult sciences. An interesting drawing appears in this treatise, a man with symbols of the Moon and the Sun at the center of his figure and whose extremities—i.e., head, arms, and feet—correspond to the five vertices of a pentagram (Fig. 11.4). At each extreme, a sign which corresponds to one of the five planets in astrology is traced. The man, in this way, is placed directly in relationship with the pentagram and, through it, with the golden ratio. At the same time, he is inscribed into a circle where the five cosmic elements appear. Aside from the references to the occult and its relationship to the golden ratio, the similarity to the various versions of the *Vitruvian Man* which we discussed in Chap. 9 is striking. At first glance, the only difference seems to be the missing square.

[8] *Three Books of Occult Philosophy Book One: A Modern Translation*, written by Henry Cornelius Agrippa, Translated by Eric Purdue (2012). Renaissance Astrology Press.

It is clear that, during the Renaissance, geometry exerted a fascination that went beyond practical purposes and that it was often used as a bridge to transcendent representations of reality and the divine relationship of man to God and the universe. It is important to stress, however, that Leonardo, in contrast to so many of his contemporaries, did not share this mystical approach and tendency toward magic interpretations, his mind always concentrated on practical problems and always trying to use experiments and theoretical studies to support his hypotheses.

The God's Eye

The fascinating aspects of the golden ratio do not end here, and it is worth delving into some of its other properties here if only because of the innumerable attempts of scholars and enthusiasts to find traces of the ratio and its influence in Leonardo's works.

Let us first return to the golden rectangle. As we have just seen, it is possible to divide it into a square and another smaller golden rectangle. If the same procedure is applied again to this smaller rectangle, another square and another rectangle in the same proportion will be obtained. Because the golden ratio is an irrational number, this iterative process can go on forever.

However, the process of building successively smaller golden rectangles eventually converges to a precise point as smaller and smaller squares are built. From the first square, a quarter-circle arc (blue curve in Fig. 11.5) can be

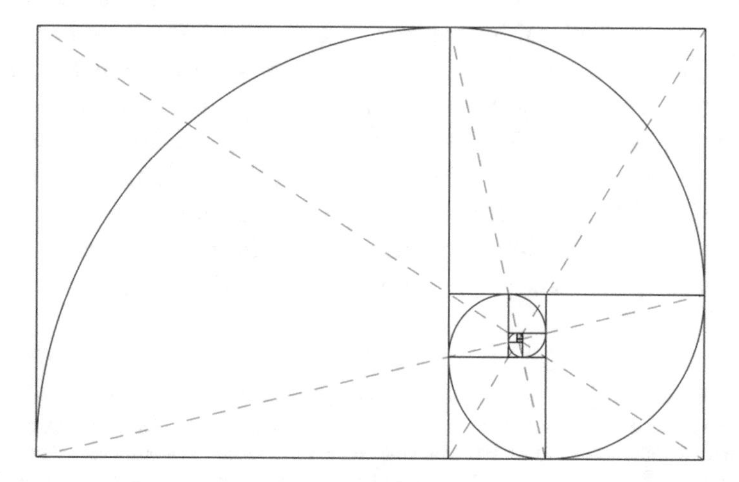

Fig. 11.5 The spiral of Fibonacci and the God's eye

drawn that can be connected with another quarter arc in the second square, and then in the third, and so on. The union of these arcs forms a spiral, the so-called Fibonacci spiral,[9] that converges toward a specific point in the rectangle, the so-called eye of God.

This spiral has another peculiarity: the point of convergence lies at the intersection of the diagonals of the golden rectangles (orange segments in the figure). There is yet another exciting feature. Apart from the diagonal and the Fibonacci spiral, other segments also converge at this same point (in green in the figure): those connecting the vertices of one golden rectangle with those of the one two sizes smaller.

Why is the spiral thus formed called Fibonacci and who was this guy? Leonardo Pisano, also known as Fibonacci (c. 1170–1240), was an Italian mathematician who had a key role in the history of science: In his *Liber Abaci* (*The Book of Calculation*), he introduced in Europe for the first time the decimal number system with nine "Indian" digits and zero,[10] in addition to describing the fundamentals of algebra and arithmetic. He is best known for the sequence of numbers that bears his name, the so-called Fibonacci sequence which was also introduced in *Liber Abaci.*

This sequence formed as follows: Starting with 0 and 1, you then get the next term by adding the previous two terms: 0 + 1 = 1; 1 + 1 = 2; 1 + 2 = 3; 2 + 3 = 5; 3 + 5 = 8, and so on, giving you the sequence 0, 1, 1, 2, 3, 5, 8, 13, 21, 34, 55, 89, 144, 233….

This sequence was discovered as a solution to a famous problem of calculating the total offspring after a year starting from a pair of rabbits and assuming that each couple is fertile from the second month.

However, by chance, the Fibonacci sequence has a direct correlation to the golden ratio, a relationship that was unknown to Luca Pacioli and which would only be discovered by Kepler: The ratio between two consecutive numbers in the Fibonacci sequence approximates the golden ratio with increasing accuracy: 55/34 = 1,617,647; 89/55 = 1,618,182; 144/89 = 1,617,978; 233/144 = 1,618,056; 377/233 = 1,6,118,026; 610/377 = 1,618,037; 987/610 = 1,618,033…. That is, the ratio between any given number in the sequence and the preceding one tends to approximate more and more closely the irrational golden ratio as one goes farther and farther along in the sequence.

[9] The spiral of Fibonacci is not a real mathematical spiral, because it has instead a logarithmic form.

[10] "Novem figure indorum he sunt 9 8 7 6 5 4 3 2 1 Cum his itaque novem figuris, et cum hoc signo 0, quod arabice zephirum appellatur, scribitur quilibet numerus, ut inferius demonstratur. (The nine Indian figures are these: 9 8 7 6 5 4 3 2 1. With such nine figures and the sign 0, which in Arabic is called zephiro, any number can be written, as would be later demonstrated)."

The Fibonacci spiral has been widely used as a popular means of proving Leonardo's use of the golden ratio in works such as the *Mona Lisa* and the *Vitruvian Man*, always with the same use of a rather arbitrary choice of reference points that would make this exercise applicable to a huge number of paintings and architectural works. As we will see, the golden ratio did hold a great fascination for Leonardo but more for his scientific and mathematical pursuits than for his artistic ones.

Divina Proportione

Leonardo's friendship with Luca Pacioli is a good indication that he was well aware of the golden ratio,[11] given the Tuscan friar's great passion for the topic. As discussed above, the knowledge of the golden ratio dates back to the Greek mathematicians, but, with time, it fell into oblivion in Europe until Luca Pacioli brought it back into the spotlight and renamed it "divine." The importance and fame of Pacioli's *Divina Proportione*,[12] however, are also in part due to Leonardo's contribution to the book's illustrations. On the first page, Luca Pacioli presented his treatise as a work necessary to all men of genius in the various fields of knowledge, who sought to be initiated into a secret science: "Work necessary to all the intelligent and curious minds. Every scholar of philosophy, perspective, painting, sculpture, architecture, music and other mathematics will acquire a fundamental knowledge about different aspects of this very secret science."[13]

As we can see from this bold declaration, Pacioli's book was a rather self-confident and self-promoting "manifesto." Pacioli tries to attract the readers with the intriguing reference to the "secret" knowledge of the golden ratio, making implicit reference to the "divine" nature of this number. The term is also used to underline that the knowledge of the intimate details of the technique and its applications will be disclosed.

Pacioli wrote the *Divina Proportione* in Milan between 1496 and 1497,[14] when Leonardo was also there at the court of Ludovico il Moro. Three

---

[11] Contin D, Odifreddi P, Pieretti A (2010) Antologia della Divina Proporzione di Luca Pacioli, Piero della Francesca e Leonardo da Vinci. Aboca Museum Edizioni.

[12] Luca Pacioli, *Divina Proportione*. Luca Paganinem de Paganinus de Brescia (Antonio Capella) 1509, Venezia.

[13] Original text: "Opera a tutti glingegni perspicaci e curiosi necessaria. Ove ciascun studioso di philosophia, Prospectiva, Pictura, Scultura: Architectura, Musica e altre Mathematice: suavissima: sottile: e admirabile doctrina consequira: e delectarassi: co' varie questione de secretissima scientia."

[14] At the end of Chapter 71, Luca Pacioli annotated that the work was terminated on 14 December 1497, VII year under the pontifex Alessandro VI.

manuscript copies of the treatise were made by different scribes. Two of these original copies are still well conserved—one, which was given to Galeazzo Sanseverino, Captain General of the Sforza army, is at the Biblioteca Ambrosiana[15] in Milan, and the second, which Pacioli gave to Ludovico il Moro, is at the Bibliothèque Publique et Universitaire of Geneva in Switzerland.[16] The third copy was offered to Pier Soderini, Gonfaloniere (a communal high officer) of Florence and has been lost.

Eventually, in 1509, the *Divina Proportione* was printed in Venice, by P. Paganini,[17] over 10 years after Pacioli had first completed the work. The printed version is actually made up of three parts, the first devoted to philosophy, music, and other arts in relation to the golden ratio; the second being a treatise on architecture, mostly based on the theory of Vitruvius; and the third a discussion of the five regular solids, the so-called Platonic solids. This last part is actually a translation into the vernacular of Piero della Francesca's Latin treatise *Libellus de quinque corporibus regularibus*, ("small book about the five regular bodies"). As we mentioned in the previous chapter, Pacioli's contemporaries accused him of plagiarism because of the inclusion of this text in his book. At least in this specific case, the accusations were unfounded, since Pacioli clearly stated the source of the text, writing: "I hope that you will be fully satisfied about me and also I promise to give you plenty of information about the perspective by using the documents of our fellow citizen and contemporary Maestro Petro de Franceschi [Piero della Francesca] who is a king of such a topic and I already did a good compendium about it."[18]

As the acknowledgment and reference to Piero della Francesca make clear, Pacioli's intention was to share this important work in a language which was understood by more people. This is an important merit of the friar, and he should be credited with having greatly contributed to the diffusion of knowledge and interest in mathematics during the Renaissance. In addition to his mathematical contributions (in particular, his work in financial accounting),

[15] *Manuscript 170 sup.* BAM.

[16] L. Pacioli, *Divina Proportione*. Reproduction of the copy at the Biblioteca Ambrosiana of Milan. Silvana Ed., Milano, (2010); L. Pacioli, *Divina Proportione*. Reproduction of the copy at the Bibliothèque Publique et Universitaire of Geneve. Aboca Edizioni, Sansepolcro.

[17] In the frontispiece of the *Divina Proportione*: *A. Paganius Paganinus characteribus elegantissimis accuratissime imprimebat* (Divine Proportion: A. Paganius Paganinus printed with very elegant and accurate characters).

[18] Luca Pacioli, *Divina Proportione*, p. 23r. Original text in vernacular: "Spero in breve sirete apieno da me satisfactit e anco con quella prometto darve piena notitia de prospectiva mediante li documenti del nostro conterraneo e contemporale di tal faculta ali nostri tempi monarcha Maestro Petro de Franceschi della qual gia feci dignitissime compendio."

Fig. 11.6 The letter M of Luca Pacioli in *Divina Proportione*, 1509 printed version (*left*); former logo of the Metropolitan Museum of Art in New York (*right*)

Pacioli's great insight was to have understood the importance of printing and making his work accessible to a large number of people.

Two sections of *Divina Proportione* were devoted to illustrations.[19] In the first of these, Pacioli himself drew all of the capital letters of the alphabet using only a compass and a ruler. The graphics of these letters would have a profound influence on encouraging the use of capital letters for printing, including being used from 1971 to 2016 as the letter "M" logo of the Metropolitan Museum of Art in New York (Fig. 11.6).

The Platonic Solids

Now let us return to the Platonic solids that were at the center of both Pacioli and Leonardo's attention, as we shall see.[20] The golden ratio is common in three-dimensional geometry as well and, in particular, can be found in the Platonic solids the dodecahedron and the icosahedron, both of which were the subject of drawings by Leonardo.

According to Plato, writing in the *Timaeus*,[21] there are five geometrical solids that constitute the fundamental structures of the universe.

[19] The *Divina Proportione* contains altogether 87 illustrations. 23 of them show the construction of the capital letters, 59 are hollow and solid geometrical forms, 3 are illustrations of architecture, and 1 is the profile of a human head. The treatise also contains a table related to the construction of the tree of proportions. In addition, the text contains another 185 small drawings in the margins.

[20] Ciocci A. I poliedri regolari tra arte e geometria. Pacioli e Leonardo tra Euclide e Platone. Available online:

http://www.centrostudimariopancrazi.it/pdf/quaderno2/pagg%20,153–184.pdf. Accessed 8 August 2016

[21] Cornford Macdonald F (1997). Plato's Cosmology: The Timaeus of Plato, Translated with a Running Commentary. Indianapolis: Hackett Publishing Company, Inc.

The *Timaeus*

The *Timaeus*, written c. 360 BC, consists of a dialogue between the Greek philosophers Socrates, Timaeus of Locri, Hermocrates, and Critias. Three problems are addressed in the dialogue: the origin of the universe (cosmological problem), its physical structure (physical problem), and human nature (eschatological problem).

The text includes a prologue, followed by three main sections, each dedicated to one of these topics. The Platonic Cosmos consists of four elements—Earth, Air, Water, and Fire—to which correspond precise geometrical shapes: the cube, the octahedron, the icosahedron, and the tetrahedron, respectively. A fifth element, representing the universe in its entirety, must also be added, whose definition and nature is more elusive and to which is connected the dodecahedron, because, with its twelve faces, it approaches the shape of a sphere.

But what was so special about these Platonic solids that they came to have such a profound influence on Renaissance thought?

Suppose we want to build some solids whose faces are all identical geometrical figures, for example, only equilateral triangles, or squares, or pentagons. To satisfy this requirement, the same number of faces must meet at each vertex. The only solids which can do this are those consisting of 4, 8, or 20 triangles, or 6 squares, or 12 pentagons—just 5 combinations in total—the 5 Platonic solids (Fig. 11.7).

The characteristics of the Platonic solids are briefly summarized in Table 11.1.

The name of each Platonic solid is derived from the number of faces, F, which are 4, 6, 8, 12, and 20, respectively. This composition was well understood by Leonardo, who described it on folio 7v of the *Codex Forster I.* He ordered the solids as a function of the number of faces: 4 equilateral triangles (*corpo di 4 base triangolari equilatere* "*body of 4 equilateral triangular bases*"), 6

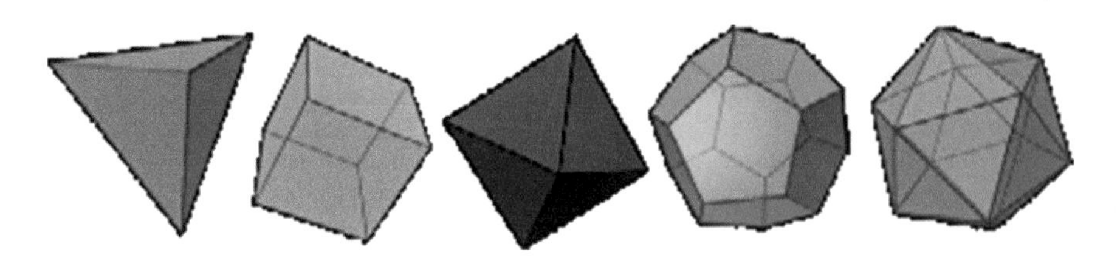

Fig. 11.7 The five Platonic solids from left to right: tetrahedron, hexahedron, octahedron, dodecahedron, and icosahedron. (This work reproduced under the attribution-ShareAlike 3.0 Unported (CC BY-SA 3.0) license. https://en.wikipedia.org/wiki/Regular_polytope)

Table 11.1 The properties of the Platonic solids

Platonic solid	Number of faces (F)	Number of edges (E)	Number of vertexes (V)	Figure for faces
Tetrahedron	4	6	4	Equilateral triangle
Hexahedron (cube)	6	12	8	Square
Octahedron	8	12	6	Equilateral triangle
Dodecahedron	12	30	20	Pentagon
Icosahedron	20	30	12	Equilateral triangle

equilateral squares, 8 equilateral triangles, 12 equilateral pentagons, and 20 equilateral triangles.

If we use the *polyhedral formula*,[22] developed after Leonardo's death, which relates the number of vertices *V*, faces *F*, and edges *E*, of a connected polyhedron, it can be easily verified that for each Platonic solid, the value 2 is always obtained:

$$V - E + F = 2$$

These geometrical properties make the Platonic solids particularly evocative, and it is no wonder that in the Renaissance, when the Platonic philosophy became a fundamental basis for the culture of the time and the use of symbolism in art a common practice, these solids attracted so much attention.

The five Platonic solids are then the foundation of a divine perfection which is expressed through the golden ratio, φ. There is, in fact, a strict connection between these solids and φ. If we consider, for instance, a dodecahedron and an icosahedron whose sides are each 1 unit long, their volume can be expressed in terms of φ via the formulas:

$$V_{\text{dodecahedron}} = 5\varphi^2 / (6 - 2\varphi) \qquad V_{\text{icosahedron}} = 5\varphi^5 / 6$$

It is possible to derive additional solids or empty polyhedral shapes from the five Platonic solids, a subject which occupies chapters 48–55 of the *Divina Proportione* and the 60 drawings by Leonardo which are printed at the end of

[22] The formula was independently discovered by Leonhard Euler (1752) and Descartes and is also known as the Euler-Descartes formula. It can be also applied to some non-convex polyhedra.

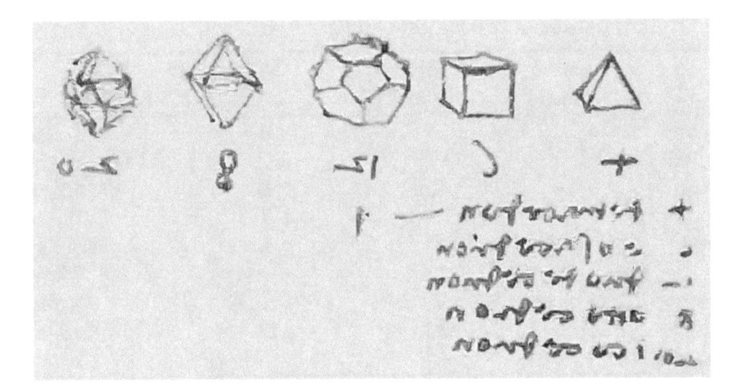

Fig. 11.8 The Platonic Solids. Leonardo da Vinci. Manuscript M, folio 80v, detail. BIF

the book, immediately after the section devoted to the graphics of capital letters (Fig. 11.10). These drawings are the only artwork by Leonardo printed during his lifetime.

The basic properties of the Platonic solids were quite clear to Leonardo. On folio 80v of *Manuscript M*, there is a complete representation of the five solids, including their names (although with some orthographic errors) and the number of faces: 4 tetracedron, 6 eusacedron, 12 duodecedron, 8 ottocedron, and 20 icocedron (Fig. 11.8). On this same folio, Leonardo also transcribed, albeit in a rather approximate way, the closing sentence of the *Divina Proportione*.[23]

The Platonic Solids Before Pacioli and Leonardo

As we will see, Leonardo's images for his friend's book were realized with an unprecedented technical skill. This is yet another demonstration of how his skills as a painter were enhanced by his remarkable ability to read and interpret space, combining the aesthetic sense of the artist with the geometrical precision of a mathematician. Leonardo was not, however, unique in this: As we have remarked again and again, before him other leading figures of the Renaissance, such as Piero della Francesca, shared the passion for art and mathematics and were able to make important contributions to both fields.

Another such figure is Paolo Uccello (1397–1475), one of the most famous Florentine painters, who was greatly intrigued by the geometry of solids.

[23] The original of Luca Pacioli is: "El dolce frutto vago e sì diletto costrinse già i Philosophi cercare causa de che pasci l'intelletto." Corpora ad lectorem. Divina proportione.

Fig. 11.9 Mosaic attributed to Paolo Uccello in the Basilica di San Marco in Venice showing a stellated dodecahedron (c. 1425–1430). (This work is in the public domain in its country of origin and other countries and areas where the copyright term is the author's life plus 70 years or less. The image has been modified with respect to the original to reduce the noise. https://upload.wikimedia.org/wikipedia/commons/9/95/Marble_floor_mosaic_Basilica_of_St_Mark_Vencice.jpg)

As we will see in Chap. 12, in several of his paintings, he explored the expressive potentialities of these bodies in art. A surprising trace of his interest in geometry can be found the Basilica di San Marco in Venice. A marble mosaic by Uccello in the church floor shows the reproduction of a solid body, which is an almost perfect stellated dodecahedron (Fig. 11.9). The work likely was realized between 1425 and 1430, some time before the publication of *Divina Proportione* (1509) and the "official" mathematical discovery and description of this solid by Johannes von Kepler in his 1596 *Mysterium Cosmographicum* ("Cosmographic Mystery").

Leonardo's Illustrations for the Divina Proportione

It is likely that not only his personal friendship with Pacioli but also the charm of the argument moved Leonardo to contribute a series of panels illustrating various geometrical figures to the *Divina Proportione* (Fig. 11.10), panels

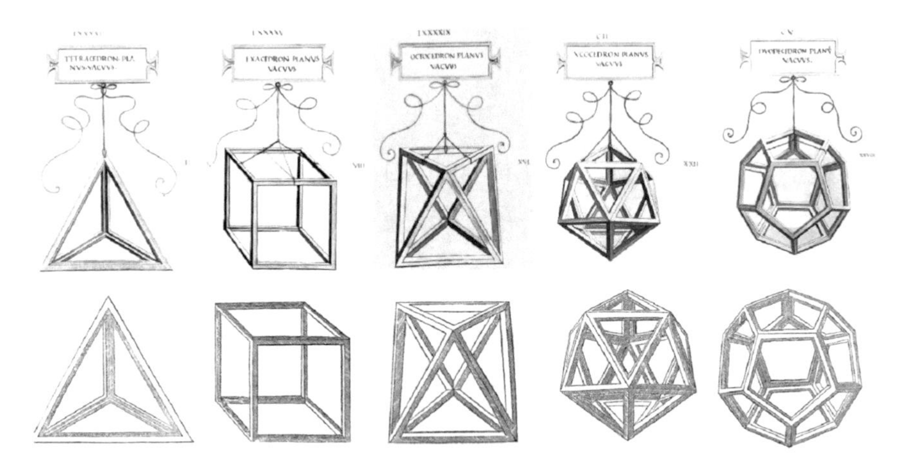

Fig. 11.10 The five Platonic solids of Leonardo for the *Divina Proportione* of Luca Pacioli in the colored manuscript copy in Milan (*top row*) and in the printed copy of 1509 (*bottom row*). From *left*: tetracedron (tetrahedron), exaedron (hexahedron), octaedron (octahedron), icocedron (icosahedron), duodecedron (dodecahedron)

which Pacioli would praise as being "*Made and formed by the ineffable left hand, well versed in all mathematical disciplines, of the man who now is the Prince among mortals, our Leonardo da Vinci.*"[24] Leonardo's illustrations, especially the manuscript versions, still have a great deal of visual appeal. Luca Pacioli had previously produced small wooden models of these solids that were delivered to the Duke of Milan and for which he also received a payment. Leonardo probably used the friar's wood solids as the starting point for his own renderings.[25]

The images in the printed version (Fig. 11.10, *bottom row*) slightly lose some of the charm of the colored illustrations in the manuscripts (Fig. 11.10, *top row*). For each shape, Leonardo realized two versions—full and hollow. In the handwritten versions, the figures are shown hanging from a thread attached to a label containing the solid's description.

The illustrations, from the three available copies (the two handwritten versions and the single printed one), are not exactly identical and exhibit some

[24] Original text: "Facte et formate per quella ineffabile senistra mano, a tutte le discipline mathematici acomodatissima del principe oggi fra' mortali, Lionardo nostro da Venci." *De viribus quantitatis.* Manuscript 250 of the Biblioteca Universitaria in Bologna. Folio 237r.

[25] "Questi son quelli Magnanimo Duca di quali le forme materiali con asai adornezze nelle proprie mani di V.D.S. nel sublime palazzo del Reverendissimo cardinale nostro protectore Monsignor de San Pietro in vincula quando quella venne ala visitatione del Summo pontefice Innocentio 8°, negli anni della nostra salute 1489, del mese de aprile, che già sonno 5 anni elapsi. E insiemi con quelli vi foron molti altri da ditti regulari dependenti. Quali fabricai per lo Reverendo Monsegnor meser Pietro de Valetarij de Genoa, dignissimo vescovo de Carpentras, al cui obsequio alora foi deputato in casa de la felicissima memoria del R.mo Cardinale de Fois, nel palazzo ursino in campo de fiore." L. Pacioli, *Summa,* second part f. 68v.

Fig. 11.11 The geometrical bodies obtained from the Platonic solids: hexahedron (cube), truncated hexahedron, elevated hexahedron, truncated stellated hexahedron (elevated). (Images from the 1509 printed version of *Divina Proportione*)

differences.[26] In the printed version, the figures are simpler, losing a little of their visual appeal and are drawn in the middle of the page and without any attached thread and label. The originals of the drawings were jealously guarded by Luca Pacioli, who wrote "*Leonardo made by his hands the mathematical bodies, which we took with us.*"[27] Unfortunately, however, no trace of these originals has remained.

Starting with the series of five Platonic or regular solids, it is possible to derive a second family of semi-regular solids, the *Archimedean solids*. This family is composed of 13 geometrical figures obtained from the Platonic solids by cutting off the edges in order to produce regular faces even if not all of the faces are identical (Fig. 11.11).[28]

For example, if we take a tetrahedron and cut it at the top, what results is a truncated tetrahedron, a solid made up of four triangles together with four additional hexagons.

Leonardo tried to demonstrate this transformation using a proper and effective visualization of the different solids, but it was not as easy a task as it might seem. Before Leonardo, there were very few attempts to produce three-dimensional renderings of geometrical solids, so Leonardo needed to rely on his own ingenuity and creativity. His drawings probably represent the first real 3D images of solids, long before the advent of computer graphics.

The illustrations of the Archimedean solids in Pacioli's book (Fig. 11.11) immediately follow those of the Platonic ones, underscoring their close relationship.

[26] A detailed study of the differences in the illustrations in the different manuscript versions can be found in the article: Dirk Huylebrouck. *Observations about Leonardo's drawings for Luca Pacioli*. http://arxiv.org/ftp/arxiv/papers/1311/1311.2855.pdf.

[27] *De viribus quantitatis*. Folio 106r, manuscript 250, Biblioteca Universitaria, Bologna

[28] An Archimedean solid should not be a prism or an antiprism.

The sketches in Fig. 11.11 show the construction of possible forms derived from the Platonic solids. Throughout, the Latin term *Vacuum* is used to indicate that the shapes are hollow. An Archimedean—or semi-regular—solid is obtained by cutting the edges of the cube (the *Hexaedron Planuum Vacuum*) to produce the truncated hexahedron (*Hexaedrum Ascissum Vacuum*). The faces of the latter solid consist of only two geometric figures—six equilateral triangles and eight squares. Leonardo trims the corners of the cube using triangles that cut the side of the square in half; this is a special case—otherwise, the operation would give a truncated hexahedron formed by six octagons and eight triangles.

The third solid in Fig. 11.11 is also derived from a Platonic solid, from the simplest one—i.e., the cube. It is the so-called high or *stellata* form (*Hexaedron Elevatum Vacuum*) that is obtained by building on each face of the polyhedron a pyramid formed of equilateral triangles. The same procedure can be performed using the semi-regular, truncated form instead of the regular one to obtain an elevated truncated solid where the pyramids are built on the truncated form of the Platonic solid (*Hexaedron Ascissum Elevatum Vacuum*).

While the finished results may look simple, in actuality, it is not—especially if you were one of the first to play with these types of graphics to represent such complex geometrical figures. Indeed, it is very easy to make some errors, and Leonardo himself was not immune.[29] For example, let us look at one of the most famous illustrations in the series for the *Divina Proportione*, the rhombicuboctahedron ("Vigintisex Basium Planum Vacuum," as it is called in the treatise itself). Starting from this, a truncated stellated form (defined by Pacioli as elevated), the *Vigintisex Ascissum Elevatum Vacuum* (trunked star-shaped rhombicuboctahedron), can be built (Fig. 11.12).

However, Leonardo's three-dimensional drawing contains some minor errors. The triangular pyramids in fact must always be surrounded by six pyramids with a square base. However, if we look at the base of the pyramid at the bottom of the figure, it looks to be four-sided when it should be triangular. The opposite problem is observed in another section of the figure. We should note, however, that, since the extant images are copies rather than Leonardo's originals, it is always possible that these errors were introduced by later copiers rather than by Leonardo himself.

Given the prolific and rich nature of Leonardo's notebooks—over 6000 pages' worth—the question begs itself as to whether similar studies of these or related figures can be found there as well. The answer is that yes, on a few

[29] Huylebrouck D (2012) Lost in Enumeration: Leonardo da Vinci's Slip-Ups in Arithmetic and Mechanics. The Mathematical Intelligencer, 34: 15.

Fig. 11.12 The figure in *Divina Proportione* with the errors circled, of the rhombicuboctahedron truncated and stellated. (Images from the 1509 printed version of *Divina Proportione*)

folios, uncommented illustrations of similar type can be found, and they are very similar to those in Pacioli's treatise. Some of these figures are part of more complex studies such as the transformation of a dodecahedron into a cube (folio 7r, *Codex Forster I.* See Chap. 14), but, in general, they are isolated passing sketches, scattered across different codices, as we can see in the various dodecahedra shown in Fig. 11.13.[30].

Another example (Fig. 11.14) is the hollow solid that appears on folio 518r of *Codex Atlanticus*, which is basically identical to the icosahedron in the printed copy of *Divina Proportione.*

Another interesting trace of Leonardo's work in drawing the Platonic and Archimedean solids remains on folio 1061v of the *Codex Atlanticus*, a simple study on how to build more complex three-dimensional geometric figures starting from Platonic solids (Fig. 11.15).

As we can see from these and the many other similar sketches in the notebooks, the figures for the *Divina Proportione* were the result of careful studies and reflections.

Pacioli's treatise ends with an illustration showing the tree of proportionalities (*Arbor proportio et proportionalitatis*) representing the theory of proportions and the various properties of a number and its multiples (Fig. 11.16, *left*).

[30] Studies on the dodecahedron are also reported on folio 533r of the *Codex Atlanticus*.

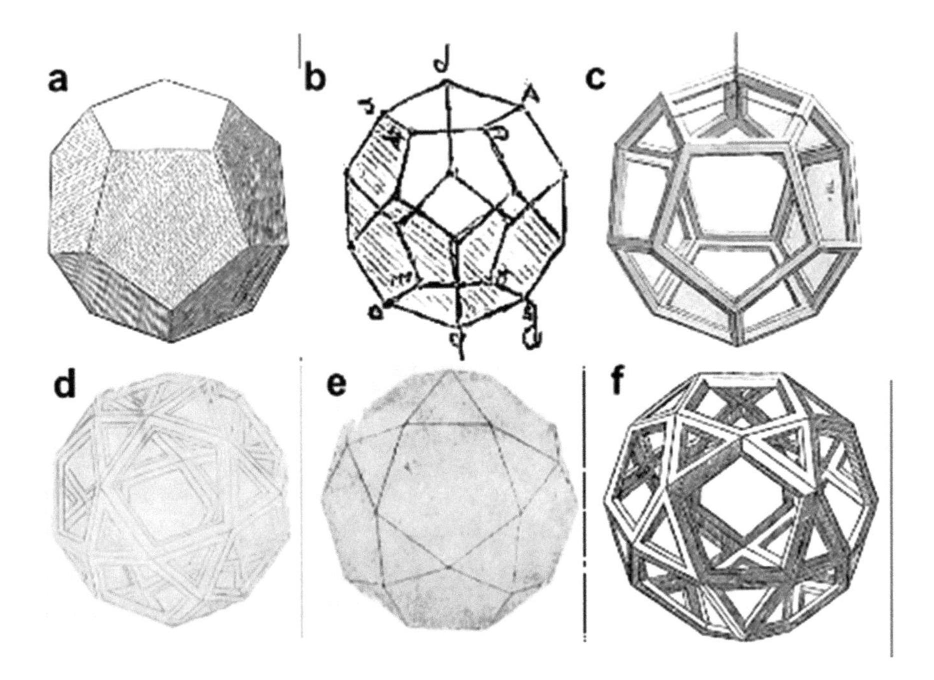

Fig. 11.13 Comparison between the images of the solids in the *Divina Proportione* and in Leonardo's notebooks. First row: (**a**) dodecahedron, 1509 printed version; (**b**) hollow dodecahedron, folio 72, Codex Forster I, detail; (**c**) hollow dodecahedron, colored manuscript copy. Second row: (**d**) hollow truncated dodecahedron, Codex Atlanticus folio 707r; (**e**) truncated dodecahedron Codex Atlanticus folio 708r; (**f**) hollow truncated dodecahedron, *Divina Proportione*, 1509 printed version

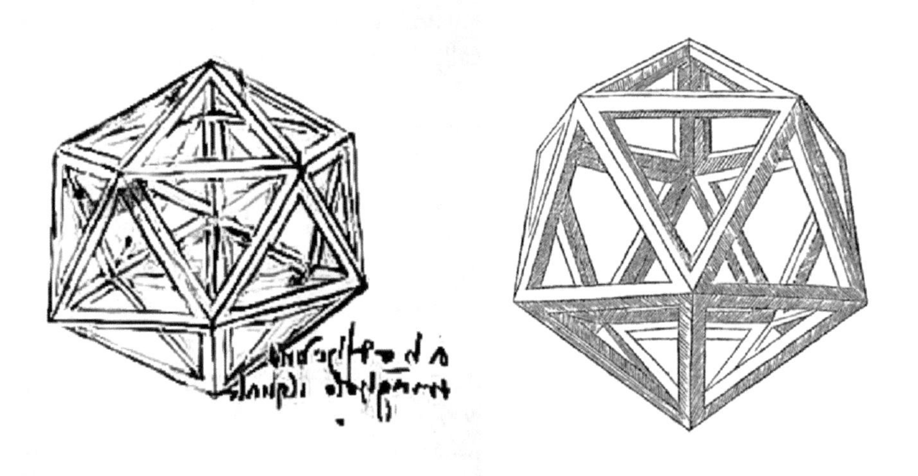

Fig. 11.14 Study for icosahedra. Leonardo da Vinci. Codex Atlanticus, (*left*) folio 518r (detail), BAM; icosahedron, printed version of the *Divina Proportione*, 1509 (*right*)

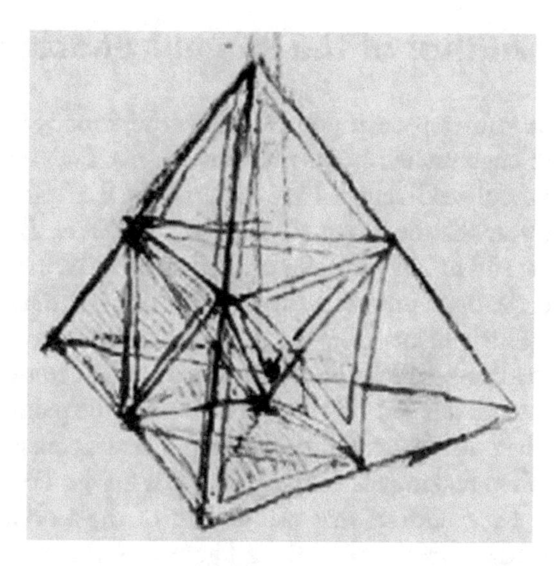

Fig. 11.15 Study of a pyramid. Leonardo da Vinci. Codex Atlanticus, folio 1061v, detail. BAM

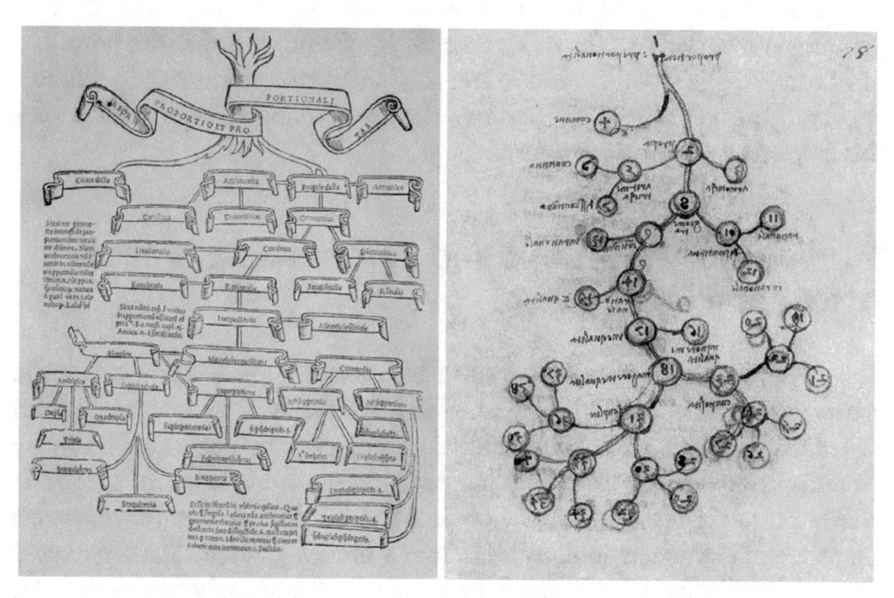

Fig. 11.16 The "Arbor proportio et proportionalitatis" in Luca Pacioli's *Divina Proportione* (*left*). 1509 printed version. Leonardo's version of the tree (*right*). Codex of Madrid II, folio 78r. BNE

Leonardo made a simple draft reproduction of the tree on folio 78r of *Codex Madrid II*, probably as a form of reminder (Fig. 11.16, *right*). It is the testimony of his great attention to the work and lessons of his mathematician friend.

Who Is the Author of the Rhombicuboctahedron?

At this point we can step back and return to the famous portrait of Luca Pacioli which we encountered in the previous chapter. Let's recap some dates: The painting was realized likely in 1495, prior to the friar's arrival in Milan in 1496, the same year that he began to work on the *Divina Proportione*, completed toward the end of 1497 but not printed until 1509. Two solids in three dimensions are clearly represented in the painting—a Platonic solid (the dodecahedron) which hangs from a thread and another convex solid (the rhombicuboctahedron), which appears on top of a book to the friar's left.

When we compare this second solid to the rest of the painting, something doesn't seem to be quite right. One possibility is that Leonardo was inspired by the portrait when making his drawings of solids for the *Divina Proportione*. However, many have noticed that the quality of the workmanship of the rhombicuboctahedron in the portrait, a figure of great technical difficulty, could suggest a different hand than the rest of the picture and have suggested that perhaps Leonardo himself or someone from his workshop added it to the painting at a later date. The other possibility, of course, is that the portraitist saw Luca Pacioli's wooden models and used them to reproduce the solids that appear in the painting. As we can see, the mystery and intrigue surrounding this enigmatic painting continues.

Studies of the Platonic and Archimedean Solids After Leonardo

The Platonic and Archimedean solids have exerted a great charm also in the artistic world; there are many examples from the Renaissance up to more recent times of the use of these solids in a variety of contexts to introduce a symbolic element, or, more simply as a decorative motif. In Verona in Northern Italy, famous as the city of Shakespeare's Romeo and Juliet, there is a small, not particular famous and almost hidden church, Santa Maria in Organo, which contains a small masterpiece of a Benedictine monk, Fra Giovanni da Verona (1457–1525), who was probably the most talented and famous maker of inlays of the Renaissance.[31] The monk perfected the use of perspective in wooden inlays and was able to recreate a fabulous and abstract

[31] Bagatin PL (2000) Preghiere di legno. Tarsie ed. intagli di fra Giovanni da Verona. Catalogo. Centro Editoriale Toscano.

Fig. 11.17 Fra Giovanni da Verona's wooden inlays in the sacristy of Santa Maria in Organo in Verona. The inlays reproduce some of the solids drawn by Leonardo for Luca Pacioli's *Divina Proportione* (1520)

world which appears to us incredibly modern.[32] The inlays of the sacristy, made in 1520, offer an exceptional surprise: They reproduce without any doubt exactly the solids designed by Leonardo for the *Divina Proportione* (Fig. 11.17).

In the inlays, the hollow solids are rendered with great skill and extraordinary use of perspective that gives an impressive visual impact. The two inlays depict a pair of open shelves that display a variety of objects alongside the solids. Starting from the top left and moving clockwise, the inlays depict the following solids: a truncated icosahedron, an icosahedron and a geode, a star-like dodecahedron, a cuboctahedron, and a rhombic dodecahedron. It is a vivid and precise representation of the geometrical figures that captures the attention of the observers.

The inlays of Santa Maria in Organo are not the only place where Fra Giovanni produced solids inspired by Leonardo.[33] In the monastery of Monte

[32]Tormey A, Farr J (1982) Renaissance Intarsia: The Art of Geometry. Vol. 247, issue 1, pp. 136–143.

[33]Bagatin PL (2016) Fra Giovanni Da Verona e la Scuola Olivetana dall'Intarsio Ligneo. Antilia, Treviso.

Fig. 11.18 Fra Damiano Zambelli's wood inlay. Museo della Basilica di San Domenico, Bologna. (1541–1549). (Permission is granted to copy, distribute, and/or modify this document under the terms of the Gnu Free Documentation License (https://commons.wikimedia.org/wiki/Commons:GNU_Free_Documentation_License), Version 1.2 or any later version published by the Free Software Foundation; with no invariant sections, no front-cover texts, and no back-cover texts. A copy of the license is included in the section entitled Gnu Free Documentation (https://commons.wikimedia.org/wiki/Commons:GNU_Free_Documentation_License), Version 1.2. This file is licensed under the Creative Commons Attribution-Share Alike 3.0 Unreported (https://creativecommons.org/licenses/by/3.0/deed.en) license. Courtesy of Georges Jansoone)

Oliveto Maggiore in Tuscany, a similar inlay appears. Again, Fra Giovanni used these geometrical figures as a decorative pattern.

Fra Giovanni was not alone in this passion for Leonardo's solids: Another friar and master of inlays, Damiano Zambelli from Bergamo (1490–1549), reproduced solids in Leonardo's style in the choir of his church in Bologna, now in the collection of the Museum of San Domenico (Fig. 11.18). These inlays are also eye-catching, even if they do not reach the high artistic level of Fra Giovanni.

It is hard to say whether the two friars were somehow influenced by each other or if they were independently inspired by Leonardo's figures in the *Divina Proportione*. Their work in any case testifies to the diffusion of Pacioli and Leonardo's work and its impact on the contemporary artistic world.

Now let us return to another great man of the Renaissance, the German artist Albrecht Dürer, whom we have already met in the previous chapter. Dürer is somehow intertwined with the fates of Leonardo and Luca Pacioli.

Fig. 11.19 Albrecht Dürer. Melencolia I (1514). Engraving. (This work is in the public domain in its country of origin and other countries and areas where the copyright term is the author's life plus 70 years or less. The image has been modified with respect to the original to reduce the noise. https://upload.wikimedia.org/wikipedia/commons/1/18/D%C3%BCrer_Melancholia_I.jpg)

We have already mentioned one of Dürer's most famous and enigmatic works, *Melencolia I*, an engraving realized in 1514, 5 years after the printing of the *Divina Proportione*, where several allegorical elements overlap to form a picture with many allusions and interpretations (Fig. 11.19).

The engraving is part of a triptych, whose other two panels are *Knight, Death and the Devil* (1513), which represents the moral sphere; and *Saint Jerome in his Study* (1514), which represents the sphere of meditation. *Melencolia I*, in turn, focuses on the intellectual sphere dominated by the planet Saturn, which is associated in astrology with melancholy. The engraving is full of symbols, and its interpretation has been the subject of lengthy treatises. The citations in this work to Luca Pacioli are, however, what interests us—the magic square, which we discussed in the previous chapter, and the strange large solid at the engraving's center left. This solid is a truncated rhombohedron, and, together with the sphere, it contributes to the work's complex symbolism.

The Perspectiva Corporum Regularium

Luca Pacioli's book enjoyed a large diffusion and spread through the scholar community of Europe. An echo of his work can be found in the treatise *Perspectiva Corporum Regularium* of Jamnitzer Wenzel (c. 1507–1585), a famous German goldsmith who also devoted his attention to the geometry of solids. Wenzel dedicated his work to the study of the five Platonic solids, realizing 24 variations on each of the regular solids, for a total of 120 engraved figures which cover truncated, stellated, and faceted regular solids.[34] The images in the *Perspectiva Corporum Regularium* were realized by the engraver Jost Amman (1539–1591) from original drawings by Wenzel, and the treatise was published in 1568.

Wenzel's study of the Platonic solids is systematic, even if a little pedantic, and the images were clearly inspired by Leonardo's. Other images in the book, however, are original. Wenzel, even if was not a formally trained mathematician, was able to explore new possibilities for the geometry of solids with creativity and intelligence. His solids were realized by combining the Platonic solids in a variety of ways and using both hollow and solids drawings, as had also been the practice of Leonardo. Some of Wenzel's solids are placed on pedestals, which are themselves highly geometrical in nature, which gives the whole image a very attractive futuristic appearance.

An example of the variations realized by Wenzel in the geometry of the regular solids is given in Fig. 11.20. The almost perfect rhombic triacontahedron, a convex polyhedron with 30 rhombic faces, and the great stellated dodecahedron in the core of the second image are complex figures whose physical representation is not trivial and which would not be represented mathematically until the work of the German astronomer Johannes von Kepler (1571–1630) in 1619.

Kepler was also very much attracted by the regular solids, and in his *Mysterium Cosmograficum* (1595), he represented the solar system using a series of nested Platonic solids. The distances of the six planets known at the time (Saturn, Jupiter, Mars, Earth, Venus, and Mercury) were correlated to the five Platonic solids, and each body was used to fit a pair of planetary spheres (Fig. 11.21). This theory, although subsequently proven to be quite wrong, was quite appealing at the time.

It is interesting to compare, on the one hand, Kepler, who was a great scientist but very much influenced by religion and mysticism, and, on the other,

[34] Swetz FJ (2013) Mathematical treasure: Wenzel Jamnitzer's Platonic Solids. Convergence. Penn State University.

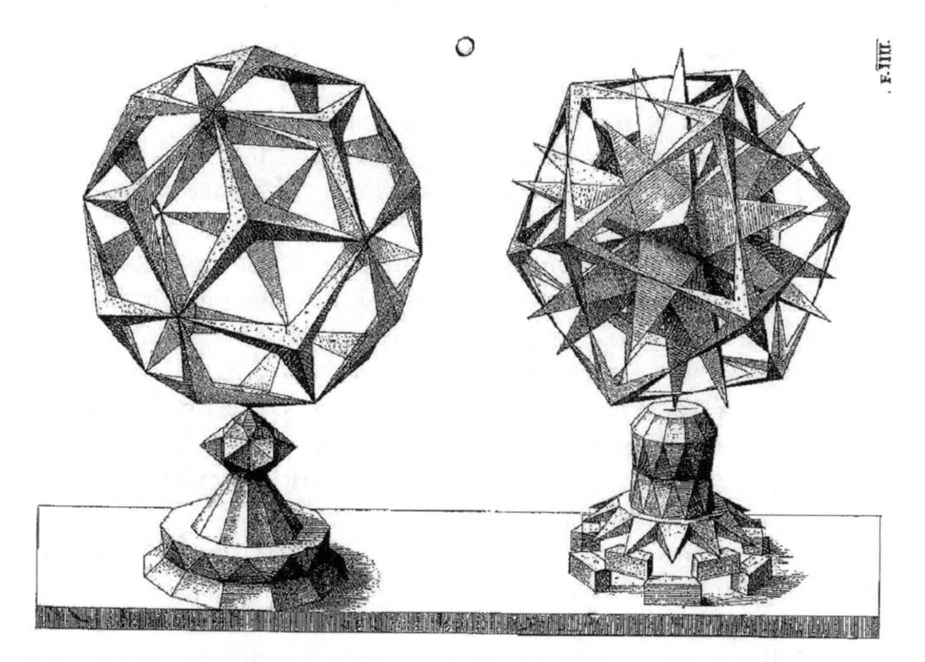

Fig. 11.20 A rhombic triacontahedron (*left*) and a great stellated dodecahedron (core of the figure on *right*). Jamnitzer Wenzel. Perspectiva Corporum Regularium. 1568

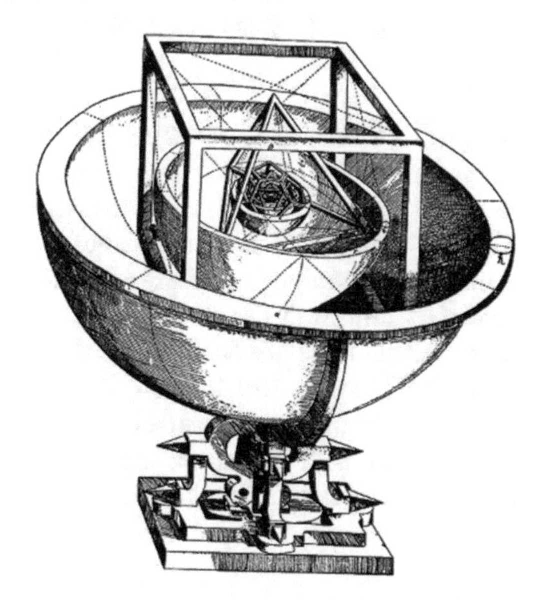

Fig. 11.21 The model of solar system from Kepler elaborated using the five Platonic solids. *Mysterium Cosmograficum* (1595). (This work is in the public domain in its country of origin and other countries and areas where the copyright term is the author's life plus 70 years or less. The image has been modified with respect to the original to reduce the noise. https://upload.wikimedia.org/wikipedia/commons/1/19/Kepler-solar-system-1.png)

Piero della Francesca, Luca Pacioli, and Leonardo. Even if they were born several years before Kepler, the latter three were only attracted by the geometrical properties of the regular Platonic and Archimedean solids. They never exhibited any particular interest in philosophical and mystical interpretations of the figures. This was a very distinctive sign of discontinuity with the past.

Another German author who needs to be cited in our exploration of the elaboration of solids is Lorenz Stoer (c. 1537–1621), who published in 1567 the *Geometria et perspectiva: Corpora regulata et irregulata*[35] (*Geometry and perspective: regular and irregular bodies*). While this book does not contain any systematic study of geometrical solids, it offers a very fascinating collection of drawings showing decadent architectural landscapes with the fancy and extemporaneous presence of geometrical bodies to mark a hypothetical border with reality.

An explicit citation of Pacioli and Leonardo's work can be found in a famous engraving of Giovanni Jacopo Caraglio (c. 1500–1565) based on an original design of Parmigianino (Girolamo Francesco Maria Mazzola, 1503–1540, Fig. 11.22). The engraving (c. 1524–27) shows the Greek philosopher Diogenes pointing at a geometrical figure in an open book which can be easily recognized to be the *Dodecaedron Planum Solidum* (drawing XXVII) from the printed version of *Divina Proportione*.

Salvador Dalí and Maurits Cornelis Escher

We will now make a time jump of a few centuries to arrive at a Spanish surrealist artist, Salvador Dalí (1904–1989), another fan of the golden ratio and well aware of Leonardo's legacy. Among the many works of Dalí, there is one in particular that connects him to this story of golden numbers and divine proportions, and it is fatally entitled *The Sacrament of the Last Supper* (*El sacramento de la última cena*)[36] and is now in the collection of the National Gallery of Art in Washington (Fig. 11.23). Aside from the amazing iconography of the painting, the dimensions of the canvas are in the golden ratio. Of course, the reference to Leonardo's own Last Supper is evident.

[35] Stoer L (2006) Geometria et perspectiva: Corpora regulata et irregulata. Manuscript Cim 103, University Library Munich. Harald Fischer Verlach.

[36] The *Sacrament of the Last Supper* is an oil on canvas painting with dimensions of 267 cm × 166.7 cm, realized by Salvador Dalì in 1955.

Fig. 11.22 Giovanni Jacopo Caraglio. Diogenes. Engraving (c. 1524–27). (This work is in the public domain in its country of origin and other countries and areas where the copyright term is the author's life plus 70 years or less. The image has been modified with respect to the original to reduce the noise. http://www.metmuseum.org/art/collection/search/357255)

The environment in which Dalí's Last Supper takes place is a translucent dodecahedral dome which, with its transparency, confuses the exterior and interior spaces. None of the Apostles display his face, and a Jesus, radiant of light, dominates the scene. Dalí called his painting an obsession of the number 12, as the dodecahedron (the Platonic solid composed of 12 pentagonal faces) represents the mathematical perfection of the universe. The divine proportion reappears in each pentagon to recall the unfathomable link between numbers and the perfection of creation. A more beautiful tribute to the obsession of Pacioli and his *Divina Proportione* perhaps couldn't have been imagined. The surreal atmosphere of the painting leaves the door open to different interpretations and possible mathematical representations of the universe.

Fig. 11.23 The Sacrament of the Last Supper of Salvador Dalì (El sacramento de la última cena). (1955). Oil on canvas. National Gallery of Art in Washington. (This file is licensed under the Creative Commons Attribution 2.0 generic (CC BY 2.0) (https://creativecommons.org/licenses/by/2.0/). *Courtesy of Jorge Elías*)

A Dutch artist, Maurits Cornelis Escher,[37] would come to describe, perhaps better than any other artist or illustrator, the world of mathematical and geometrical forms that, inside a sort of magic universe, can become a source of games and illusions. Escher inserted in many different works geometric solids whose transformations become part of a more complex change of the dimensional borders. In the lithograph *Reptiles*, printed in 1943, a dodecahedron is placed along a path where some lizards enter in a 2D page only to escape again into a surrealistic 3D reality.[38] Other loose objects included in the image are books, a plant, a bottle, and a glass, which combine to create a random mathematical game.

The image of the lizard reappears in the woodcarving *Stars* (1948)[38], this time at the center of another Leonardo-style solid, which appears on a black background surrounded by other tiny solids of different shapes and sizes. This is yet another of Escher's illusionistic games, where lizards are caged in a sort of geometric impossibility. The similarity of Escher's drawings to the way Leonardo represented the hollow forms of the Platonic solids and their derived forms is remarkable. One last curiosity: Like Leonardo, Escher was also left-handed.

[37] Maurits Cornelis Escher (Leeuwarden, 1898–Laren, 1972) is very famous for his graphical work, which is in part inspired by geometry.

[38] The image can be found in the Escher's official website: www.mcescher.com.

12

The Geometry of Shapes and Unfashionable Headgear

As we saw in the previous chapter, Leonardo developed a very special affection for the geometry of solids which guided him into a fascinating exploration of new forms. This work was not only driven by intellectual curiosity but was also inspired by the need to couple function and form in the design of his innovative machines, such as the "automobile" described in Chap. 7 and the mechanical lion and other devices that he created for theatrical shows and installations in which he staged his devices for the delight and amazement of the Italian courts and their Lords. A famous theatrical representation based on Leonardo's machines was the Festa del Paradiso (Feast of Heaven) which we have described in Chap. 10.

In the previous chapter, we saw that Leonardo dedicated particular attention to exploring the Platonic solids and their possible graphical renderings. His playful exploration was also motivated by a scientific interest which pushed him to study several geometrical problems in depth. In Leonardo's notebooks, we find the main result of this exploration, a number of drawings which provide a vivid representation of geometrical objects. One notable example is the object obtained by assembling three square frames with a cross in the middle found on folio 709r of the *Codex Atlanticus* (Fig. 12.1). This is a very original sketch, a geometrical game realized by combining the light and shadow *chiaroscuro* technique and a 3D perspective. The object's simplicity is also part of its aesthetic attraction.

P. Innocenzi, *The Innovators Behind Leonardo*, https://doi.org/10.1007/978-3-319-90449-8_12

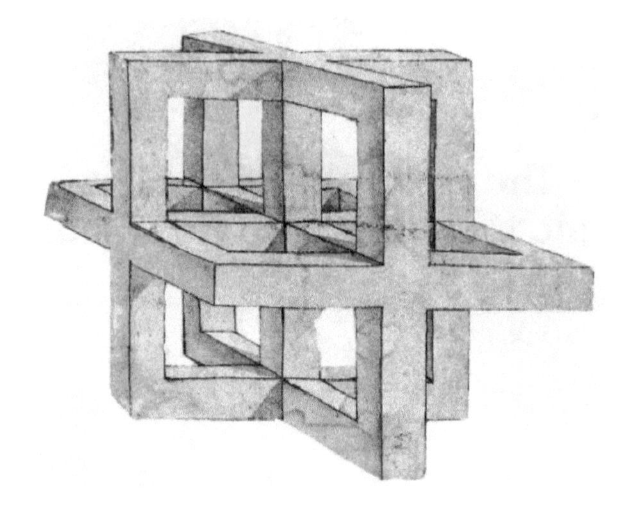

Fig. 12.1 Geometrical figure. Leonardo da Vinci. *Codex Atlanticus*, folio 709r, detail. BAM

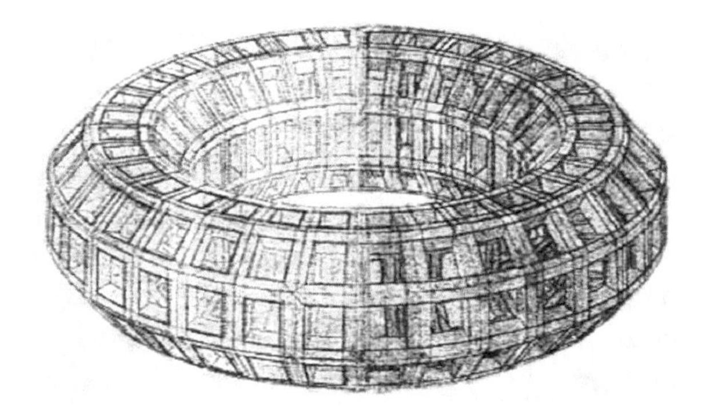

Fig. 12.2 *Mazzocchio*. Leonardo da Vinci. *Codex Atlanticus*, folio 710r, detail. BAM

The Strange Story of the Mazzocchio

Leonardo, as we have just seen, liked to play with geometrical objects using his incredible capability for graphical representation to create forms of unprecedented elegance and beauty. One of the most striking examples of this is the pencil-and-paper drawing of the popular Renaissance headgear the *mazzocchio*[1] on *Codex Atlanticus*, folio 710r (Fig. 12.2). Together with his perfect mastery of *chiaroscuro*, Leonardo's deep understanding of the intricacies of

[1] Brecher K (2011). The Mazzocchio in Perspective. Bridges: Mathematics, Music, Art, Architecture, Culture.

perspective, which were well explored and described in Piero della Francesca's *De Prospectiva Pingendi*, formed the basis of his graphical capability.

The *mazzocchio* or *mazzocco* headgear had the shape of an asymmetrical ring, with the top somewhat larger than the bottom. While enjoying its greatest popularity in Florence, it became a distinctive element for the rich merchants and bourgeoisie, and its use quickly spread to many European cities. Some versions were simply formed by a thick cloth ring, but in other cases, as in Leonardo's drawing in Fig. 12.2, the frame had a complex, faceted geometrical structure.

A great fan of the *mazzocchio* was the talented Florentine painter Paolo Uccello (1397–1475), whose mosaic in the *Basilica di San Marco in Venice* we saw in the previous chapter (Fig. 11.9). Uccello was a true master of perspective, of whom Vasari wrote: "he had no other pleasure than investigating some things of perspective difficult and impossible"[2]; he used the *perspective naturalis,* introducing multiple vanishing points in the same painting. Very likely because of its intrinsic difficulty of graphical representation and complexity of 3D perspective, the *mazzocchio* was a decorative element which Uccello included in many of his works. For example, in *The Battle of San Romano, Niccolò Mauruzi da Tolentino unseats Bernardino della Ciarda at the Battle of San Romano*[3] (1438, Fig. 12.3), now in the collection of the Galleria degli Uffizi of Florence, several of the knights wear multicolored, structured *mazzocchios*.

The *mazzocchios* depicted by Paolo Uccello contribute to the strange surrealistic atmosphere which is transmitted by the painting, with the orange and blue horses and the knights who look more like they are taking part in a tournament than in a true battle. It is difficult to imagine soldiers participating in a bloody fight wearing such elegant headgear. Thus, we see that the *mazzocchios* are there to exhibit once more the artist's capability of illustrating complex geometrical forms in the three-dimensional space as well as its unconventional character.

There is also a study dedicated to perspective usually attributed to Uccello but sometimes to Piero della Francesca where the *mazzocchio* is used as a deco-

[2] Original text: *non ebbe altro diletto che d'investigare alcune cose di prospettiva difficili e impossibili.*

[3] This painting was the central panel of a triptych by Paolo Uccello dedicated to the Battle of San Romano. The other two panels were dispersed and are now at the National Gallery of London (*Niccolò Mauruzi da Tolentino at the Battle of San Romano* (1438–1440?)) and the Musée du Louvre, Paris (*The Counterattack of Michelotto da Cotignola at the Battle of San Romano* (c. 1455)).

The Battle of San Romano (around 50 km from Florence) was fought on 1 June 1432 between the Florentine army guided by Niccolò Mauruzi da Tolentino and the Sienese army under the command of Bernardino Ubaldini della Ciarda.

Fig. 12.3 Paolo Uccello. The Battle of San Romano, Niccolò Mauruzi da Tolentino unseats Bernardino della Ciarda at the Battle of San Romano (1438). The mazzocchios in the painting are shown within the white circles. (Galleria degli Uffizi of Florence. This file is licensed under the Creative Commons Attribution-Share Alike 3.0 Unported (https://creativecommons.org/licenses/by-sa/3.0/deed.en) license. Courtesy of VivaItalia 1974. https://commons.wikimedia.org/wiki/File:Uccello_Battle_of_San_Romano_Uffizi.jpg)

rative element for the representation of a calyx[4,5] (Fig. 12.4). The drawing is intentionally schematic to highlight the geometrical details of the object, which is reproduced in 3D perspective. To our modern eyes, the calyx resembles a perfect, three-dimensional computer rendering.

As we have indicated, Paolo Uccello was not the only painter attracted by the *mazzocchio's* shape, and not by chance another great master of the Renaissance, Piero della Francesca not only used the *mazzocchio* in several of his paintings but also put its description in his *De Prospectiva Pingendi.*

The use of the *mazzocchio* as an element of decoration in the paintings of these two famous Renaissance masters very likely contributed significantly to the rise in interest and attention toward this original geometrical object.

A trace of the *mazzocchio* can be found in another masterpiece of Italian Renaissance, the *Studiolo di Guidobaldo da Montefeltro* (Studiolo di Gubbio, 1479–1482, Fig. 12.5) which was located in the Palazzo Ducale of Gubbio.

[4] Roccasecca, P (2000) Il 'Calice' degli Uffizi: da Paolo Uccello e Piero della Francesca a Evangelista Torricelli e l'Accademia del Disegno di Firenze. Ricerche di Storia dell'Arte, 70, pp. 65–78.

[5] Talbot R (2006) Construction: why is the Chalice the shape it is?. Nexus VI. Architecture and Mathematics, 6, pp. 121–134.

Fig. 12.4 Perspective study of a calyx. Paolo Uccello. Drawing on paper (34 × 24 cm), 1758A, Galleria degli Uffizi, Gabinetto dei Disegni e delle Stampe. Florence, Italy. (This work is in the public domain in its country of origin and other countries and areas where the copyright term is the author's life plus 70 years or less. The original figure has been modified to reduce the noise. https://commons.wikimedia.org/wiki/File:Paolo_uccello,_studio_di_vaso_in_prospettiva_02.jpg)

As an aside, Guidobaldo da Montefeltro was one of the possible candidates for the second individual in the portrait of Luca Pacioli which we discussed in Chap. 10.

The *studiolo* (a small room reserved for study and meditation) was realized with the wood-inlay technique (intarsia).[6] The inlays were realized by local artisans very likely based on drawings by Francesco di Giorgio, who was in the service of Guidobaldo's father, Federico da Montefeltro, the second Duke of Urbino. The inlayed cabinets, whose doors appear partially opened or closed to play with perspective, show several objects chosen to illustrate the wide cultural interests of the Duke. The *Studiolo* was dismantled in 1878, and, after changing hands several times, it eventually arrived at the Metropolitan Museum of Art in New York in 1939, where it can still be admired, even if completely out of the context for which it was originally created.

[6] Raggio O (1999) The Gubbio Studiolo and Its Conservation. Federico da Montefeltro's palace at Gubbio and its studiolo. Metropolitan Museum of Art.

Fig. 12.5 Studiolo di Gubbio. Metropolitan Museum of Art, New York, USA. (This work is in the public domain in its country of origin and other countries and areas where the copyright term is the author's life plus 70 years or less. The original image has been modified to reduce the noise. http://images.metmuseum.org/CRDImages/es/original/DT237370.jpg)

The *mazzocchio*, gently placed on a table which appears in the middle of the wood wall, enjoys pride of place at the center of the scene. The overall effect is quite astonishing, and the small object unavoidably attracts the full attention of the observer. What is the meaning of the *mazzocchio* just there—is it an allegoric representation, or is it just a nice decorative element? The *mazzocchio* does not play any real function but is able to transmit an undefined feeling: the room is empty, and the useless object is waiting for somebody, maybe its mysterious owner.

As we begin to see here, putting aside its practical function, the *mazzocchio* became a widely reproduced decorative motif in wood inlays and elsewhere because of its very attractive geometrical shape.

Another intarsia Master, Giovanni da Verona, whom we have already encountered in Chap. 11 in the context of his wood-inlay reproduction of Leonardo's images from Luca Pacioli's *Divina Proportione*, used the *mazzoc-*

Fig. 12.6 Fra Giovanni da Verona. Intarsia for the Monastery of Oliveto Maggiore. Italy. (This file is licensed under the Creative Commons Attribution 3.0 Unported (https://creativecommons.org/licenses/by/3.0/deed.en) license. Courtesy of Sailko. https://upload.wikimedia.org/wikipedia/commons/f/fd/Fra_giovanni_da_verona%2C_coro_intarsiato_di_monte_oliveto_maggiore%2C_1503–05%2C_12.JPG)

chio in another of his masterpieces, the intarsia for the Monastery of Oliveto Maggiore in Tuscany (Fig. 12.6).

Thus, the geometry of the *mazzocchio*, as we have just seen, was at the center of the attention of many Renaissance artists. They liked to explore the potentiality of complex objects for studying perspective, but they were also artists and therefore, at the same time, were attracted by decorative elements which are able to add allegorical and symbolic allusions and elusive meanings.

Leonardo's drawing of the *mazzocchio* in Fig. 12.2 exhibits, if we look very carefully, an extraordinary quality and beauty. Leonardo, however, did not constrain the drawing to a perfect general representation of the three-dimensional geometrical shape but instead, showing an unprecedented capability, reproduced the intimate details of the structure's internal facets. The *mazzocchio* is shown in axonometric projection using 32 sections with 8 faces and appears extremely detailed and complex. It represents a clear attempt to sketch geometrical objects in a 3D space with a sophisticated technical approach.

Leonardo used an interesting trick to draw the *mazzocchio*: in fact, by taking advantage of the symmetry of the figure, he sketched only half of it; then he folded the paper and used small holes to transfer the image onto the other half of the page as a basic trace.

However, Leonardo did not draw the *mazzocchio* just as an exercise in geometry. In fact, he was motivated by a very practical purpose: he had in mind to produce a real version to be used as a decorative element in the Castello Sforzesco or in some special events, such as one of the famous parties he organized for Ludovico il Moro, the Duke of Milan. On the same page, in fact, he described very much in detail how to prepare an actual *mazzocchio*[7]: "In order to make it, you must use thick glued paper; cut the pieces into two basic shapes that make up the Mazzocchio, then assemble all the pieces of paper and hold them together with some thread, then pour liquid wax into the paper shapes. The total number of pieces is 512: 256 rings and 256 joints. If you want to make pieces in lead, as it is done with printing characters, the pieces are the same as those in wax, and by making the calculations, the number of pieces required totals 2048."[8] A nice computer graphic rendering of the mazzocchio can be found in the dedicated chapter of the book *Da Vinci's Workshop in the Ideal City. Codices, Machines and Drawings* by M. Lisa, M. Taddei, and E. Zanon.[8,9]

As described by Leonardo, it is possible to produce a metallic version of the *mazzocchio* out of lead by joining a total of 2048 pieces.

[7] Original text: "Per farla di rilievo, torrai carbone in[c]ollato, e taglia li pezzi secondo le due proprie forme che in ta mazzocchio qui s'adoperò e metti insieme le carte di ciascun membro, e quelle avvolgi con filo di rete, e poi gitta dentro la cera liquefatta. Sdranno li pezzi che si debbon tagliare di carta, 512, cioè 256 né cerchi e 256 nelle cinture. E se tu li volessi gittare di piombo ovver della materia che si gittano le lettere di stampa, li pezzi sarebbero il medesimo di sopra, e perche la calcolazione fu errata, perché li pezzi saran duemila 48."

[8] Translation taken from page 219 of *Da Vinci's Workshop in the Ideal City. Codices, Machines and Drawings*. Massimiliano Lisa, Mario Taddei, Edoardo Zanon. Catalog of the exhibition: Il Laboratorio di Leonardo. 2009, Leonardo3 srl.

A very nice physical model reconstruction using computer graphics can also be found in the chapter dedicated to the *mazzocchio*, which was one of the objects shown in the exhibition. pp. 219–223.

[9] http://www.leonardo3.net/mazzocchio/index.html.

The Perfect Description of the Mazzocchio

The *mazzocchio* as headgear fell out of fashion soon after the Renaissance, and it was almost completely forgotten in a relatively short time. It encountered a similar destiny also in the artistic field, after being at the center of the attention of some of the main masters of perspective of the Renaissance. It was apparently never reproduced again from the end of the sixteenth century up to its modern rediscovery, when it attracted again the attention of some contemporary artists (e.g., see Fig. 12.7).

Before falling into oblivion, however, the figure of *mazzocchio* enjoyed its moment of maximum glory with the Venetian Daniele Barbaro[10] (1514–1570), author of a famous treatise *Della Perspettiva di Monsignor Daniel Barbaro, Eletto Patriarca d'Aquileia. Opera molto utile a Pittori, a scultori & ad Architetti* ("About perspective by Daniel Barbaro elected Bishop of Aquileia. Work very

Fig. 12.7 Artistic reproduction of a mazzocchio by Ben Jakober and Yannick Vu (1994). Prato (Florence), Italy. (This file is licensed under the Creative Commons Attribution 3.0 Unreported (https://creativecommons.org/liceses/by/3.0/deed.en) license. Courtesy of Sansa55. http://commons.wikimedia.org/wiki/File:Antiche_mura_di_Prato,_Toscana,_Italia_03.jpg)

[10] Vacca G (1930). Entry on Daniele Barbaro. In: Enciclopedia Italiana (1930)

helpful to painters, sculptors and architects"), printed in Venice in 1569. This treatise is very useful in providing definite clarification about the use of the *mazzocchio* by the Renaissance artists: Barbaro writes that it was an act of daring to exhibit their skill in mastering perspective.

Even in the case of *mazzocchio*, it has been open season for esotericism among Leonardo scholars and enthusiasts, and several recondite messages have been attributed to it. However, as we have seen, for Leonardo, the *mazzocchio* basically represented a geometry study, an exercise in perspective, and an attempt to produce an object of original and complex shape that could attract the imagination and attention of his contemporaries.

Daniele Barbaro

Daniele Barbaro was a cultivated person with several interests. He studied philosophy, mathematics, and astronomy at the University of Padua and had an important career as Ambassador for the Republic of Venice and as a high-ranking priest. Beside these public activities, he was also very much devoted to the study of science and the humanities. His translation with comments of Vitruvius's *De Architectura* was reprinted several times, and, furthermore, he introduced an important innovation in the dark room—the diaphragm—showing noticeable skill as technological innovator.

Della Perspettiva (About Perspective) was his more original work. The introduction itself is a small masterpiece because Barbaro was able to transmit to his readers his passion for knowledge. Barbaro explained how he started his studies for pure amusement, but soon after, he realized that he could achieve something more challenging and of practical use, the knowledge of perspective.

At the beginning of every chapter in the *Della Perspettiva*, at the top of the page, there appears a different version of the *mazzocchio* (Fig. 12.8). Moreover, Barbaro was clearly so deeply fascinated and attracted by this geometrical

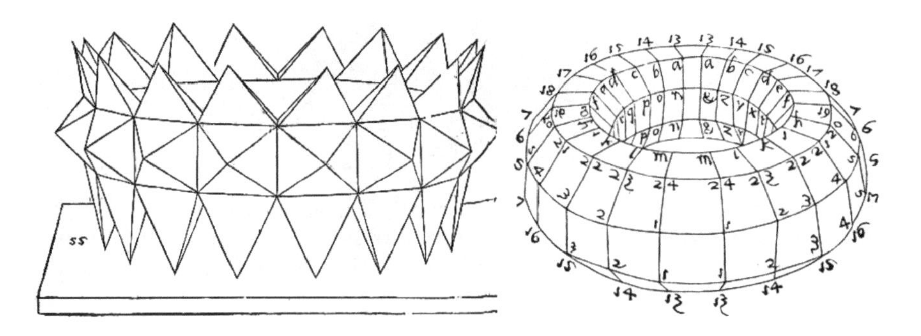

Fig. 12.8 Star *Mazzocchio* (left), detail, pag. 114; perspective study of a *mazzocchio* (right), detail, pag. 123. Daniele Barbaro, Della Perspettiva

shape that he dedicated to it two specific chapters: 38, *Descrizione del torchio, ovvero mazzocco*[11] (*"Description of mazzocchio"*) and 39, *La perfetta descrizione del Mazzocco* (*"The perfect description of mazzocchio"*).

Barbaro's study on the *mazzocchio* represents a kind of ultimate word on the subject: all the geometrical potentialities of the object were exploited up to the smallest detail, including several types of related structures.

The Gentleman Lorenzo Sirigatti

The mania for perspective studies and *mazzocchios* does not end here, however, and we have to introduce one final individual whose variations on the same subject are of some interest, especially for the quality of the drawings. He is Lorenzo Sirigatti (1596–1625), author of *La pratica di prospettiva del cavaliere Lorenzo Sirigatti (The practice of perspective of gentleman Lorenzo Sirigatti*), printed in Venice in 1596.[12]

The treatise is divided into two books, the first containing some scholastic exercises on perspective, including the *mazzocchio*, and the second clearly inspired by Luca Pacioli's *Divina Proportione* or, more precisely, by the figures which Leonardo realized for his friend's work. Beside some canonical drawings of architecture, Sirigatti indulges in a series of variations on the theme of solids similar to those of Leonardo. He added, however, an original section which is precisely the one related to the *mazzocchio* (Fig. 12.9). It is very likely that Sirigatti ignored Leonardo's work on the *mazzocchio*, but used the images in the *Divina Proportione* for drawing several new figures, some of them obtained by combining different geometrical solids. Leonardo's lesson was somehow internalized, and the work of Sirigatti contains one of the few tangible traces left by the master in his close successors, excluding of course his paintings.

This fact is worth some particular consideration. As we have noted before, the only printed work of Leonardo not only during his lifetime but also for quite a long time after his death is the graphical representation of the solids in Luca Pacioli's *Divina Proportione*. In its influence on Sirigatti's book and other works, we witness that printing a manuscript was able to change the destiny of a work—the difference between oblivion and fame. Only printing could allow an immediate and wide diffusion of a manuscript. Unfortunately, Leonardo was not yet very well aware of this revolution or was too much distracted by the many activities in which he was daily involved, and the notebooks remained unpublished for centuries.

[11]The Italian word used for indicating the headgear is not written in a standard way, and several words with small changes were used, such as mazzocchio and mazzocco.

[12]The book of Sirigatti is available online at https://archive.org/details/gri_33125008485225.

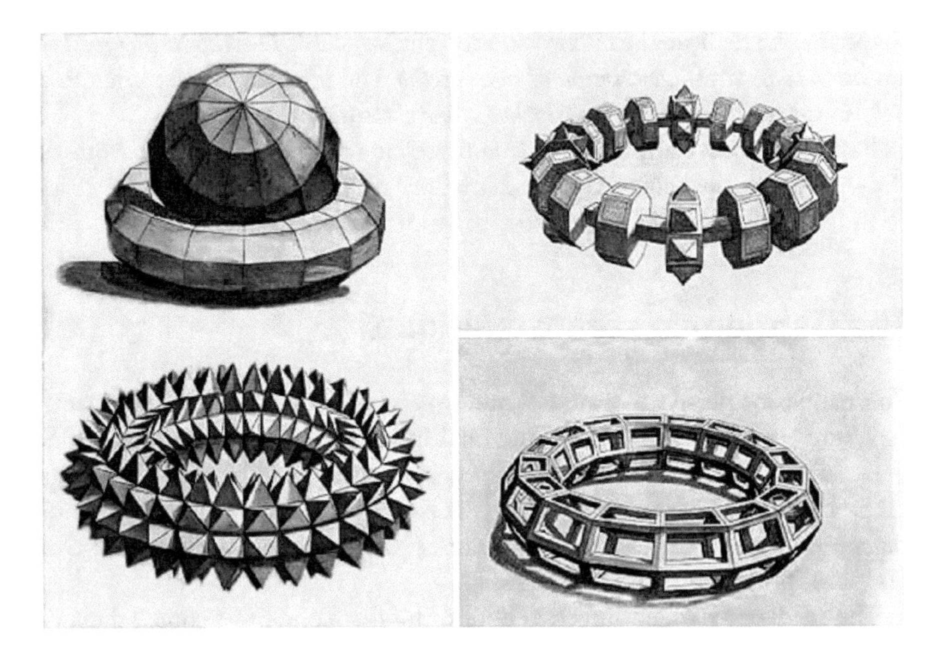

Fig. 12.9 Exercises on the theme of the *mazzocchio. La pratica di prospettiva del cavaliere Lorenzo Sirigatti*, second book

Beyond the Mazzocchio

The example of the *mazzocchio* is very interesting, and it attracted so much attention not only because of its peculiar geometry but also because of its playful ability to allow for so much variation and experimentation. In time, this last aspect of the *mazzocchio* in particular was lost, but there still remained the pleasure of exploring new shapes and geometrical forms.

Leonardo went beyond this game by mixing this pure amusement with the scientific exploration of geometry. There are several examples in his notebooks of this effort, such as a spiral version of the *mazzocchio* which is cut to form a coil of intertwined rings (Fig. 12.10). This *mazzocchio* appears almost alive and full of movement, a geometrical monster out of another fantastic universe, forerunner of modern topology. It is a product of Leonardo's imagination and a good example of his incredible capability for visualizing complex objects. It is very likely that, with the materials available at the time, a real version of his topological creation could hardly be realized. Leonardo however was very satisfied with his innovation, and, in fact, he wrote in the notes accompanying the spiral *mazzocchio*: "Body born from the perspective of

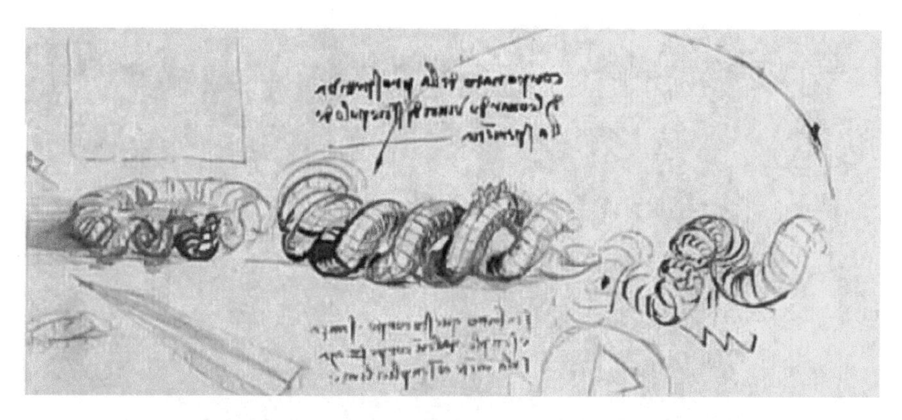

Fig. 12.10 Spiral version of *Mazzocchio*. Leonardo da Vinci. *Codex Atlanticus*, folio 520r, detail. BAM

Leonardo da Vinci, pupil of experiments. This body should be done without any example of any other body, but only with simple lines."[13]

There is yet another example of great interest which is also linked to the idea of the *mazzocchio* in Leonardo's notebooks, this time strictly related to a practical application—

the development of new mechanical devices. On folio 1106v of the *Codex Atlanticus*, Leonardo shows a project for a device to be used as part of a mechanical clock (Fig. 12.11). It appears to be an endless screw for the transmission of motion, and just below it is a second sketch showing how to seal the device. It is interesting to observe the correlation that exists between the form and function of the device.

Thus we see how the exploration of new geometrical forms and the capability to visualize and draw them became in Leonardo a source of inspiration and a practical tool for making innovative mechanical projects. This is the beginning of the modern field of industrial design.

Another impressive study of geometrical shapes from Leonardo can be found on folio 706r of the *Codex Atlanticus*. It is the skeleton of a new type of solid, where zigzagging interconnected open frames form a ring (Fig. 12.12).

Some of Leonardo's ideas for practical applications of the geometry of the *mazzocchio*, which was clearly a very fascinating shape for him, become rather clear if we examine another small drawing on folio 849v of the *Codex Atlanticus* (Fig. 12.13).

[13] Original text: "*Corpo nato della prospettiva di Leonardo Vinci, discepolo della sperienza. Sia fatto questo corpo sanza esempio d'alcun corpo, ma solamente con semplici linie.*"

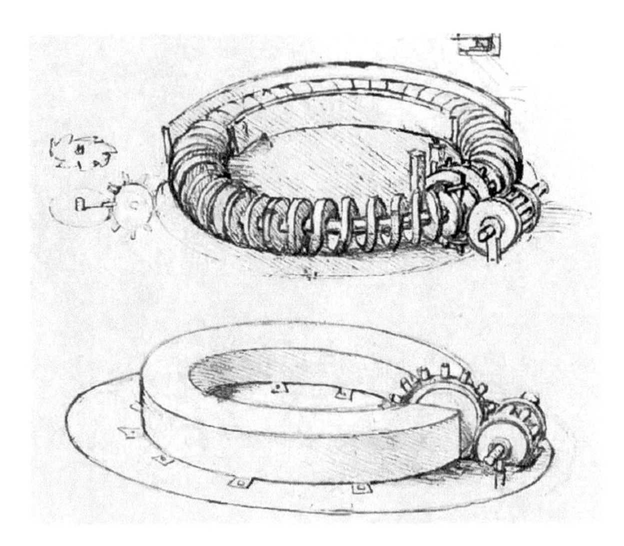

Fig. 12.11 Circular endless screw. Leonardo da Vinci. *Codex Atlanticus*, folio 1106v, detail. BAM

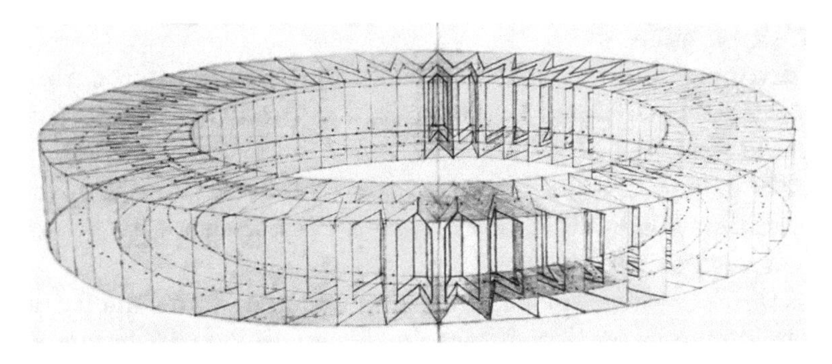

Fig. 12.12 Geometrical study for a figure similar to a *mazzocchio*. Leonardo da Vinci. *Codex Atlanticus*, folio 706r, detail. BAM

A small portion of a *mazzocchio* is used as the inspiration for an architectural project. It is just a simple sketch where Leonardo puts in relationship the circular shape of a *mazzocchio* with its regular "windows" and a classical Roman amphitheater, with its typical concentric structure of regularly alternating arcs at different levels.

Fig. 12.13 From *mazzocchio* to architecture. Leonardo da Vinci. *Codex Atlanticus*, folio 849v, detail. BAM

More Mazzocchios

Before we close this fanciful detour into the world of Renaissance headgear, it is worth pausing to note a strange connection which exists between this figure of Leonardo and an intarsia of Damiano Zambelli, whose wooden inlay of the Platonic solids in the wooden choir in the Basilica of San Domenico in Bologna, completed about 20 years after Leonardo's death, we saw in the previous chapter (Fig. 11.18). In another intarsia in this same work, we find Leonardo's spiked *mazzocchio* ruff (Fig. 12.14).

The presence in the intarsia of a solid which reproduces so closely Leonardo's drawing is quite intriguing. One wonders whether Zambelli had had an opportunity to see Leonardo's work or if they were simply both referring to something that, as in the case of *mazzocchio*, was part of the common artistic discourse. In either case, Zambelli's work attests to the ongoing Renaissance interest in complex geometric shapes and their depiction.

Fig. 12.14 Intarsia with geometrical objects, 1537. Fra Damiano Zambelli da Bergamo. Museum of Basilica of San Domenico, Bologna. (Permission is granted to copy, distribute, and/or modify this document under the terms of the GNU Free Documentation License. Courtesy of Georges Jansoone. http://commons.wikimedia.org/wiki/File:Antiche_mura_di_Prato,_Toscana,_Italia_03.jpg. (https://en.wikipedia.org/wiki/GNU_Free_Documentation_License), Version 1.2 or any later version published by the Free Software Foundation; with no Invariant Sections, no Front-Cover Texts, and no Back-Cover Texts. A copy of the license is included in the section entitled GNU Free Documentation License (https://commons.wikimedia.org/wiki/Commons:GNU_Free_Documentation_License,_version_1.2). This file is licensed under the Creative Commons Attribution-Share Alike 3.0 Unported (https://creativecommons.org/licenses/by-sa/3.0/deed.en) license)

13

The Measure of Time

Measuring time as accurately as possible has always represented a significant technological challenge. Rather simple time measurement systems had been developed since antiquity, such as hourglasses, sundials, and water clocks, but it was only with the advent of the first mechanical systems in the late Middle Ages that the way would be paved for the creation of what would become the accurate watches in use today.

Medieval monks developed some simple mechanical systems, known as "monastic clocks," that served to indicate the key moments of the day. The sound of a bell marked the hours for praying, working, and eating.[1] These monastic clocks began to spread in Europe beginning around the year 1000 and can be considered the forerunners of modern watches even if they did not have a dial and only marked the passage of time with the sound of a bell.

Around 1300, the first large clocks with a dial built and housed in towers began to become popular. They were driven by weights and had fairly good accuracy up to minutes but not seconds. These systems were based on a falling weight which drove a pinion gear system for transmission of motion and a brake system to control the speed of the fall, the so-called escape wheel, without which the thrust of the weight would be depleted too quickly. However, obtaining high-precision time measurements with smaller-scale mechanical clocks activated by weights remained a challenge. Because of the complex design of the system and the large dimensions necessary to operate with weights, reducing the size of the clock without loss of precision was a difficult task to accomplish.

[1] Dohrn-van Rossum G (1996) History of the Hour: Clocks and Modern Temporal Orders, University of Chicago Press.

P. Innocenzi, *The Innovators Behind Leonardo*, https://doi.org/10.1007/978-3-319-90449-8_13

It was only from the end of the fifteenth century that more sophisticated mechanical clocks began to appear. They were significantly more advanced compared to the small monastic clocks for daily use, thanks to the introduction of a new source of power for motion: the spring.

The Clocks in Villard de Honnecourt

The first documentation of a clock tower is very likely due to an old friend of ours, Villard de Honnecourt, who drew the image of a *house of a clock* (*cest li masons don orologe*) in his *Livre de portraiture* (Fig. 13.1, left image). The drawing does not show any clock but rather the infrastructure necessary to host it. Villard was clearly more interested in architecture than in mechanics. In any case, his drawing is a clear indication that clocks of large dimensions were already widely diffused in Europe during his time. Villard shows also on another page, together with other mechanical devices, a simple system for a mechanical sundial of sorts (Fig. 13.1, right image). The system, which is basically composed of a rope with two weights at the extremes, a wheel and an axle, should allow the contraption to move an angel (not in the drawing of Villard) according to the motion of Sun: "By this system an angel keeps his

Fig. 13.1 Clock tower (*left*) and system to turn the figure of an angel (not shown) in time with the Sun (*right*). Villard de Honnecourt. Livre de Portraiture (about 1230). Manuscript Fr.19093, Planche and XLIII. BNF

finger always pointing to the sun."[2] Whether or not this system, which is still quite naïve, can be taken as a part of a more complex clock design, it is, however, a clear indication that interest in the mechanics of the clock was rising and would, as we will see, in a short time reach a very sophisticated level.

A Brief History of the Escapement and Clocks

The most important device in mechanical clocks is the escapement, which serves to free up (escape) at regular intervals the driving force, and it is exactly the advanced development of the so-called *verge-foliot* system (Fig. 13.2) that allowed the construction of the first tower clocks. The movement of the escapement is controlled via a horizontal bar which bears two regulating weights at its ends, the so-called *foliot* in French.

The operation of the mechanical clock with a *verge-foliot* escapement is relatively simple. The movement is activated via a weight connected to a

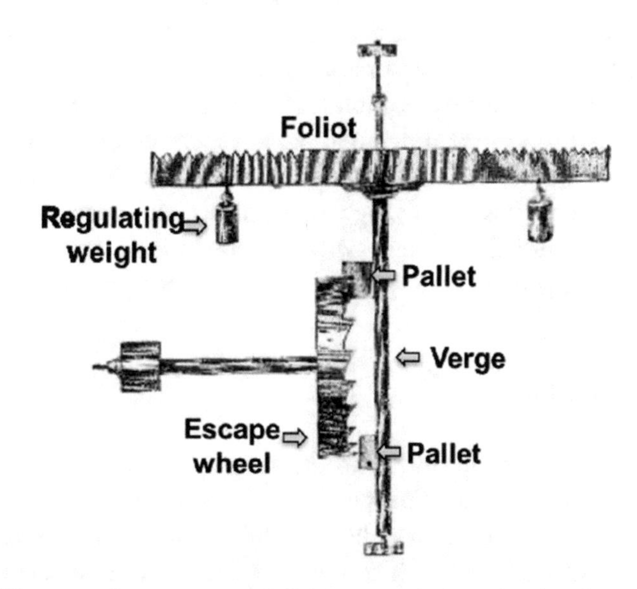

Fig. 13.2 Diagram of a verge and foliot escapement mechanism from the De Vick clock, built in 1379 in Paris by Henri De Vick. Pierre Dubois, Historie de l'Horlogerie, Paris, 1849, p. 22. (This work is in the public domain in its country of origin and other countries and areas where the copyright term is the author's life plus 70 years or less. https://commons.wikimedia.org/wiki/File:De_Vick_Clock_Verge_%26_Foliot.png)

[2] Original text: "Par chu om un angle tenir son doit ades vers le solel" (Par ce moyen fait-on qu'un ange tienne tohours son doigt tourneé vers le soleil).

crown gear wheel (the system is also known as a crown-wheel escapement). The wheel is slowly put in motion when it meets some small metallic plates (pallets) attached to a central shaft (verge rod). The two pallets are positioned with respect to the verge 90° apart from each other; in this way, the crown wheel teeth will be engaged by only one pallet at a time. The pallets block and release the crown teeth alternately, allowing a regular jerky movement. At the top of the verge, the attached *foliot* is swung now in one way and now in another when the pallets meet the teeth of the crown wheel; the weights allow adjusting the resistance to boost the crown. The accuracy of clocks with this type of escapement was not very high, and continuous regulation using, for example, sundials, was necessary. The main drawback and source of inaccuracy are the recoil effect generated by the momentum of the *foliot*. This produces a backward movement of the crown wheel before the weight is able to reverse the motion.

After the *verge-foliot*, which for several centuries in Europe was the main system for controlling the movement of clocks, many other escapement systems were designed. The *verge-foliot*, however, was definitely the most widely used system at the time of Leonardo, even though, as we have just seen, it had major limitations in terms of accuracy.

Leonardo developed a deep passion for clocks, and several projects not only of entire clocks but also of individual parts, especially escapements, can be found in his manuscripts. It is clear that the wide interest of Leonardo for mechanics was at the basis of his high attention also on mechanical clocks. As we will see, he studied different types of escapements and explored with enthusiasm the new possibilities offered by springs as a power source.

The Astrarium

Around the middle of the fourteenth century, when the large clock towers were widely spreading across Europe, there was realized in Padua, the university city close to Venice, a great little wonder: the *Astrarium* of Giovanni Dondi dall'Orologio,[3] a truly incomparable masterpiece of mechanics and ingenuity (Fig. 13.3). Begun in 1348 and completed 16 years later, the *Astrarium*[4] was an astronomical clock, with seven faces and a very complex mechanical system comprising 107 gears, capable of showing the positions of

[3] Pesenti T (1992) Entry on Giovanni Dondi dall'Orologio. In: Dizionario Biografico degli Italiani. Enciclopedia Treccani. Volume 41.

[4] Bedini SA, Maddison FR (1966) Mechanical Universe: The Astrarium of Giovanni de' Dondi. Transactions of the American Philosophical Society. New Series, 56:1–69.

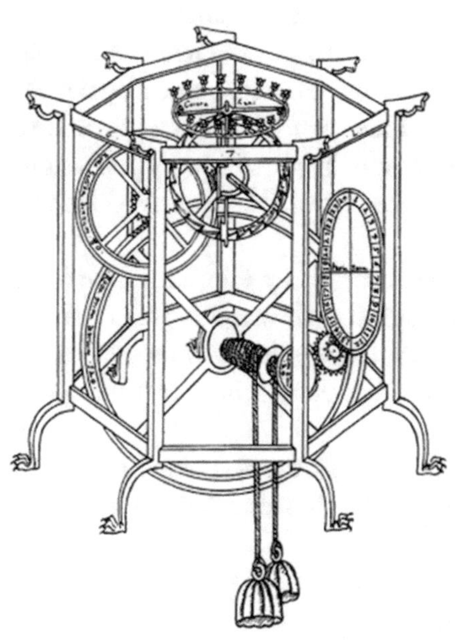

Fig. 13.3 Design of the bottom part of the astronomical clock of Giovanni Dondi. Tractatus astrarii (copy of the original manuscript dated 1461), Manuscript Laud. Misc. 620, folio 10. Bodleian Library, Oxford. (*left*). (This work is in the public domain in its country of origin and other countries and areas where the copyright term is the author's life plus 70 years or less. http://commons.wikimedia.org/wiki/File:Giovanni_Di_Dondi_clock.png). Reconstruction of the Astrarium by Luigi Pippa, 1963. Museo della Scienza e Tecnica, Milan (*right*). (This file is licensed under the Creative Commons Attribution-Share Alike 4.0 International. (https://creativecommons.org/licenses/by-sa/4.0/deed.en) license. Courtesy of Catalogo collezioni Museo nazionale della scienza e della tecnologia Leonardo da Vinci, Milano. http://www.museoscienza.org/dipartimenti/catalogo_collezioni/scheda_oggetto.asp?idk_in=ST040-00103)

the Sun, Moon, and the five planets then known (Venus, Saturn, Mercury, Jupiter, and Mars), as well as marking religious holidays. It was almost a meter tall and was built entirely by hand without the use of screws to secure the individual gears that had been molded into a variety of shapes, such as elliptical wheels, the better to simulate the irregular movement of the planets.

Giovanni Dondi (1330–1388) was actually the son of another clockmaker, Jacopo de Dondi[5] (1293–1359), who produced the astronomical clock of the clock tower of Padua with such great expertise that he was subsequently authorized to add "dall'Orologio (from clock)" to his family name. Jacopo

[5] Pesenti T (1992) Entry on Iacopo Dondi dall'Orologio. In: Dizionario Biografico degli Italiani. Enciclopedia Treccani. Volume 41

Dondi's clock was installed in 1344 but was severely damaged only a few decades later, in 1390. The clock was later rebuilt, and the new one, finished only in 1436, is still well preserved in Padua and contains some original parts of Dondi's mechanism. We see that the interest in clocks was, therefore, a family passion, even if father and son were actually physicians by profession as well as both being professors at the University of Padua in medicine and astrology.

The construction of the *Astrarium* was recorded by Giovanni Dondi with remarkable accuracy in a treatise whose original manuscript is preserved at the Biblioteca Capitolare of Padua (Manuscript D39), the *Tractatus astrarii.* This text represents one of the first extant detailed descriptions of a mechanical clock.

In 1381, 17 years after its completion, Giovanni Dondi offered the *Astrarium* to the Duke of Milan, Gian Galeazzo Visconti, who put it on display in the library of the castle of Pavia.

The original *Astrarium* has been lost. However, thanks to the detailed description of Giovanni Dondi, several beautiful and quite faithful reconstructions of the astronomical clock were recently realized,[6] one of which can be seen in Fig. 13.3. In any case, the *Astrarium* remained a fine show at least until 1495, and at this point it is natural to ask if Leonardo himself, so attentive and curious to any mechanical device, did not see and himself study this mechanical wonder.

The answer is on folio 92v of *Manuscript L*, where we can find a drawing in which Leonardo reproduced a mechanism for an astronomical clock (Fig. 13.4, left). A short explanatory note indicates quite clearly its purpose and operation: "Venus-Sun-ab-c. The carriage a takes the second wheel and the carriage c the first (wheel) and the carriage b the third (wheel)."[7] The system thus serves to regulate the movements depending on the motion of Venus and the Sun's position. It is undoubtedly part of an astronomical clock. Leonardo's drawing has 3 wheels with 110 teeth, and in addition there are 3 smaller wheels, the *carrelli (carriages)*, which move around each of these. The drawing only shows the inside of the system, perhaps the most interesting and original to Leonardo. However, this is not the only part of the *Astrarium* in which Leonardo seems to have been interested: In the same *Manuscript L,* folio 93v,

[6] See, for instance, the working model at the Observatoire de Paris inv. 404, anc. 20–36. *Giovanni de' Dondi's astrarium* reconstruction by A. Segonds, E. Poulle, J.P. Verdet, Paris. Another model realized in 1963 by the clockmaker Luigi Pippa is the Astrarium conserved at the Museo della Scienza e della Tecnica of Milan.

[7] Original text: *Venus - Sole - ab - c. Il carrello a piglia la seconda rota e 'l carrello c e la prima e 'l carrello b e la terza.*

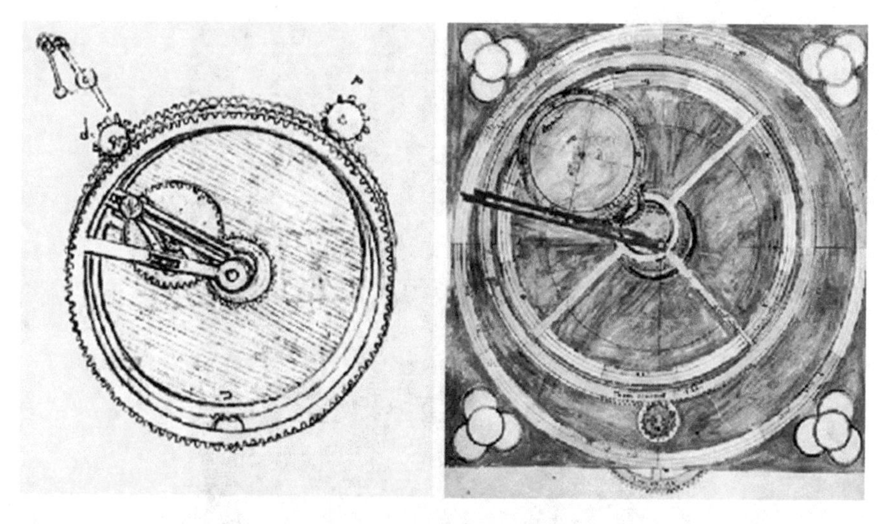

Fig. 13.4 Mechanism for an astronomical clock showing the dial of Venus and the Sun. Leonardo da Vinci. Manuscript L, folio 92v (detail), left. BIF. The dial of Venus in the Tractatus astrarii. Giovanni Dondi dell'Orologio. Ms. D.39, folio 12v. Biblioteca Capitolare, Padua, right. (This work is in the public domain in its country of origin and other countries and areas where the copyright term is the author's life plus 70 years or less. The original image has been modified by taking from the page the bottom part and by reducing the noise. https://upload.wikimedia.org/wikipedia/commons/1/15/Giovanni_Dondi_dell%27Orologio_dial_of_Venus.jpg)

in fact, there is a similar drawing, though drawn very lightly and just barely legible, which has been interpreted as a mechanism for the motion of Mars. This, coupled with Leonardo's well-documented frequent visits to the library in Padua, means that it is quite plausible that he had seen Dondi's *Astrarium* and that it inspired him in his own studies for an astronomical clock.

Another Wonder

There is, however, another interesting trace worth following: in fact, in Florence a century and a half later, another clockmaker, Lorenzo della Volpaia[8] (1446–1512), was able to create a masterpiece that could rival Dondi's clock. Commissioned by Lorenzo il Magnifico, Lorenzo della Volpaia's work, the *Orologio dei Pianeti* ("the clock of the planets"),[9] introduced an important

[8] Pagliara PN (1989) Entry on Lorenzo Della Volpaia. In: Dizionario Biografico degli Italiani. Enciclopedia Treccani. Volume 37.

[9] Brusa G (1994) L'orologio dei pianeti di Lorenzo della Volpaia. Nuncius, *IX*:645–69. https://doi.org/10.1163/182539184X00991.

Fig. 13.5 Working replica realized in 1994 of the planetary clock designed by Lorenzo della Volpaia and completed in 1510. Museo Galileo. Florence, Italy

breakthrough compared to Dondi's *Astrarium*: in Lorenzo della Volpaia's clock, it was possible to observe in a single dial, instead of the seven of the *Astrarium*, all of the planetary motions and the other measures of time (Fig. 13.5). Della Volpaia's clock was able to show the movements of Mercury, Venus, Mars, Jupiter, and Saturn, the phases of the Moon, and the average motion and the position of the Sun in addition to indicating the hour, the day, and the month. The clock was not designed, however, with the main purpose to keep the measure of the time but rather to show the exact positions of the astral bodies with respect to the Earth. This could allow an effective determination of the astrological influences.

Lorenzo della Volpaia was yet another multifaceted protagonist of the Renaissance. He is mentioned as an architect, goldsmith, mathematician, and clockmaker, but unfortunately, no trace remains of his original astronomical clock. It was brought in 1510 to the Palazzo Vecchio in Florence, to the *Sala dei Gigli* (Hall of Lilies), but after a while it stopped working and was no longer restored. It was definitely lost in the seventeenth century. The sons of Lorenzo della Volpaia, however, gathered diligently such detailed and precise notes of the clock's construction that it has been possible to realize modern copies. These are quite faithful and reproduce well this mechanical masterpiece of the Renaissance.[10]

Lorenzo della Volpaia and his heir Bernardo both had a great passion for mechanics, and several records of projects of machines remain—not only for clocks but also for winches, mills, and hydraulic saws, among others. Among the drawings in the *Codice Marciano* in Venice compiled by Lorenzo della Volpaia's son Bernardo,[11] there are at least three studies that are clearly based on models of Leonardo, who is explicitly mentioned.[12,13] So there is definitely a direct link between the two.[14] Indeed, they even met in person in Florence in 1504, when they participated in a meeting to decide the placement of Donatello's sculpture of David.

The Chiaravalle Clock Tower

The evidence of Leonardo's direct knowledge of mechanical clocks doesn't stop here, however. In fact, there was yet another astronomical clock tower, also built with extraordinary mechanical expertise, the clock of the Abbazia Cistercense of Chiaravalle near Milan. No traces have been found of this clock up to recent times, besides a vague historical memory, and the clock currently at the site only dates to 1862. The most important information about this clock is actually in the pages of Leonardo's *Codex Atlanticus*. On folio 1111r there is drawn a clock mechanism which is quite intriguing for its technical characteristics. The sketch is accompanied by Leonardo's note: "Clock Tower of C[h]iaravalle which shows the Moon, Sun, hours and minutes."[15] The mechanism is original, and its peculiarity is due to the orthogonal position of the shafts for transmitting motion, in contrast to the standard parallel placement (Fig. 13.6).

[10] Codice Marciano 5363 (Venice, Bibl. Naz. Marciana, Mss. Ital., cl. IV, 41 [= 5363]).

[11] The codex includes 96 numbered pages and was written, according to Carlo Pedretti, between 1520 and 1524 by Bernardo della Volpaia, aside from some pages of his brother Girolamo Camillo.

[12] Pedretti C (1957) Studi Vinciani. Librairie E. Droz, Geneve.

[13] The references to Leonardo regard a hydraulic machine, how to split a line, and a geometry drawing.

[14] Tagliagamba S (2011) Lorenzo Della Volpaia e Leonardo da Vinci. Scambi e ricambi: il caso del contatore d'acqua per Bernardo Rucellai. Microstoria XII:35–38.

[15] Original text: *Oriolo della torre di C[h]iaravalle il quale mostra luna, sole, ore e minuti.*

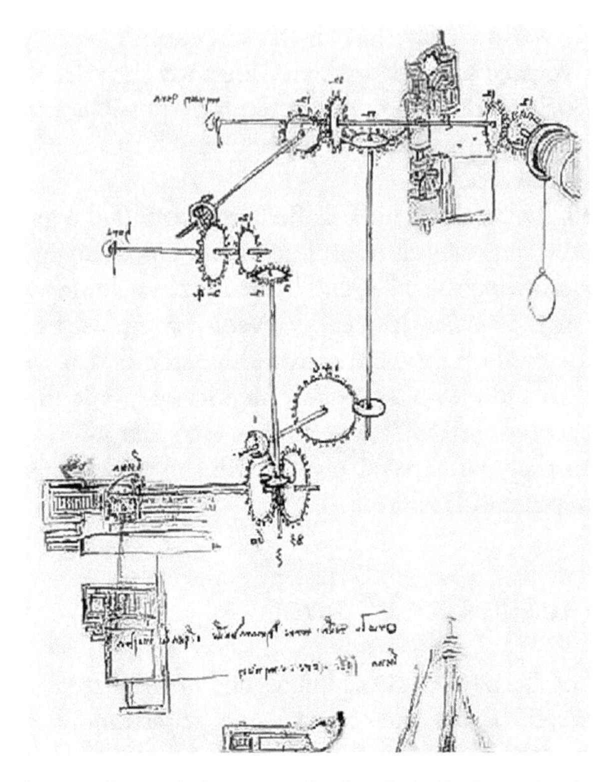

Fig. 13.6 Mechanism for a clock. Leonardo da Vinci. *Codex Atlanticus*, folio 1111v. BAM

Initially, Leonardo's description suggested to scholars the hypothesis that he himself had designed a clock for the Abbey of Chiaravalle. Given the sketchy and imprecise nature of the drawing, it was supposed that this was a rough draft to which he intended to return at a later date.

Very recently, however, the discovery of a manuscript in the Chiaravalle Abbey library[16] describing in sufficient detail the story and mechanism of the clock[17,18] has made it clear that, on the contrary, this drawing was a study of an existing mechanism in the Chiaravalle clock tower. The text reports that the astronomical clock was built in 1368 and was composed of three dials: one for minutes, one for hours, and a third—the biggest—to mark monthly the position of the Sun and the related zodiac signs as well as the lunar rotation: a very complex mechanism with an original structure, "And it is made in

[16] The manuscript was written by Benedetto de Blachis, a monk of the abbey.

[17] Addomine M (2014) *Leonardo e l'orologio di Chiaravalle*. La Clessidra.

[18] Addomine M (2007) *Ancora sull'orologio astronomico di Chiaravalle*. La Voce di Hora, n. 23.

the shape of a cross not as the others which are made out of four columns,"[19] as is also seen from the drawing by Leonardo. This is probably the first mechanical clock which is documented as having a dial for minutes.

There are also other folios—10v, 11r, and 11 s in *Codex of Madrid I*—where it is possible to find some studies of the mechanism of Chiaravalle's clock. These probably contain some variations introduced by Leonardo himself. The clock tracked the different phases of the Moon through a rotating hollow globe that contained a partially gold-plated second globe. This is the mechanism that probably attracted the attention of Leonardo. On folio 11v of the *Codex of Madrid* I, Leonardo explicitly refers to the clock of the Abbey, writing "poles in poles as at Chiaravalle,"[20] namely, axles to axles, to describe this system of concentric spheres for lunar motion. The inclusion of these and many of the other sketches that we will encounter in this chapter in the *Codex of Madrid I* is not surprising, given that this is the manuscript in which can be found Leonardo's more systematic and ordered studies on mechanics.

A very nice reconstruction with detailed description of the operation of Chiaravalle's clock based on Leonardo's drawings and the Abbey library manuscript can be found in the article of Addomine, "Leonardo e l'orologio di Chiaravalle" (*Leonardo and the clock of Chiaravalle*).

The great clock was still in place until at least the beginning of the 1800s, but was probably damaged during the Napoleonic wars and eventually replaced by a new clock in 1862. It is likely thanks to Leonardo that we owe its rediscovery, another piece in the history of mechanics that shows the great wealth of knowledge already widely spread in Europe from 1300—much earlier and much more sophisticated than the general perception.

So we see that Leonardo had direct knowledge of Dondi's *Astrarium* and Della Volpaia's *Orologio dei Pianeti* as well as having studied in detail the astronomical clock of the Abbey of Chiaravalle. This was more than enough to whet his imagination and push him to grapple enthusiastically with the study of mechanical clocks.

The Sienese Engineers' Clocks

However, before going to see in more detail some of Leonardo's ideas on the mechanics of clocks, it is interesting to wonder if the two Sienese engineers, Taccola and Francesco di Giorgio, also devoted their attention to mechanical

[19] Original text: *Et è fabricato in forma di croce no(n) come son li altri quali son fatti su quatro colo(n)ne.*

[20] Original text: *Poli in poli come a Chiaravalle.*

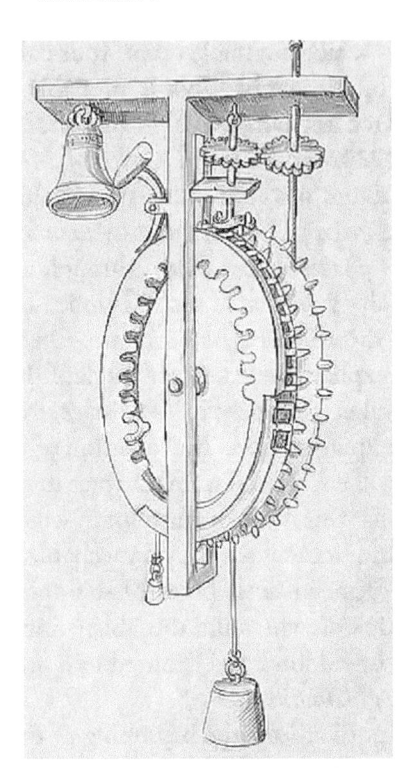

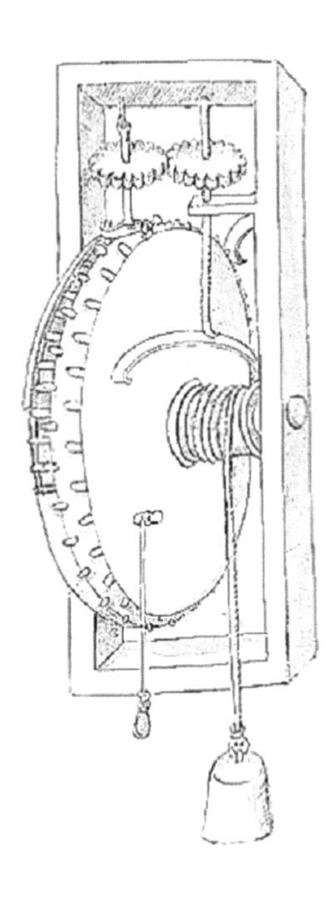

Fig. 13.7 Clocks by Taccola and Francesco di Giorgio. Anonymous. Manuscript Palatino 767, folios 43 and 44. BNCF (late fifteenth century)

clocks. The answer in this case is quite obvious: indeed, as we have seen again and again, what connects them to Leonardo is the shared passion for machines and mechanics. Clocks, with their complex mechanisms, undoubtedly represented an irresistible attraction for all of them.

In the Florentine *Manuscript Palatino 767*, which shows projects from both Taccola and Francesco di Giorgio, there are two drawings of simple clocks, mostly derived from the monastic models with their basic mechanics and with time marked by a bell (Fig. 13.7). They are still unsophisticated systems, with the descent of the weight controlled by a simple escapement. As we will see, Leonardo would develop the mechanics of clocks in a more innovative way through a detailed analysis of different mechanical solutions, especially with regard to the mechanics of the escapements.

More interesting is the drawing on folio 156r of the *Manuscript Additional 34113* in the British Library in London, where a mechanical clock with a small quadrant for marking the hours is shown (Fig. 13.8). The interesting

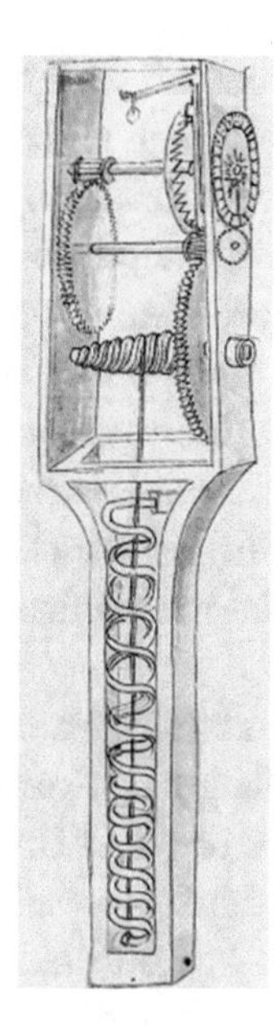

Fig. 13.8 Mechanical model of an early alarm clock design driven by a coil spring. Anonymous Sienese engineer (possibly Taccola?). Additional Manuscript 34113, folio 156r. BL

part is that the mechanism is driven by a coil spring loaded via a crank. These studies are among the first to explore the possibility of an alternative source of motion with respect to the weight system used until then, marking the beginning of a new season of mechanical clocks. The attribution of this drawing is unclear, but it has been suggested that it may be from Taccola.

The application of springs as a source of motion in clocks represented a clear advantage with respect to systems driven by weights and rope. The spring, in fact, is able to release the energy in any position, allowing for a more flexible design. At the same time, the spring could also work in movement, opening the door to the possibility of fabricating portable clocks and watches. The spring system presents, however, a main drawback which is the uneven release of the force with the unwinding of the spring.

An Unexpected Clockmaker

The first documented use of springs to move a clock, however, is, perhaps surprisingly, due to Filippo Brunelleschi, who, among other things, was also a skilled clockmaker. Giorgio Vasari in his *Vita di Filippo Brunelleschi scultore ed architetto* (Life of Filippo Brunelleschi, sculptor and architect) described with enthusiasm Brunelleschi's capabilities in this regard: "… and he realized by his own hand some very good and very beautiful clocks." Vasari goes on to write: "Having engaged with pleasure in the past in making some clocks and alarm clocks in which there are many different generations of springs, which are multiplied by great numbers of ingenious devices … this was extremely helpful to him in being able to imagine various machines used for hauling and hoisting and drawing." This comment is extremely interesting because it puts in relationship Brunelleschi's abilities as clockmaker with the skill he acquired in designing and realizing special mechanical devices to be used in his construction sites.

Some traces of his work as clockmaker are still to be found in the vicinity of Florence, especially in the villages of Scarperia and Castelfiorentino. The construction of the clock for the tower of the Palazzo dei Vicari in Scarperia by Brunelleschi in 1445, in particular, is well documented, and a part of the original clock still survives and is exhibited in the palace (Fig. 13.9). The clock is formed of two separate mechanisms, one for the time and another for the tolling of the bell. Each of these parts has three wheels. The escapement, originally a *verge-foliot*, was replaced by a pendulum during a restoration. Brunelleschi's clock worked well for several centuries before being replaced.

Fig. 13.9 Part of Brunelleschi's clock for the tower of Palazzo dei Capitani in Scarperia (Florence), right. Detail of the clock's mechanism (*right*)

Leonardo's Studies of Clocks

Leonardo was also rather fast to catch the novelty represented by the use of springs for different types of mechanical applications and devoted specific studies to the various possible models of springs and their manufacturing techniques.[21] Among the many examples, there are mechanisms and levers for automatic ignition of firearms (folio 18v, *Codex of Madrid I*); spring locks (folios 50r and 99v, *Codex of Madrid I*); a tension spring (folio 13v, *Codex of Madrid I*); and springs with progressive charge (folio 14r, *Codex of Madrid I*), a really large variety of devices in which Leonardo introduced a wide range of innovative ideas. Meanwhile he began to worry about the technology necessary to manufacture high quality springs, a challenge to which the project on folio 14v of *Codex of Madrid I* is dedicated (Fig. 13.10).

Leonardo's description of the process is particularly interesting: "This tool is to fabricate a spring for a clock. And should be done in this way, that is a piece of steel should be done in the form as shown in **n b**. and such instrument should be put in the way you see. And with the screw m should be bonded to the steel c, in a way that will be strong. ..."[22]

A spring-driven clock model designed by Leonardo can be found on folio 863r of the *Codex Atlanticus* (Fig. 13.11). It is a little sketch at the bottom of a page where the project for a large flying machine attracts most of the attention. At the center of the small sketch, a short note reads: "Foundation of motion."

The clock moves thanks to a pair of leaf springs, which are springs consisting of flexible metal foils. If strained downward, the two springs stretch, and this creates a controlled release of the force that sets the whole system in motion. The spring drives a pair of weights placed on an axle and, through a

[21] Reti L, Bedini SA (1974) Horology. In: The unknown Leonardo. Reti L (ed.). Hutchinson, London.

[22] *Codex Atlanticus*, folio 14v. Original text: "Questo strumento è da ffare una molla da dare moto a uno orilogio. E ffassi in questo modo, cioà sia fatto un pezzo d'acciaro nella forma che si dimostra n b. E ssia messo in tale strumento nel modo che tu vedi. E colla vite m sia serato coll'acciaro c, in modo forte. E poi sia essa verga d'acciaro tirata per forza di vite, a panca, nel modo che ssi tira il ferro trafilato. E ppoi sia ricotta e rimessa nella trafila c, più stretta alquanto che prima, e di novo ritrafilata. E così, senpre d'una grossa carta. Di poi l'avolta, e mettu nella sua cassa nel modo che ssai. Di poi la tenpera e lla rinvieni al coloro azuro nel pi[o]nbo fonduto, o voi 'n fornello, a uso dove si lavora il vetro, acciò rinvenga equalmente per tutto. E poi adopera come ti accade. E perché l'acciaro è dove duro e dove tenero, si debbe ciascun pezo di verga ricalcare per testa, e ssaldarle in croce, quadrate che ll'hai a uso di dado. E bollile bene insieme, e poi distendi, e tirala come detto per trafila. Po' l'avolta con un filo di ferro in mezo, tra volta e volta, aciò ch'e sua spati fieno equali. Di poi tenpera e rinvieni nel pionbo fonduto, aciò che vada per tutto."

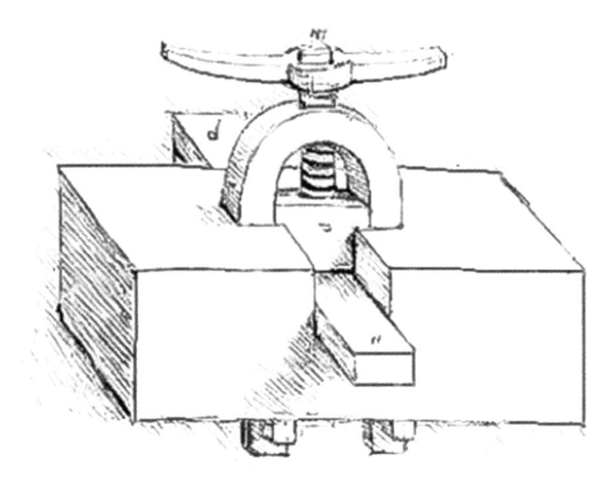

Fig. 13.10 Spring-making machine. Leonardo da Vinci. Codex of Madrid I, folio 14v (detail). BNE. (ca. 1493–1497)

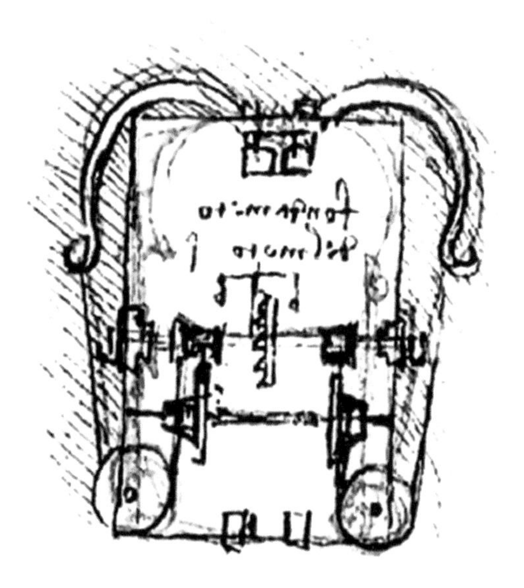

Fig. 13.11 Design of a clock with a leaf spring mechanism. Leonardo da Vinci. *Codex Atlanticus*, folio 863r (detail). BAM

system of gearwheels, transmits the movement to a verge escapement.[23] In the middle of the drawing, it is possible to observe the double weights of the *foliot*, the crown wheel, and the rod verge of the escapement system. In its

[23] It is possible to admire a beautiful and working reproduction of Leonardo's clock at the Istituto e Museo di Storia della Scienza in Florence.

essence, this mechanism looks very similar to the one drawn by Leonardo for the *automobile* described in Fig. 7.16, although in that case the real engine system was formed not by leaf springs but by spiral-like springs.

The interest in the new spring-powered devices for the clocks' gears is evident in Leonardo. This system could have, in fact, a universal use, and he quickly envisaged the possibility of applying it even to move more sophisticated mechanisms, such as robots and automata. The use of springs for putting mechanical clocks in motion presents, however, a problem with respect to the weight system used up to that point. The advantage of the system is evident, as it allows the fabrication of much smaller clocks; however the force exerted by the spring decreases with time, while that generated by the weight remains constant. This problem represented a big technical limitation for the construction of accurate spring clocks. Leonardo understood the problem and was likely one of the first to find the solution which would eventually be universally adopted: the *fusee* (Fig. 13.12). The *fusee* consists of a cone-shaped pulley whose

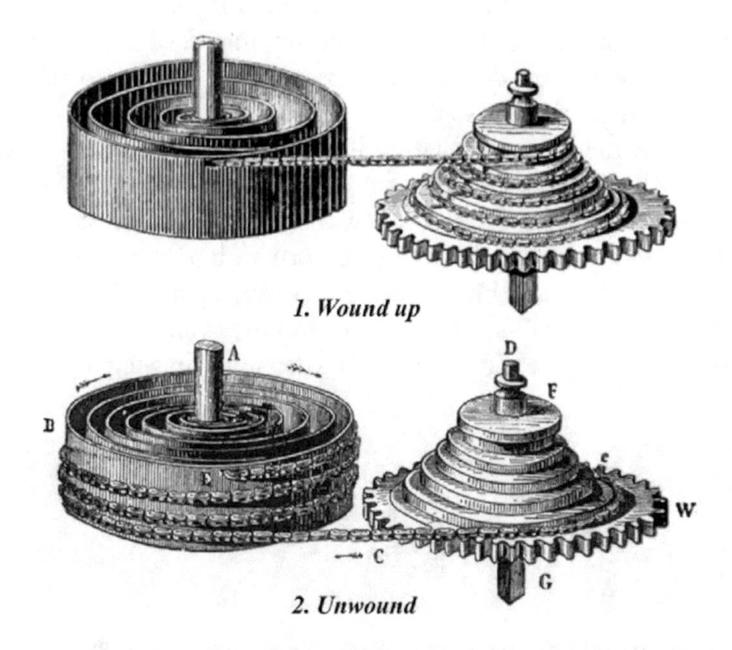

Fig. 13.12 Fusee for springs. (A) mainspring arbor, (B) barrel, (C) chain, (E) attachment of chain to barrel, (e) attachment of chain to fusee, (F) fusee, (G) winding arbor, (W) output gear. Dionysius Lardner, The Museum of Science and Art, Vol. 6, Walton & Maberly, London, 1855, Figs. 15 and 14, p. 24–25. (This work is in the public domain in its country of origin and other countries and areas where the copyright term is the author's life plus 70 years or less. The original image has been modified to reduce the noise. https://upload.wikimedia.org/wikipedia/commons/5/5c/Fusee.png)

surface has been modified to have a helical groove with a wound cord, or small chain, connected to a main spring barrel. This system allows one to adjust the force exerted by the spring and to keep it as constant as possible to compensate in this manner for the uneven pull as it runs down. The capability of the *fusee* to be an effective spring compensator is due to its conical shape, with the radius of the groves gradually increasing from the top to the bottom. When the main spring unwinds, the released force also decreases over time. The weaker pull of the spring is, however, compensated for by the larger radius of the bottom part of the *fusee*, which, in this way, keeps the driving torque constant. This simple and elegant system, whose real inventor is unknown, remained in use for quite a long time, from the 15th up until the beginning of the twentieth century, before being replaced by more precise spring and escapement mechanisms.

Leonardo's intuition about the potential of the *fusee* is really impressive. The proof is found on folio 85r of the *Codex of Madrid I*, where he listed different types of springs and finally the *fuse* (Fig. 13.13). This appears virtually identical as a conception to the one that would become a common system for watches. It has to be remembered, however, that the *fusee* was already in use much before Leonardo, and he discovered nothing new. He was, however, able to catch the possible developments of the system and tried to integrate the *fusee* in more precise and sophisticated devices.

Just beneath the figure, Leonardo described very clearly the problem of using springs for generating movement in mechanical systems: "If the spring has the same dimension, for sure when is losing the charge, its power will always decrease. Because of this we can say that such power has the nature of a pyramid, because it starts big and decreases to end up in nothing. It is necessary to face such pyramid-like power, with another pyramidal power, with an opposite decrease of power".[24] Comparing Leonardo's drawing with a *fusee* for

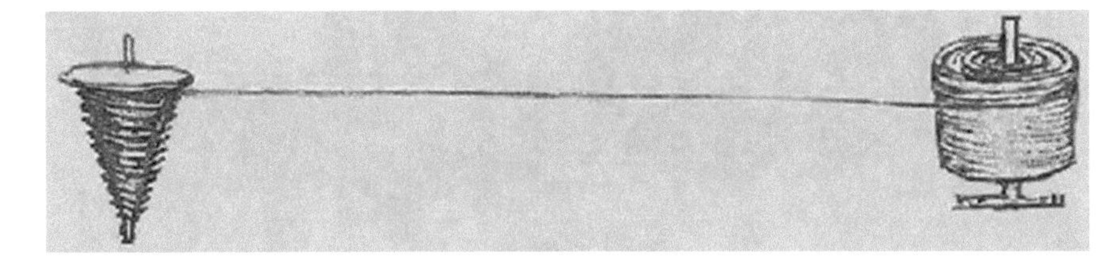

Fig. 13.13 Fusee for clocks. Leonardo da Vinci. Codex Madrid I, folio 85r, (detail). BNE. (ca. 1493–1497)

[24] Original text: "Se lla molla fia d'equal grosseza, cierto nel suo disscaricarsi, la sua potentia senpre va diminuendo. Onde diremo tal potentia essere di natura di piramide, perché comincia grande e diminuisscie i' niente. Onde è neciessario a tal potentia piramidale contradire, over contrastare con potentia piramidale, con contraria diminutione di resistentia."

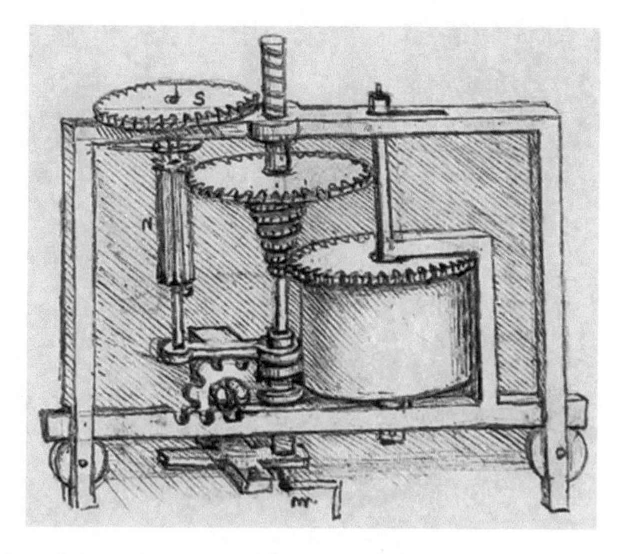

Fig. 13.14 Device to regulate the force released by a spring. Leonardo da Vinci. Folio 14r, (detail). Codex of Madrid I. BNE. (ca. 1493–1497)

watches is really impressive, and the similarity of the two concepts is immediately apparent.

Leonardo understood, therefore, very well the potentiality of the *fusee* system and used it to design some advanced devices for controlling the release of the spring force. An example of such a device can be found on folio 14r of the *Codex of Madrid I* (Fig. 13.14). It consists of a spring inside a cylindrical drum with a crown wheel at the top which drives the movement of a winding cone-shaped gear. The top part of the vertical shaft is a worm screw which, in turn, transmits the movement to a toothed wheel that drives a rack. The gear system was designed to have a controlled horizontal shift to the right of the bar where the cylindrical drum is attached. This is, in fact, necessary to mesh the crown wheel of the drum with the conical toothed wheel.

Starting from this intuition, Leonardo developed a series of very innovative and brilliant systems for spring-based mechanical devices. One such example can be found on folio 4r of the *Codex of Madrid I*, where Leonardo designed a spring system to be used for a clock. At the center of the drawing, which has a strong visual impact, he wrote: "Here there is the spring that moves (the system),"[25] indicating that a coiled spring was to be located inside the central cylindrical drum (Fig. 13.15).

[25] Original text: "Qui sta la molla che move."

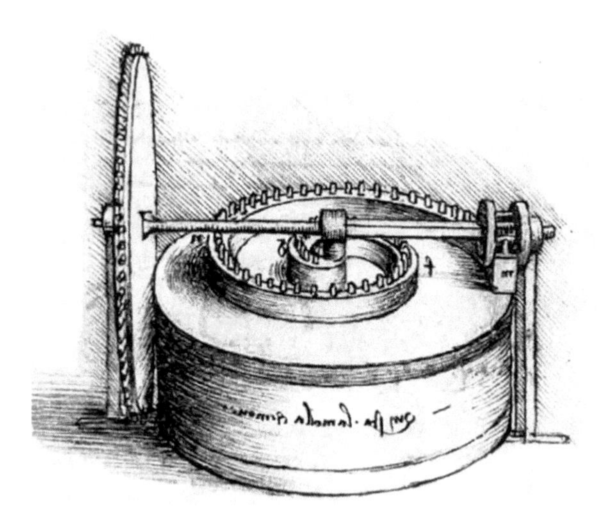

Fig. 13.15 Spring system for a clock. Leonardo da Vinci. Codex of Madrid I, folio 4r, (detail). BNE. (ca. 1493–1497)

The system is simple but shows a brilliant solution. The spring is stuck in the middle of the drum and attached with a rod to an external axis. When the spring begins to unwind, it would initiate a clockwise motion of the case, which, however, would move following the path directed by the helical gear. In turn, the movement would be transmitted to the external gear through the shaft. This system allows for a controlled unwinding of the spring, a variation on the theme of the *fusee*.

This idea underwent a number of interesting variations in Leonardo's notebooks. A different version of this model is to be found on folio 45r of the *Codex of Madrid I* (Fig. 13.16, *left image*). Leonardo's notes only specify that the spring drum should be made of *tempered spring* (i.e., of hardened steel). No other details for manufacturing the device are included. An intermediate version of the model can also be found on folio 16r (Fig. 13.16, *right image*).

The system shown in Fig. 13.16 is a bit more complex than the previous one and increases the unwinding time of the spring that puts in motion the drum, which, this time, is not fixed. As soon as the drum is moved by the spring, a grooved spool hinged to a vertical external rod begins to climb the spiral-toothed staircase-like path, moving, in turn, the gear toward the top. This system is able to do about two and a half laps with respect to the previous one.

Leonardo studied clock mechanisms in great detail. In particular, as we have seen, he was clearly attracted by the problem of controlling the spring

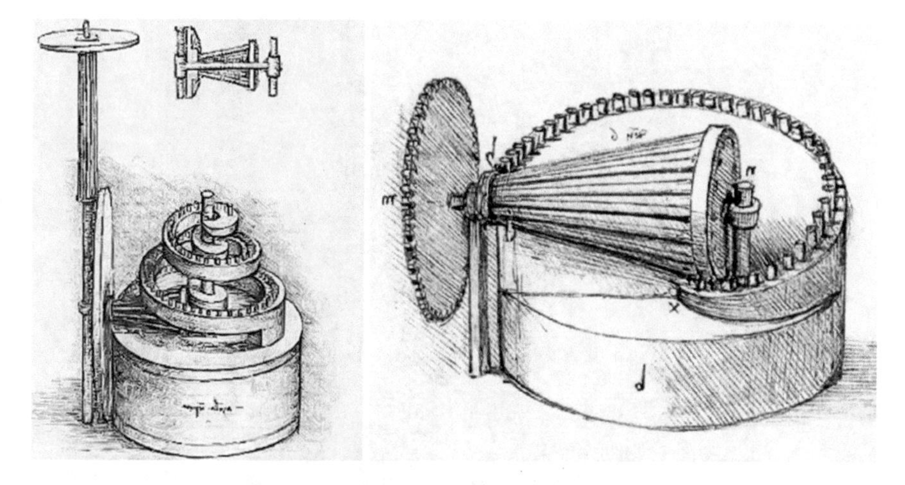

Fig. 13.16 Micro-motor for clock (*left image*). Leonardo da Vinci. Codex of Madrid I, folio 45r. BNE. Study for controlled unwinding of the power of a spring (*right image*). Codex of Madrid I, folio 16r. BNE. (ca. 1493–1497)

force release. He also explored different solutions for other parts of a clock. Again in the *Codex of Madrid I*, several studies for clock parts are shown. For example, on folio 15r, there is another interesting sketch of a device, probably a striking mechanism for a clock and detailed instructions for using it (Fig. 13.17).

Other examples are on folio 13r of the *Codex Forster II* and on folio 81r of the *Manuscript M*, where can be found a helical cam for clocks.

Leonardo also showed numerous examples of possible mechanisms for clocks with escapements and moved by a weight, one of which is on folio 964r of the *Codex Atlanticus*, where the scheme for of a "*tempo d'orilogio* (literally 'time of a watch,' i.e., escapement)," is reported in detail (Fig. 13.18). As Leonardo's term indicates, the escapement precisely regulates the "time" to be measured.

Leonardo, even if he was also exploring different and innovative solutions for the escapement, continued to study how to improve the traditional system by introducing small changes in the design of the *verge-foliot* escapement.

An example of these projects is on folio 115v of the *Codex of Madrid I*, where Leonardo represented a *verge-foliot* escapement adding also a long and exhaustive description of the system's operation (Fig. 13.19): "…It is custom to oppose the violent motions of the wheels of the clock by their counter-weights with certain devices called escapements (tempo in Italian), as they keep the timing of the wheels which move it. They regulate the motion

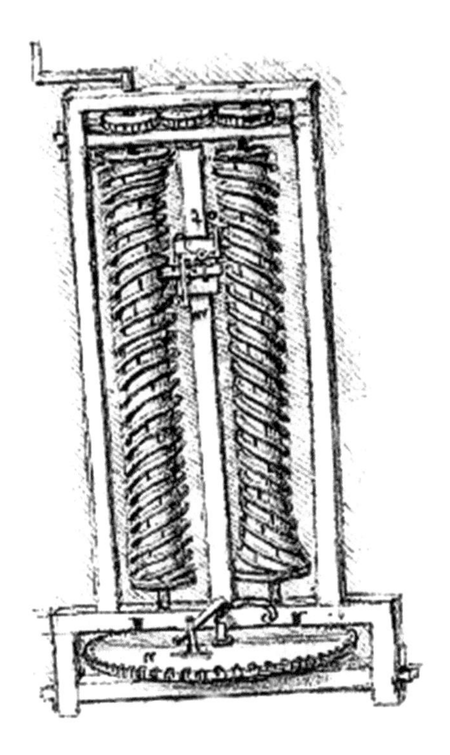

Fig. 13.17 Striking mechanism for clock. Leonardo da Vinci. Codex of Madrid I, folio 15r, (detail). BNE

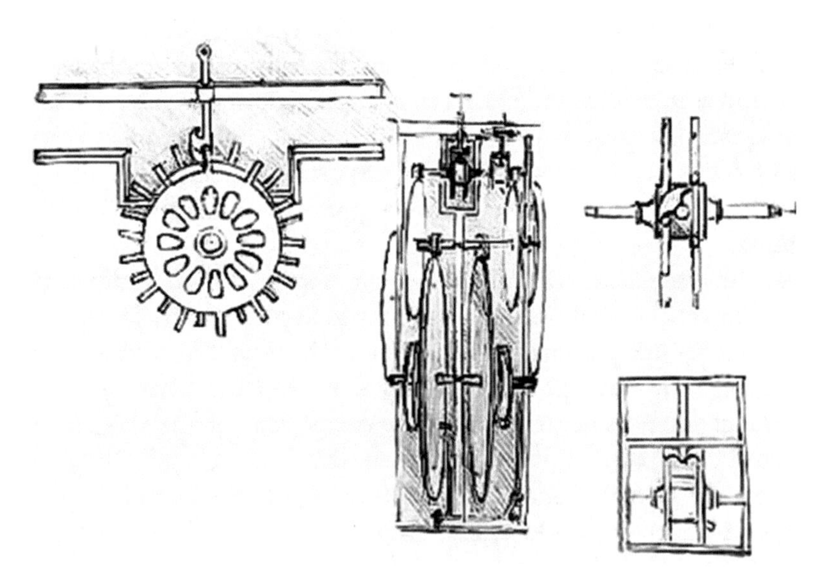

Fig. 13.18 Clock escapement mechanism. Leonardo da Vinci. Folio 964r. *Codex Atlanticus*. BAM

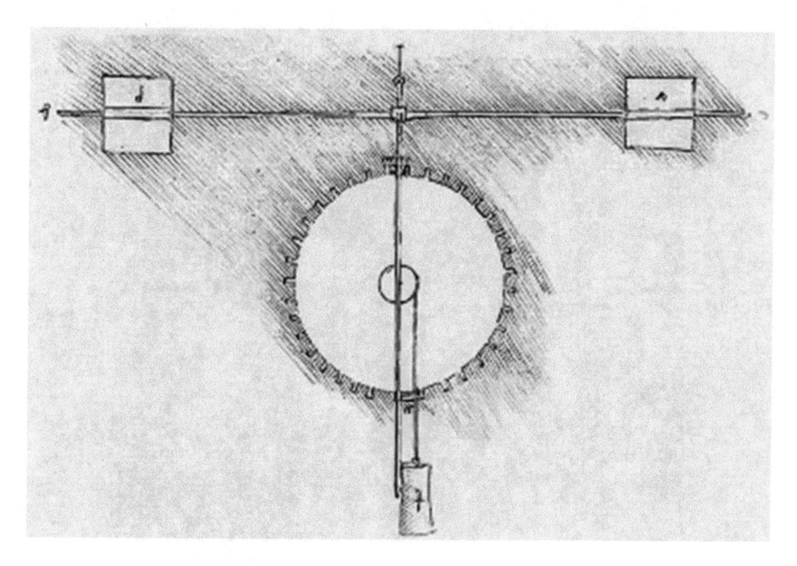

Fig. 13.19 Project for a verge-foliot escapement. Leonardo da Vinci. Codex Madrid I, folio 115v, detail. BNE

according to the required slowness and the length of the hours. The purpose of the device is to lengthen the time, a most useful thing."[26,27]

Leonardo is also looking for simple and effective solutions for the mechanics; this is true in particular for devices such as clocks where the possibility of using a compact and reliable mechanism is very important: "This wheel should allow making a clock showing the hours, which is simple and good" (folio 115v, *Codex Madrid I*).[28]

Another interesting study can be found on folio 27r of the *Codex of Madrid I* (Fig. 13.20). In this case, Leonardo approached the problem of slowing down the weight which moves the clock gears via a pulley system. To ensure that the movement of the weights does not terminate too soon, it was necessary to design the system to extend as much as possible the descent time. Hence, the difficulty in building portable clocks based on weights. In this

[26]Translation taken from: Moon FC (2007) The Machines of Leonardo Da Vinci and Franz Reuleaux: Kinematics of Machines from the Renaissance to the twentieth Century. History of Mechanism and Original text: Machine Science, Volume 2. Springer Science & Business Media.

[27] *Codex Madrid I*, folio 115v. Original text: "Usano alcuni contra l'inpito de' moti i quali spesse volte possono acadere a' contrapesi che muovono le rote degli orilogi, alcuni strumenti i quali dimandano tenpo, perchè tiene in tenpo il moto delle ruote che le movano, e va ritardato secondo che ssi richiede alla tardità e llunghezza dell'ore. Questo strumento è in tale di prolungare il tempo, cosa molto utilissima. ...".

[28] *Codex Madrid I*, folio 115v. Original text: "Puossi questa rota solo fare un orilogio che mostri l'ora, che sia semplice e bbuo[no]".

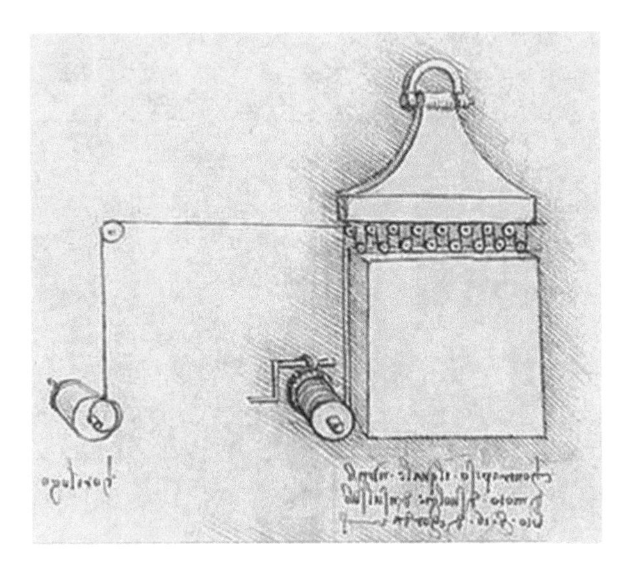

Fig. 13.20 Pulley system for weight-driven clock gear. Leonardo da Vinci. Codex of Madrid I, folio 27r, detail. BNE. (ca. 1493–1497)

project, Leonardo conceived an innovative solution with a system of pulleys to increase the descent time while minimizing the space required for housing the weights.

Leonardo's investigations into clocks fit into his more general interest in the transmission of movement through various types of gears. An interesting example of such studies is the transformation of different types of movements, as from vertical to rotary motion.

Did Leonardo ever translate these various studies into an integrated project of a clock? The answer is on folio 27v of the *Codex of Madrid I*, where Leonardo realized a very intriguing sketch with a complete project for a mechanical clock (Fig. 13.21). This drawing has been used to produce a working copy of "Leonardo's clock," and several of them are even sold on the market in kits.

The previous examples are only a small part of the many studies performed by Leonardo on the mechanics of clocks in the *Codex of Madrid I*. Several pages contain studies of specific mechanical parts, new ideas such as a pendulum escapement mechanism (*Codex of Madrid I*, folio 8r), and incremental improvements for existing devices. These studies bear witness to Leonardo's special interest in the subject, which was most likely more technologically than philosophically driven.

Leonardo's work on the mechanics of clocks is particularly interesting because it allows us to get a better understanding of his general approach to

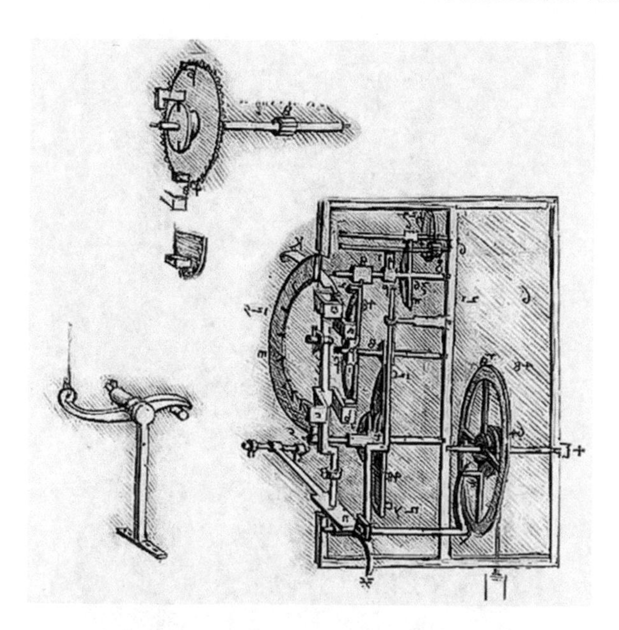

Fig. 13.21 Project of mechanical clock. Leonardo da Vinci. Codex of Madrid I, folio 27v, detail. BNE. (ca. 1493–1497)

engineering problems and his working method. Most of the studies on clocks and their parts are reported in the *Codex of Madrid I*, and they partially belong to an organic and systematic study. In other manuscripts, several drawings and notes related to mechanical clocks can be also found, but in most cases, they are studies of parts of existing clocks, such as the sketches of the *Astrarium* and the tower clock of the Abbey of Chiaravalle which we saw earlier in this chapter.

The first step for Leonardo was therefore to study and understand the mechanics of clocks using what had already been done in the field, which is also the current common approach in science. In the second phase, he introduced incremental innovations, and the studies on the *verge-foliot* escapements are a typical example. At the same time, he did not hesitate to explore new solutions, and again we have the example of the spring system for clock motion.

It is clear, however, that he was not alone and was not starting from scratch. If we consider what had already been done in the field, his work appears to have had only a limited impact. Leonardo clearly was not able to realize or even plan a complicated clock such as those of Lorenzo della Volpaia or Dondi's *Astrarium*. We should not forget that also Brunelleschi a century before Leonardo had realized big clocks and already introduced an important

Fig. 13.22 Table clock. Anonymous. Italy sixteenth century. Museo Galileo. Inv. 3821. Florence, Italy

innovation in the *fusee*. Even the use of springs for clocks was well known at the time of Leonardo. In Fig. 13.22 is shown a table clock realized by an anonymous in the sixteenth century with a mechanism that appears similar to Leonardo's device shown in Fig. 13.14.

Thus, when seen in their proper context, the final projects of clocks from Leonardo are nice but not at all comparable to the mastery shown by the chief Italian clockmakers, who realized wonders with an incredible skill not only on paper but also in working models and masterpieces. Leonardo's drawings are very attractive—they look so beautiful, and we are so much drawn to them for their aesthetic appeal—but we should not forget to put everything within the right historical perspective.

From Clocks to Mechanics of Motion

Leonardo's studies on the mechanic of clocks, which form the basis for his ideas of automata, such as the "car" and the robot, became also an important opportunity to reflect on the nature of motion and the forces that generate it. As we have seen, the two possible sources of motion known to Leonardo to activate the mechanism of a clock were weights and springs. As he progressed in his studies, it soon became clear to him that these forces were very different in nature.

On folio 96r of the *Codex Atlanticus*, Leonardo drew another small project (Fig. 13.23), similar to those for clocks and his car. However, he used this study to make some observations about the force of a spring and a weight: "What is the difference between a spring and a weight. Among the spring and the weight of equal power the spring has a higher value, its power is pyramidal because of its type of motion. The weight is composed by two powers, one which is columnar and the other pyramidal. The columnar (power) is that the

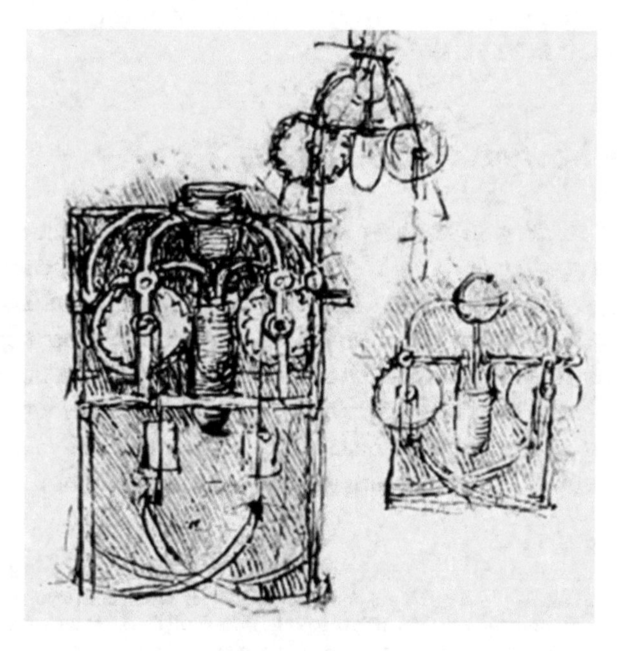

Fig. 13.23 Study of a mechanism with weights and springs. Leonardo da Vinci. *Codex Atlanticus*, folio 96r. BAM

weight is always the same and has always the same force at the beginning and the end of motion, but the pyramidal is zero at the beginning and high at the end."[29]

Leonardo, in accordance to his notes, understood the different nature of the two forces, which today we know as the elastic force and the force of gravity. His description of the motion generated by these two forces is pretty realistic; in fact, the mathematical laws used to describe them are:

$$\mathbf{F_e} = K\mathbf{x}(\text{Elastic force}) \qquad \mathbf{F_g} = m\mathbf{g}(\text{Force of gravity})$$

where F is the force, K the elastic constant which is characteristic of the spring, m the mass, and g the acceleration of gravity (9.8 m^2 s^{-1}). For Leonardo, the elastic force is "pyramidal (piramidale)" and the force of gravity is "columnar (colunnale)." The elastic force, $\mathbf{F_e}$, depends on the distance the spring is extended, and when the spring is at rest, it is zero; $\mathbf{F_e}$ increases with the increase of x. This type of motion is defined by Leonardo as pyramidal (linear) and is opposed to the force exercised by a weight, which depends only on its own mass. For the same reasons, however, Leonardo was not able to understand the nature of the motion generated by the force of gravity, and he mixed the two contributions.

How the Story of Clocks Ends

The fifteenth century, as we have seen, was a time when the mechanics of clocks experienced an important evolution.[30] The motion generated by the fall of a weight was gradually replaced by a spring system, and Leonardo captured with great foresight this transition.[31] Some of his insights were extremely innovative solutions which would be adopted in general practice only several years later; others would be forgotten along with his codices for a long time.

Dondi's *Astrarium*, Lorenzo della Volpaia's *Orologio dei Pianeti*, and the clock of the Abbey of Chiaravalle well represent the extraordinary technical

[29] Original text: "Che differenzia è da molla a contra peso. De la molla e 'l contra peso di potenzie equali sempre la molla vale di più, con ciò sia la sua potenzia è piramidale e ha la sua somma potenzia nel principio del moto suo. Il contrappeso ha potenzie composte, delle quali l'una è colunnale e l'altra è piramidale. La colunnale p che sempre il peso è in sè equale e tira con equal potenza così ne rpincipio del moto con nel fin[e], ma la piramidale è nulla nel principio e assai nel fine."

[30] Brusa G (1993) Entry on Orologio. In: Enciclopedia Italiana, Treccani. V Appendice XXV:588.

[31] Kingston Derry T, Trevor W (1993) A Short History of Technology from the Earliest Times to A.D. 1900. Dover Publications.

level achieved at the time of Leonardo, who perhaps not being a specialist was unable to fabricate such complex mechanical systems. His work in the field, in fact, was mainly limited to the study of some isolated parts and the design of alternative solutions for clock motion and escapements.

These three highly complex astronomical clocks were likely the main source of inspiration for Leonardo, but they were not the only ones in Europe. Since the first mechanical clocks in the early 1300s, in Europe there had begun an extraordinary race to build ever more precise and more sophisticated devices. This was only possible thanks to a thorough mechanical knowledge and very advanced craftsmanship. Leonardo's skill in designing such kinds of systems should be understood in this context: amazing as it may seem, we must not decontextualize his work, and the example of clocks serves very well to bring it back into its correct historical perspective.

There is another interesting observation to make: the first mechanical systems to measure time—from the monastic clocks to tower clocks—all had a practical function. They were used to mark the time necessary to conduct daily activities. For this reason, including a sound system to warn of the passage of time was very important—for example, a bell to announce hours for prayer in a monastery. Leonardo limited his interest to the mechanical part of the clocks, not really having in mind any practical application. He could not compete with the masters who made the complex astronomical clocks and was only able to study them carefully. However, the vivid imagination of Leonardo allowed him to foresee the future evolution of clock mechanics, and this was, as we have seen, his original contribution.

A beautiful reproduction of small clocks realized by Fra Giovanni da Verona in 1457–1525 (Fig. 13.24, *left*) can be admired in the Church of Santa Maria in Organo in Verona, the same work in which he also realized the copies of Leonardo's solids which we saw earlier. An early alarm clock, an instrument that we have seen had an important role in organizing the life of a monastery, can be admired among the inlays of the choir. Another version of these clocks can be found among the inlays realized by Fra Damiano da Bergamo for the Church of San Domenico in Bologna in 1551 (Fig. 13.24, *right*). As in his depictions of Leonardo's solids, he was again inspired by the master of Verona. The two clocks reproduced in the inlays show that, at the time of Leonardo, the use of a dial, even for small clocks, was already well developed. The inlays clearly show that the dial showed the time with 24 h.

These models represent somehow the natural development from the first simple monastic clocks that slowly evolved to a true system to measure time, not just to mark it with a sound. While the precision of these clocks was not so high, they nonetheless served well for their purpose, regulating the daily life of the monks.

Fig. 13.24 Detail of the monastic choir inlays with a copy of an early alarm clock, *left image*. Fra Giovanni da Verona. Church of Santa Maria in Organo, Verona (1499). Copy of a monastic clock with dial, *right image*. Inlay for the Church of San Domenico in Bologna, (1551). Fra Damiano da Bergamo

A new impulse to the development of innovative mechanical devices for clocks came from the 1581 discovery of the isochronism of the pendulum by Galileo Galilei, who always had in mind to develop a clock whose motion could be regulated by a pendulum. Galileo gave detailed instructions to his son on how to build a clock using a pendulum and escapement, but, unfortunately, this new clock was never realized. Christiaan Huygens (1629–1695) would be the first to develop pendulum clocks, although he still used verge escapements, ignoring Galileo Galilei's innovation. Huygens's treatise *Horologium Oscillatorum: sive de motu pendulorum ad horologia aptato demonstrationes geometricae (The Pendulum Clock: or geometrical demonstrations concerning the motion of pendual as applied clocks)* is not only an important text for the history of clocks and watches but would also lay the ground, together with Isaac Newton and Galileo Galilei's work, of modern mechanics.

It was seven centuries before mechanical clocks would permanently lose their supremacy in favor of the very precise quartz watches. The discovery that tiny quartz crystals are piezoelectric allowed the use of their oscillations to measure time. However, there is a little revenge of history: in fact, somehow, the great applied knowledge to develop the mechanics of clocks is not lost, and the majority of portable clocks on the market today show the hours using a dial with hands. The signal produced by the quartz is amplified and used to drive an electric motor that moves the mechanical system and then the hands. A habit of centuries is difficult to change, and the dance of the hours continues like at the time of Leonardo.

14

Mission Impossible: Squaring the Circle

To indicate the inability to achieve an enterprise or the futility of some theoretical disquisitions, a range of idiomatic expressions are frequently used in our daily conversation, such as discussing "how many angels can stand on the point of a pin" This saying comes from the long disputes of medieval scholars about angelology, which now appear of modest if not totally irrelevant interest. Another of these expressions, "squaring the circle," is used as a metaphor for an impossible task, for example, the search for perpetual motion which kept busy so many inventors for centuries. For quite a long time, some of the brightest minds dedicated intense efforts to the attempt to *square the circle*, i.e., to the construction of a square with the same area as a circle, using only a compass and straightedge and a finite number of operations. This seemingly simple problem challenged generations of mathematicians, professional, as well as amateur, until in 1882 the impossibility of such a construction was finally proven.

The Origin of the Squaring the Circle Problem

The beginning of the interest in squaring the circle is connected to Greek mathematics, especially to Hippocrates of Chios (470 BC–410 BC) (not to be confused with Hippocrates of Cos the famous physician), who had also devoted himself with great dedication, well before Leonardo, to finding a solution to the problem. Hippocrates of Chios was also unable to sort it out but somehow went a little closer because first he did manage to achieve the

P. Innocenzi, *The Innovators Behind Leonardo*, https://doi.org/10.1007/978-3-319-90449-8_14

squaring of a curved region indicated as a *lunula* (small moon).[1] The latter is a concave plane figure formed by two arcs of circumferences of different radius. The diameter of the smaller circumference corresponds to a chord forming a right angle to the larger circumference. The name derives from the fact that the surface defined by two arcs has the form of a small moon, the *lunula*. The Greek mathematician owes part of his fame to this discovery, and his solution to this particular problem triggered for a long time the illusion that squaring a circle was also possible. Leonardo himself, as we will see, also became embroiled in these frustrating attempts, in good company in any case, with intellectual luminaries both before and after him.

The First Lunula of Hippocrates

Three solutions for the squaring of *lunulae* are attributed to Hippocrates, with the first of these being the one that Leonardo likely learned through the two-volume *De Expetendis, et fugiendis rebus opus* of Giorgio Valla[2,3] (1447–1500, Venice).

To square the first *lunula* (Fig. 14.1), it is necessary to draw an isosceles right triangle ABC inscribed inside a semicircle. This can, in turn, be split again into two equilateral isosceles triangles ABD and BCD. The side AB is the chord of a quarter-circumference arc of the original circle as well as being

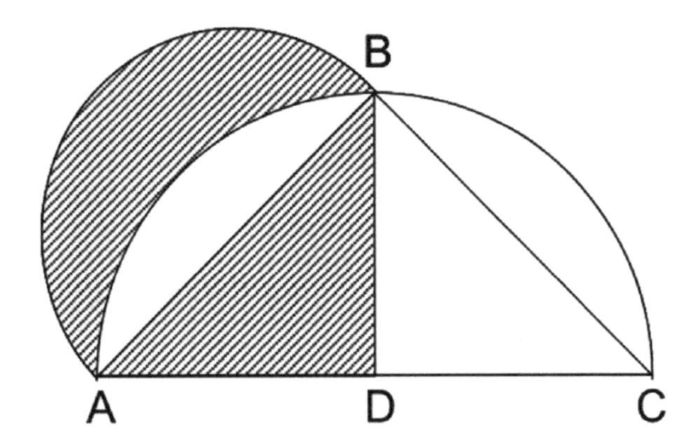

Fig. 14.1 The first lunula of Hippocrates

[1] Leonardo used the expressions of *bisangoli* and *falcata* o *lunole falcate*.

[2] A website is dedicated to Giorgio Valla: http://www.dm.unipi.it/~tucci.

[3] Gardenal G, Landucci Ruffo P, Vasoli PC (1981) Giorgio Valla tra scienza e sapienza, A cura di V. Branca. Firenze, Olschki.

the diameter of a semicircle plotted in this same section. This procedure allows the construction of a *lunula*, which has an area equal to the isosceles right triangle ABD, inscribed into the quarter circle (i.e., the two shaded regions of Fig. 14.1). This is the first known attempt at "squaring" a curved figure.

Leonardo's Interest in the Squaring the Circle Problem

As we have indicated, even Leonardo da Vinci was fascinated by the challenge of squaring the circle and worked very hard for over 12 years to search for a solution, until he ended with the conviction that he had finally found it! It was a time of great excitement, and Leonardo in fact carefully annotated with emphasis this event in his notebook. At the end of the night of Saint Andrew (30 November), having consumed the candle and filled his whole paper, he finally claimed to have found the solution which he had been seeking for so long: "The night of Saint Andrew, I finally found the solution of squaring the circle; at the end of the night, the candle and the paper where I was writing, it was concluded; at the end of the hour."[4]

The way Leonardo described his discovery of the solution of squaring the circle is certainly emphatic and naïve but is also the proof of his passion for the problems of geometry and mathematics. He devoted so much time to them, until the end of the night....

The solution, however, as we know, was a blunder, and later Leonardo himself realized that he was wrong.

What was the origin of all this interest in the squaring the circle problem? Leonardo always had a great passion for mathematics, even if his contributions to the field were not so relevant and, in the end, remained relatively overshadowed compared to his work in mechanics and physics.[5,6,7,8,9] Certainly his interest and passion in this regard had risen by reading his friend Luca Pacioli's *Summa*. We can see the great appreciation that Leonardo had for the

---

[4] Original text: "*La notte di Sancto Andre' trovai il fine della quadratura del cerchio; e 'nfine del lume e della notte e della carta dove scrivevo, fu concluso; al fine dell'ora*". *Codex Madrid II*, folio 112r.

[5] Bagni GT, D'Amore B (2006). *Leonardo e la matematica*. Giunti, Firenze.

[6] Fenyo IS (1985). *Leonardo da Vinci e la matematica*. Rendiconti del Seminario Matematico e Fisico di Milano, 54:101–125.

[7] Marcolongo R (1939). Leonardo da Vinci. Artista scienziato. Hoepli, Milano.

[8] Marcolongo R (1937). Studi vinciani. Memorie sulla geometria e la meccanica di Leonardo da Vinci. Napoli. Stabilimento Industrie Editoriali Meridionali.

[9] Marinoni A (1982) *La matematica di Leonardo da Vinci*. Philips-Arcadia, Milano.

knowledge and study of mathematics from the *Trattato della Pittura* (Treatise on Painting) where he wrote: "No investigation can be defined real science, if it does not pass through mathematical demonstrations; .... who is not a mathematician should not read my considerations; no certainty there is where you can't apply one of the mathematical sciences where they are not combined with them."[10]

How did Leonardo know the geometry of ancient Greece so well as to turn his attention to squaring the circle and, above all, as we shall see, to the study of *lunulae*? His friendship with mathematicians was not limited to Luca Pacioli but also included many other famous scholars, among whom the most important are Giorgio Valla and Fazio Cardano (1444–1524), father of the famous Girolamo Cardano (1501–1576) who himself would make fundamental contributions to mathematics and mechanics. Valla, in particular, was the author of an encyclopedic treatise on mathematics in two thick volumes, the (*De Expetendis, et fugiendis rebus opus*),[11] published posthumously in Venice in 1501 by Aldo Manuzio. This work included a summary of the first five books of Euclid's elements, the mathematics of Archimedes, and the physics of Aristotle.[12] It is in a commentary on physics that the first squaring of *lunulae* of Hippocrates is reported, and it is possible that Leonardo became aware of the results of the Greek mathematician either directly or indirectly from this citation.[13,14]

Leonardo's Lunula

Hippocrates's *lunula* is a special case, while the more general solution has at least two independent discovers. One was the Arab mathematician Ibn Al Haitam (Alhazen, c.965–c.1040), and the second was Leonardo himself. For

[10] Original text: "*Nessuna investigazione si può dimandare vera scienza, s'essa non passa per le matematiche dimostrazioni; .... non mi legga chi non è matematico nelli mia principia; nessuna certezza è dove non si può applicare una delle scienze matematiche ove che non sono unite con esse matematiche.*"

[11] See the chapter: *Il libro di Giorgio Valla*, in *Leonardo e Io*. Pedretti C (2008) Mondadori.

[12] D'Amore B (2007). Leonardo e la matematica. Giacardi L, Mosca M, Robutti O (eds.) (2007). *Conferenze e Seminari 2006–2007.* Associazione Subalpina Mathesis – Seminario di Storia delle Matematiche "T. Viola". Torino: Kim Williams Books.

[13] Another possible source, considering that the treatise of Valla was published in Latin in which Leonardo was not exactly at ease despite studies in mature age, is *De lunarum squareness* of Leon Battista Alberti. The lunulae of Hippocrates can also be found in the book, *Euclidis megarensis philosophi acutissimi matermaticorumque omnium fine controversia principis*, published in 1509 by Paganino de Paganini in Venice.

[14] It is known from a list present in the *Codex of Madrid I* that Leonardo possessed a copy of a book by Giorgio Valla, although probably it was not his encyclopedia but the treatise *Nicephoris de arte disserendi* (See: c. Sacker, *Per la biblioteca di Leonardo: «Libro di Giorgio Valla»*. Aevum. Year 74, Fasc. 3 (2000), pages 669–673.

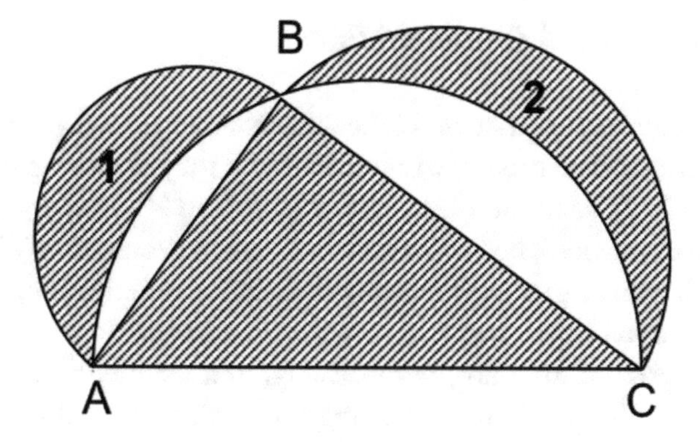

Fig. 14.2 The lunula

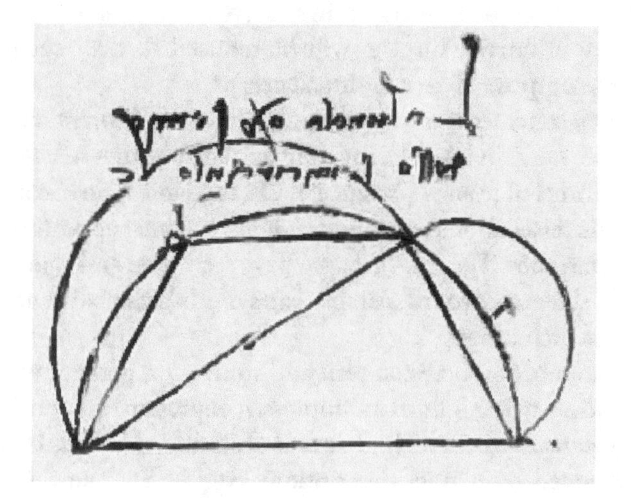

Fig. 14.3 Generalization of squaring of lunulae. Leonardo da Vinci. *Codex Atlanticus*, folio 389r, detail. BAM

this reason, these geometrical figures are referred to variously as the *lunulae of Alhazen* or the *lunulae of Leonardo*.

The general case extends the possibility of squaring a *lunula* to a generic right triangle ABC inscribed into a semicircle of diameter AC, where AC is the hypotenuse (Fig. 14.2). The two shorter sides of the triangle AB and BC are used to trace two semicircles which allow one to identify two major moons, 1 and 2 in the figure, the sum of whose areas is equal to that of the triangle.

Leonardo described very well his discovery (Fig. 14.3), which perhaps helped reinforce his conviction that he was on the right track toward squaring the circle.

Leonardo's Ludo Geometrico

Leonardo's attempts to square the circle and his other related geometric games and investigations were probably intended as the subject of a separate treatise. In fact, on folio 272v of the *Codex Atlanticus*, there is a header which is perhaps the title of the book he had in mind: "De ludo geometrico (On the game of geometry), which gives the process of infinite varieties [of] squaring of surface with curved sides." Right next to the title, Leonardo described in more detail the purpose of his study, emphasizing, in a manner that even seems a touch excessive, the problem of squaring the circle as central in the study of geometry: "The square is the end of all works of geometric surfaces. Every surface waits for its squaring, so surrounded by curved lines as such as straight lines. ... as it will be demonstrated in the process of the work, in which an infinite variety of curved surfaces will be reduced to their squaring, which squaring is the purpose of geometric science."[15]

These strong statements are softened, however, by Leonardo's inclusion of the term game (*ludo*) in his title, indicative of the pleasure one gets in exploring different forms of *scienza* ("science"). On the same folio, Leonardo played with geometric tessellations to produce *lunulae* for what appears to be primarily aesthetic purposes (Fig. 14.4). As we have seen time and again, the artistic and scientific sides are two inseparable parts of his personality, and they both feed and chase each other.

The close connection between play and study—or, perhaps we should say, the pleasure in studying—finds an impressive representation on folio 455r of the *Codex Atlanticus*, where we find dozens of studies of *lunulae* in tiny squares, each accompanied by a brief description (Fig. 14.5). The folio includes a rather large number—176, to be precise—of what, in today's mathematical terminology, would be called "topological transformations."

Modern Topology

The modern mathematical field of topology studies the properties of space which are maintained through continuous deformations without tearing, overlapping, or gluing. A cube and a dodecahedron, for example, are topologically equivalent figures because you can switch from one to the other via a continuous transformation. Another important topological object is the torus, which looks like a doughnut with a hole. Topologically, a doughnut is equivalent to a cup of coffee via a simple deformation, but you cannot obtain a torus from a sphere without creating a hole, which, in topological terms, would require a lift.

---

[15] *Codex Atlanticus*, folio 272v. Original text (): "*... E perché le circuite da linie curve ci sono in poca notizia, io mi sono affaticato co' nuova scienza e darne notizia con varie regole, le quali hanno scoperto nuove notizie,...*."

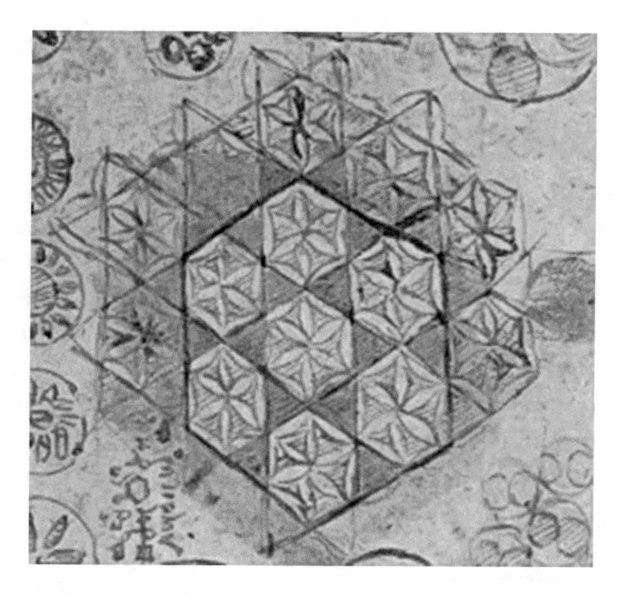

Fig. 14.4 Tessellation with hexagons and lunulae. Leonardo da Vinci. *Codex Atlanticus*, folio 272v. BAM

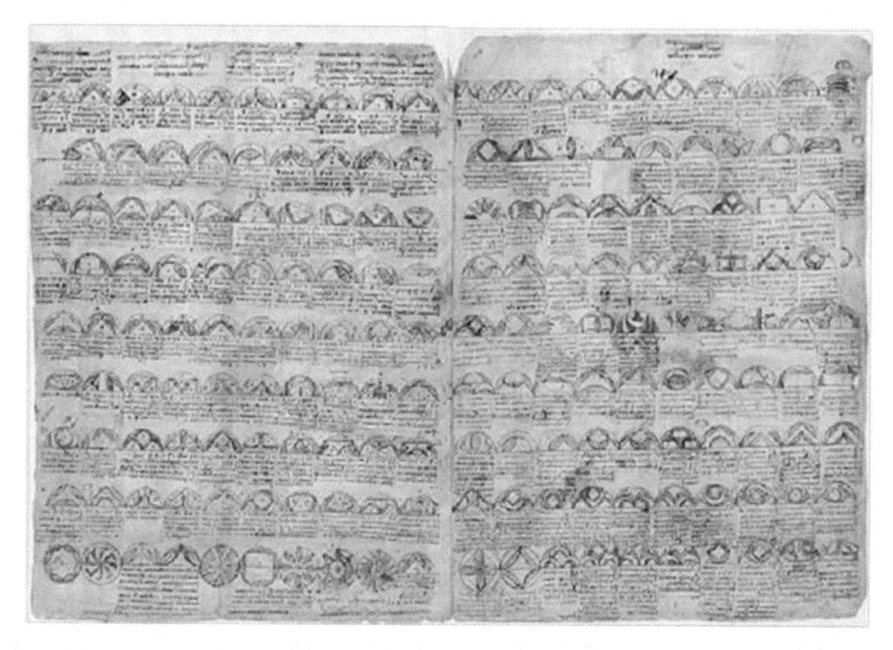

Fig. 14.5 Studies of lunulae. Leonardo da Vinci. *Codex Atlanticus*, folio 455r. BAM

It is perhaps not surprising that Leonardo came so close to foreseeing the basis of topology. He was, in fact, always attracted by natural phenomena and their transformations. Geometry was, for him, the main tool for describing and representing these transformations.

The collection of figures that can be found in different parts of the *Codex Atlanticus* is not merely a game—although Leonardo's joy and playful exploration is certainly very apparent in them—but, in fact, forms part of a careful and thorough study of geometric transformations closely linked to the principles of topology. These figures were drawn employing a wide range of different geometrical elements, including circles, triangles, squares, and *lunulae*, the same forms that are used to explore the possible transformations from one figure to another.

Figure 14.6 shows the construction of one of Leonardo's figures using only a compass and straightedge. First a circle is drawn; then the inscribed square

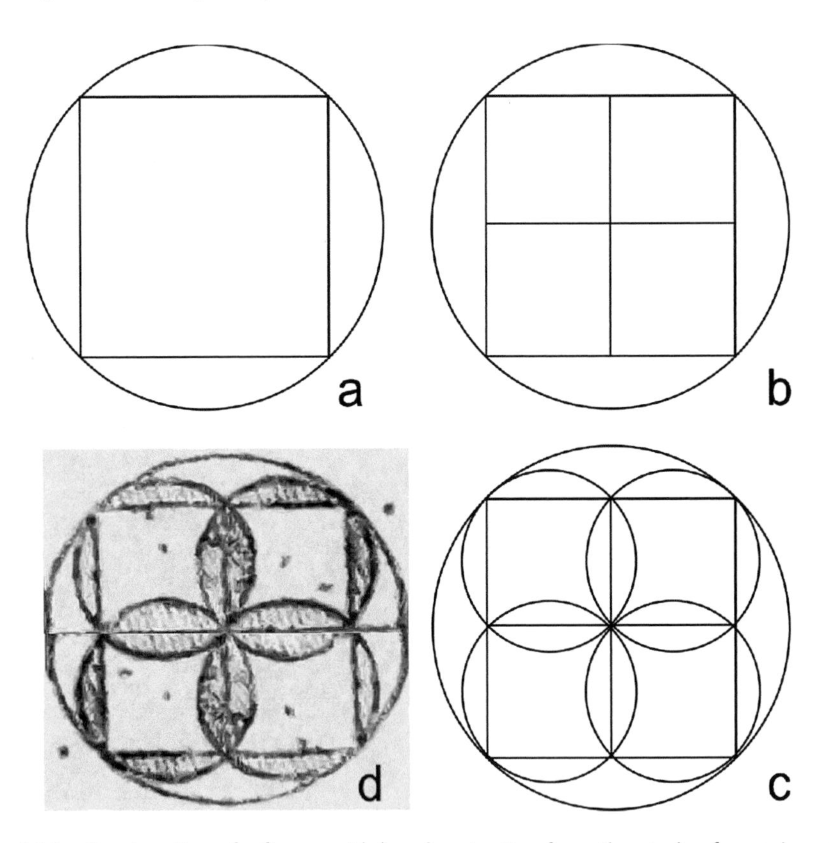

Fig. 14.6 Construction of a figure with lunulae starting from the study of squaring. **a**) a square is inscribed in a circle; **b**) the square is divided into other four squares; **c**) each square is inscribed into a circle, this process is forming the lunulae; image **d**) is the original figure from Leonardo, the shaded parts represent the lunulae. The image **d**) is the right side of the folio 455r, fifth row, seventh figure, in *Codex Atlanticus*. BAM

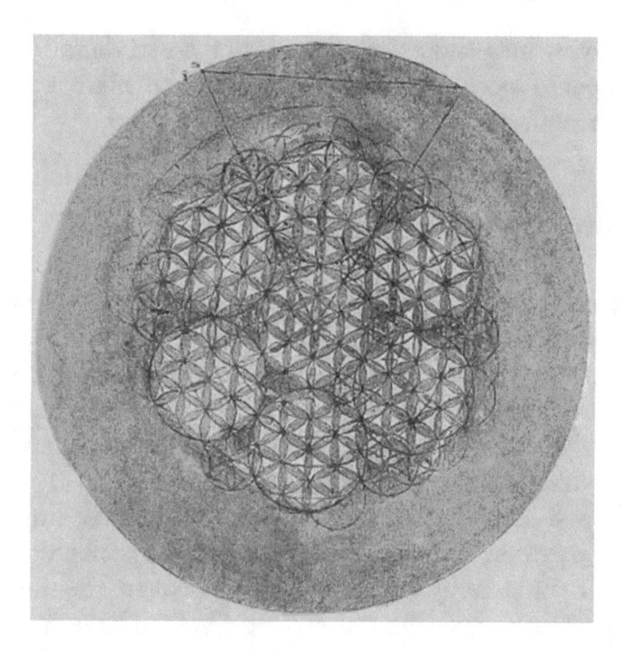

Fig. 14.7 Tessellation of a circumference with lunulae. Leonardo da Vinci. *Codex Atlanticus,* folio 307v, detail. BAM

is divided into four smaller squares. These, in turn, are each inscribed into another smaller circle. This operation produces some half *lunulae* (*falcate* in the Italian of Leonardo), squaring the figure as described by Leonardo: "Squaring putting inside the quadrilateral what is outside of it, namely the four 'falcate'…."[16]

This procedure allows one to build a large number of figures whose attraction is not merely geometric. On the folio, Leonardo resolves real geometric equations, but he also excels in the aesthetic beauty generated by the symmetry of the forms.

This interchange between game (*ludo*) and rigorous mathematical studies is even more evident in another figure, which occupies almost the entire folio 307v of the *Codex Atlanticus* (Fig. 14.7). It is a figure composed of seven circles in which *lunulae* inscribed. Leonardo described the figure in detail: Each of the seven circles contains 42 *lunulae* (*bisangoli* in Leonardo's terminology), all drawn in black. Each *lunula* is itself divided into two parts, giving a total of 84 sections (42 × 2). Since there are seven circumferences and each one contains 84 sections of a circle, there is a total of 588 sections. This calculation

[16]Original text: "*Quadrasi mettendo dentro al quadrilatero ciò che è fuori di lui, cioè le quattro falcate: vaglian le 4 porzioni.*"

allows us to know how many circles can form these sections, because each of these, as Leonardo specifies, is a sixth of a whole, so 588/6 = 98 is the total number of circumferences in the figure.

The Geometry of Transformations

We have seen that Leonardo clearly intended to turn some of his notebooks into real books dedicated to specific subjects, such as the *Ludo Geometrico*. Another one, with a similar topic, is the *De strasformazione*[17] ("On transformation"), which basically corresponds to the first 40 folios of the *Codex Forster I*. This book was intended to be dedicated to "transformation of a body into another one without decrease or increase of the matter."[18]

Leonardo well explored the transformations of bodies from a practical point of view based on a simple experimental trick: a geometrical body made in wax which could easily be molded to observe its transformation into another shape. For example, on *Codex Atlanticus* folio 820v, he writes: "Take some amount of wax and with that make a square, on it should be made a 4 side pyramid. What should be the height of the pyramid (?), Transform the wax into a cube and take 2 parts of wax, and make a cylinder with the same base and height, then take 2/3 of that and you will get the wax for the proposed pyramid."[19]

Regarding the geometry of these transformations, Leonardo had a rather systematic approach, studying at first two-dimensional planar figures with the support of his knowledge of Euclidean geometry and then extending his observations to three-dimensional bodies. He explored several of these transformations, such as turning a cube into a cylinder (*Codex Forster I*, folio 9r) and a cube into a pyramid (*Codex Forster I*, folio 15v), just to cite a few examples.

One of the most intriguing of his transformations concerns two Platonic solids: transforming an octahedron into a cube (*Codex Forster I*, folios 7r and 16r). Leonardo described in great detail how to realize the transformation. The description on folio 7r, in particular, is quite accurate, and it is worth

[17] *Codex Forster I*, folio 3r. The correct Italian word should be *Trasformazione*; likely this was an orthographic error of Leonardo.

[18] *Codex Forster I*, folio 3r. Original text: "*Libro titolato 'De Strasformazione' cioè d'un corpo in un altro sanza diminuzione o accrescimento di materia.*"

[19] *Codex Atlanticus*, folio 820v. Original text: "*....Sia data una quantità di cera e con quella si dia un quadrato, sopra il quale si facia la piramide di 4 lati. Riduci la cera in tavola e quella cuba, e prestale 2 tanti cera, e fanne cilindro tal base e poi hanno il 2/3 prestati e rimarrà la terza base per la piramide proposta.*"

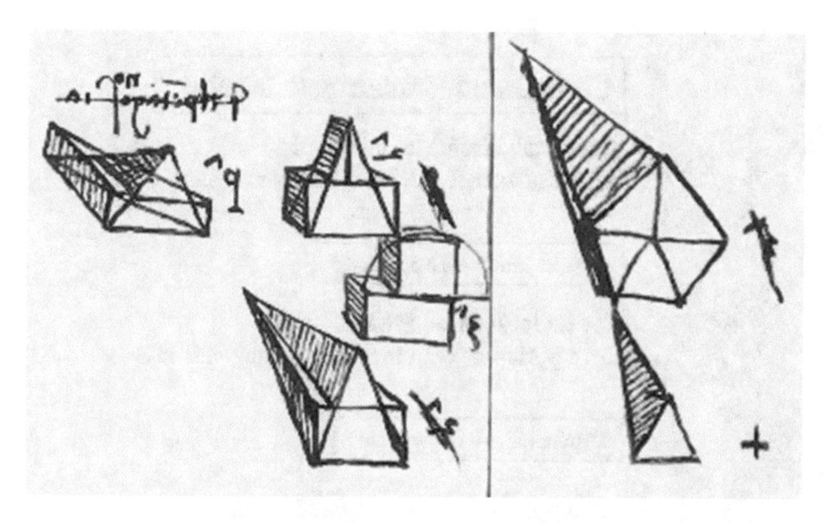

Fig. 14.8 Study of a transformation of a regular solid into a cube. Leonardo da Vinci. Codex Forster I, folio 16r, details. VAM

noting that, over time, Leonardo became more descriptive and accurate in his notes. His confidence in his level of scientific and technical knowledge grew until, in the last part of his life, he considered himself capable of reorganizing his work into thematic treatises.

On folio 16r of the *Codex Forster I*, Leonardo started describing the transformation of a pentagon into a cube, but his description remained incomplete, even if the figures are quite clear (Fig. 14.8).

On the same folio, he also discussed the transformation of a dodecahedron into a cube, writing: "The pentagonal body is composed of 12 [faces]. The surface of the pentagonal body is formed by identical pentagonal faces from which faces born 12 pyramids that end with their tip into the center of such body. Transform the pentagonal base of one of such p[yramids] in five triangles, which extend on the five sides of the pentagon and end at the center of such face."[20]

On folio 7r, all of the steps of the transformation are clearly indicated (Fig. 14.9). At first, the dodecahedron is disassembled into 12 pyramids with pentagonal bases and their points located at the center of the Platonic body. The pentagonal pyramids are then broken into 60 pyramids with triangular

[20] *Codex Forster I*, folio 16 r. Original text: "*Il corpo pentagonale è composto da 12 b[ase]. La superficie del corpo pentagonale è composta di equali base pentagonali, delle quali base nasce 12 piramide che terminano colle lor punte nel centro del predetto corpo. Risulti adunque la base pentagonale d'una coppia d'esse pira[mide] in cinque triangoli, che s'estendino da 5 lati del pentagolo e terminino nel centro di tal base. Fatto questo, colle prima di sopra...*"

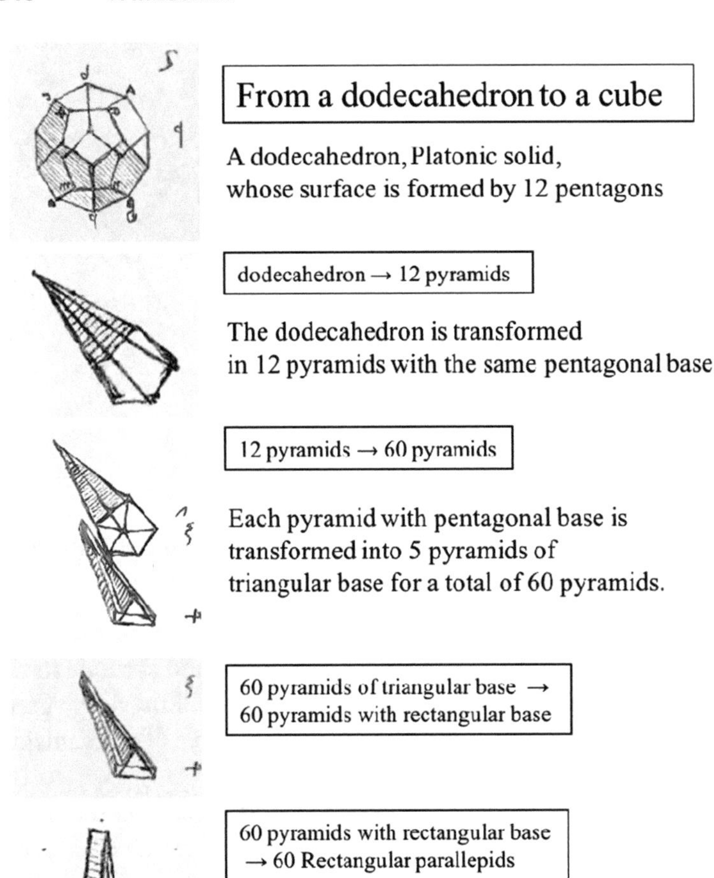

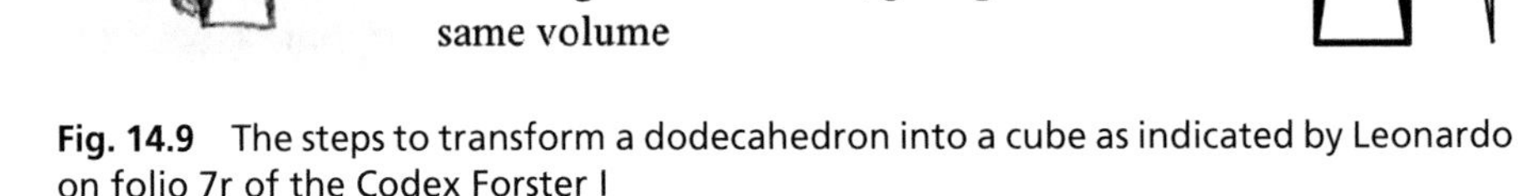

Fig. 14.9 The steps to transform a dodecahedron into a cube as indicated by Leonardo on folio 7r of the Codex Forster I

bases by dividing the pentagonal base into 5 identical triangles. In the third step, the triangular base is transformed into a rectangle so that the original octahedron is now disassembled into 60 pyramids with rectangular bases. At this point, each of the 60 bodies can be transformed into a rectangular parallelepiped or prism, which can then be stacked to form a cube. On the other hand, once a rectangular parallelepiped had been obtained, Leonardo knew well how to transform it into a cube, as demonstrated by the transformation he annotated on folio 820v of the *Codex Atlanticus* (Fig. 14.10).

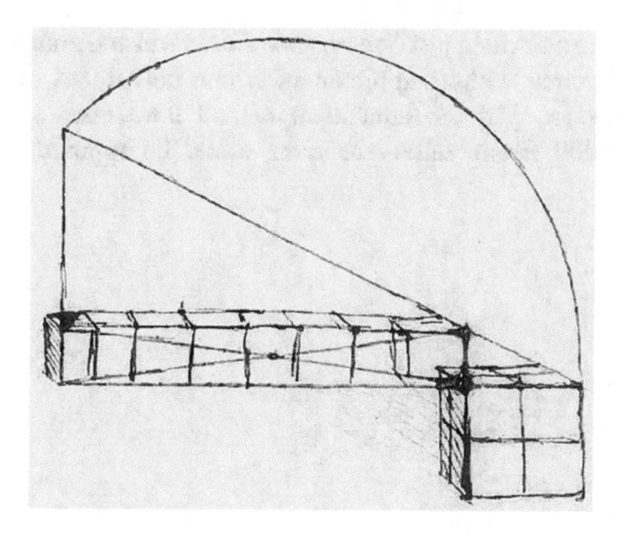

Fig. 14.10 Study of the transformation of a parallelepiped into a cube. Leonardo da Vinci. *Codex Atlanticus*, folio 820v, detail. BAM

The transformation of the octahedron into a cube shows the extraordinary ability reached by Leonardo to visualize geometrical objects and how helpful his friendship with Luca Pacioli was to Leonardo's mathematical development and level of sophistication. His drawings of the Platonic and Archimedean solids for his friend's book were a fundamental stimulus for his studies of geometry, which demonstrate Leonardo's innovative topological intuition.

Really Impossible?

As we come to this chapter's close, let us go back to the problem of squaring the circle with which we began. After centuries of attempts, the final word on this classic problem came in 1882 when the German mathematician Carl Louis Ferdinand von Lindemann (1852–1939) demonstrated that π is a transcendental number, that is, it cannot be expressed as a ratio of two integers because it is irrational and is not the solution of any polynomial equation. The transcendence of π then implies that a construction with compass and straightedge to square a circle is not possible.

All over then? Not really, because, if it is mathematically impossible to get a perfect squaring of the circumference, we can nonetheless get very close. In 1913 the Indian mathematician Srinivasa Aiyangar Ramanujan (1887–1920)

published an article titled just *Squaring the Circle*, which demonstrated a way to square the circle with an approximation that only differs to the seventh decimal place of π. With the Ramanujan method, if we square a circle with an area of 140,000 square miles, the error would be approximately 1 inch (2.54 cm)!

15

A Lost Industrial Revolution and Leonardo's Gun

The "Architronito" or Steam Cannon

Leonardo da Vinci dedicated a full page, folio 33r in the Manuscript B, to illustrating a strange type of gun whose invention he attributed to Archimedes (c. 287 BC–212 BC). Leonardo described how the system should work in detail and accompanied the note with two drawings, one showing the instrument's operation (Fig. 15.1) and the other presenting a design for a cart to transport the weapon to a battlefield (Fig. 15.2). Leonardo appears quite sure when he describes the effect of firing the gun: "The Architronito is a machine of fine copper, invention of Archimedes, and throws iron balls with great noise and rage." The name "architronito" (Archimedes' thunder) is a clear homage to the inventor of Syracuse.

The steam cannon works on a simple principle, creating enough pressure from vapor within the gun's chamber to throw the ball. The cannon's operation is controlled by the insertion of water through a valve. When the breech behind the bullet is hot enough, water immediately evaporates into the steam which gives the thrust necessary to eject the projectile.

Leonardo gave a detailed description of the cannon's mechanism: "And has to be used in this way. The third part of the instrument is placed above an intense fire produced by coal and when will be hot enough, close the screw d which is above the water vase abc, and when the screw is closed the lower part will be opened and the water will fall down on the hottest part of the instrument and then will be immediately converted into vapor, which will appear as a wonder, and look at the fury and listen the noise. This will throw an iron ball, heavy as 1 talent, [to a distance] of 6 stadia."

P. Innocenzi, *The Innovators Behind Leonardo*, https://doi.org/10.1007/978-3-319-90449-8_15

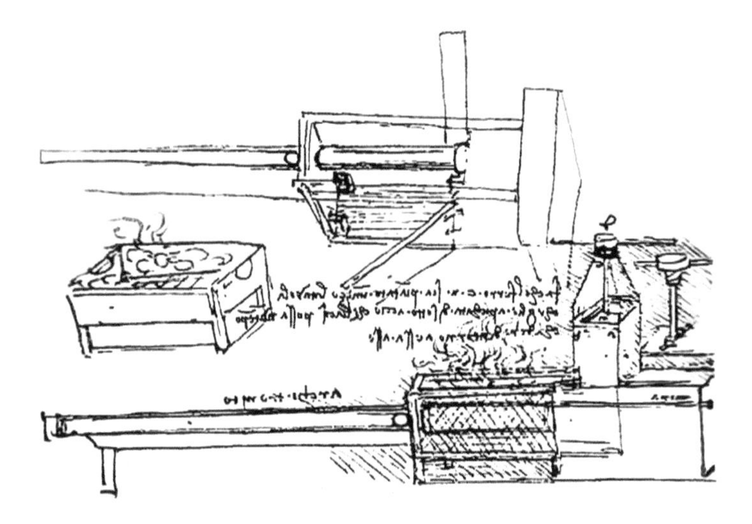

Fig. 15.1 Project for a steam gun, the "architronito". Leonardo da Vinci. Manuscript B, folio 33r, detail. BIF (about 1482)

Fig. 15.2 Project for transporting the "architronito". Leonardo da Vinci. Manuscript B, folio 33r, detail. BIF (about 1482)

If we take for one stadium the Roman value of about 185 m, it means that the cannon would be able to throw an iron ball weighing approximately 32.3 kg (the metric equivalent of a Roman talent weight) the remarkable distance of 1.11 kilometers.

Even if Leonardo described the cannon's operation quite well, as usual, the devil is in the details, which, in Leonardo's projects, are frequently not disclosed. It is not clear, for instance, how to place the projectile inside the breech, nor are the cannon's dimensions—critical for its ability to work—specified.

So, could the *architronito* actually work? In 1981, a first attempt to reproduce a steam cannon using Leonardo's design was made by Greek engineer Ioannis Sakkas. Sakkas produced a scale model which was able to throw an iron ball weighing 3 kg a distance of 50 m.[1] Not such a brilliant performance. Sakkas's primary goal, however, had been to demonstrate that the true inventor of the steam cannon had been Archimedes. His thesis (rather weak) was that if the cannon, constructed only using materials available at the time of Archimedes, could successfully operate, then the invention must be ascribed to him.

A more sophisticated reconstruction was done by researchers at the Massachusetts Institute of Technology (MIT) in 2006. They started from the scientific consideration that the vague indications of Leonardo could give room for a variety of different interpretations. The main limiting factor was the direct injection of water that, quickly poured onto the hot breech, was supposed to produce a flash of steam with enough power to launch the projectile. The rough calculation done by the MIT team was: "A 540°C copper breech can provide a heat flux of roughly 100000 kW per square meter to water that is in direct contact with its surface. And if water can be sprayed onto the breech in a uniform 1 mm thick layer, 2000 kJ of energy per square meter of surface will produce 400 PSI steam—a pressure that would generate a reasonable projectile velocity. The pressurized steam must be generated quickly, before the projectile exits the cannon, perhaps in 0.005 s. But at 100000 kW per square meter, the breech can only provide around 500 kJ per square meter in such a short time, not nearly enough energy to produce high-pressure steam." The conclusion is that, if direct injection of water is used, the steam cannon would not produce "much more than a burp."[2] The solution was found by MIT to overcome this limit but was not disclosed for safety reasons. It was assured, however, that it was compatible with Leonardo's design. The final result was astonishing: The steam cannon was able to fire a bullet at a speed of over 300 m/s with a kinetic energy 1.3–1.8 times the energy of a 0.50 BMG caliber bullet from a M2 machine gun.

But was Archimedes really the inventor of the steam cannon? Even if today several sources list Archimedes as the father of the *architronito*, there are no documents to prove it.[3] There are some vague references in Italian authors, such as Francesco Petrarca (1304–1374), who attributed to Archimedes a device capable of imitating the thunder of Zeus.[4] It is, however, a very generic

[1] Reeves Flores J, Paardekooper R (2014) Experiments Past: Histories of Experimental Archaeology. Sidestone Press.

[2] A detailed description of the MIT steam cannon from Leonardo can be found online: http://web.mit.edu/2.009/www/experiments/steamCannon/ArchimedesSteamCannon.html.

[3] Rossi C, Russo F (2017), Ancient Engineers' Inventions: Precursors of the Present. Springer.

[4] De Remediis Utriusque Fortunae, a collection of dialogues in Latin written by Francesco Petrarca between 1360 and 1366.

citation, and no original sources are known about the supposed Archimedes' steam cannon, nothing that could be used as definitive evidence. Nor does the reconstruction of a working device do much to indicate the invention's lineage.

What, therefore, was the source of Leonardo's *architronito*? Two main hypotheses have been put forth. One posits that the steam cannon is an invention of Leonardo, and he simply put the name of Archimedes to underline the importance of his invention. The second is that he had access to some lost sources[5] where he found a description of the cannon. If we look carefully at the project, however, we become increasingly convinced that it is much more likely to have been a full invention of Leonardo. For, as we shall see in the coming pages, it was very similar in concept and design to many of his other projects, and it seems unlikely that his note referred to a specific external source.

More Steam

The question about the paternity of the *architronito* can be better clarified if we consider it in the context of Leonardo's other studies on steam. This is a very interesting point, because he clearly understood the difference between steam and air and that vapor was potentially a powerful source of motion. What is strange is that he was not able to make the last connection with the use of such force for propelling machines. Had he done so, the industrial revolution could have been started several centuries in advance!

Leonardo was very likely aware of the possibilities of steam from a famous Greek device, the *aeolipile* or Heron's engine (Fig. 15.3). An *aeolipile* is formed by a copper sphere filled with water and suspended on a brazier for heating. The sphere is allowed to rotate on its axis, and two bent handles placed on opposite sides are used to rotate the sphere using the thrust generated by the steam.

This simple turbine was described by Heron of Alexandria (about 10–70 AD), who is generally credited with its invention. Almost 100 years before Heron, however, a similar device had already been described by Vitruvius in *De Architectura*: "Aeolipile are hollow brazen vessels, which have an opening or mouth of small size, by means of which they can be filled with water. Prior to the water being heated over the fire, but little wind is emitted.

[5] Rossi C (2010) Archimede's cannons against the Roman fleet? History of Mechanics and Machine Science 11:113–131. Springer.

Fig. 15.3 Aeolipile from Vitruvius in the version of Cesare Cesariano. Italian Edition of Vitruvius's *De Architectura*, translated and illustrated by Cesare Cesariano. 1521

As soon, however, as the water begins to boil, a violent wind issued forth."[6] It is exactly from this source, of which the several illustrated Italian translations were printed during Renaissance, that this steam device was known in Leonardo's time.

Leonardo even managed to realize an experiment to measure the force of steam, which he described in detail on folio 10r of the *Codex Leicester* (Fig. 15.4). He put a calfskin bag, half filled with water, into a square box open at the top (g, h, e, f, in the figure), and then he covered the bag with a plank large enough to close the box. The system was then heated from the bottom. The expansion due to the formation of the steam within the bag caused the rise of the plank and the fall of the counter-weight, *n*.

We know that Leonardo truly performed this experiment because on folio 15r of the same *Codex Leicester* he wrote that one ounce of evaporated water was able to fill the calfskin and that this value was the result of a direct experience: "The water generates the wind, that is when it transforms into air; about this I already made evident, one ounce of evaporated [water] filled a calfskin…."[7]

A small drawing on the left margin of the same page illustrates the working principle of the system (Fig. 15.5). This small rough sketch resembles a piston moved by steam. As we have already observed, Leonardo really got very close to inventing a steam engine.

[6] Translation taken from: The Ten Books on Architecture by Vitruvius, Chapter VI, paragraph 2.

[7] Original text: "*L'acqua è generatrice del vento, cioè quando essa si risolvè in aria; della quale già feci pruva: un'oncia vaporata m'empiè un otro e prima si tccava li dentri della pelle in ogni lato*".

Fig. 15.4 System to measure the conversion of water into vapor. Leonardo da Vinci. Codex Leicester, folio 10r, detail

Fig. 15.5 Pistons moved by vapor. Leonardo da Vinci. Codex Leicester, folio 15r, detail

The *Sufflator*

Already in the Middle Ages, a primitive version of a kind of boiler, the *sufflator*, was quite well known.[8] This device was generally in the form of a human head filled with water, which, when heated, emitted a jet of stream from the mouth. A funny version of a *sufflator* with a hairless head and distinctive appearance was sketched by Taccola (Fig. 15.6, bottom). This is typical Taccola's sense of humor, as we have already observed in Chap. 1. Leonardo copied the *sufflator* from Taccola's manuscript (Fig. 15.6, top) and described its working system in a note on folio 1112v of the *Codex Atlanticus:* "If this head is filled with water up to the mouth, boiling the water and leaving the steam only from the mouth, has the strength to light a fire."[9]

The interesting aspect of this simple system is that it had potential military applications and attracted a great deal of attention for this very reason. As Leonardo noted, when a *sufflator* is placed close to a fire, because the outgoing

Fig. 15.6 Steam bowlers (sufflator). Leonardo da Vinci. Codex Atlanticus, folio 1112v, detail (*top*). BAM. Mariano di Jacopo (Taccola), Cod. Lat. Monacensis 197 II, fol. 69r detail (*bottom*). BSBM

[8] White TL (1962) Medieval technology & social change. Clarendon Press.

[9] Original text: "*Se questa testa è piena insino alla bocca d'acqua, bollendo l'acqua e uscendo il fumo sol per la bocca, ha la forza d'accendere un foco*".

Fig. 15.7 Sufflator in bronze. Venetian area, sixteenth century. Museo Correr, Venice. Inv. cl. XI n. 790

stream is richer in air than in water-vapor, the strong flow can increase the strength of the fire.

An example of these sufflators used in Italy can be seen at the Museo Correr in Venice where a sixteenth-century steam bowler in bronze of Venetian origin is shown (Fig. 15.7).

A very nice illustration of a possible military application can be found in Konrad Kyeser's *Bellifortis* (c. 1405) where a giant *sufflator* is used to send a stream of fire onto an enemy tower (Fig. 15.8). So we see that, from the Middle Ages on, the power of steam was immediately applied to military purposes, as both a cannon and a system for burning one's enemies.

The properties of the *sufflator* were, however, also exploited to propel machines, for it was soon realized that, using the proper design, the force of the steam could be applied to rotate mechanical systems. The first to exploit this possibility was an Italian architect and engineer, Giovanni Branca (1571–1645), who designed a simple steam engine based on a *sufflator* (Fig. 15.9). Branca realized a collection of several different mechanical devices, which was published in Rome in 1629 with the title *Le machine* (The machines). The book contains 63 engravings, unfortunately of modest quality, with a short description of each in Italian and Latin.

Fig. 15.8 Use of a sufflator for putting fire on an enemy tower. Konrad Kyeser. Bellifortis. BSBM

Branca's simple scheme for his steam-boiler-powered system consists of a machine which alternatively moves a double pestle for milling. The system is propelled by the steam generated by a *sufflator* which propels a horizontal paddle gear. The movement is transmitted via a shaft to a system of toothed wheels which, in turn, are able to lift and lower the pair of pestles. The illustration is certainly not sophisticated, and the technical quality in comparison to the impressive drawings of Leonardo and Francesco di Giorgio appears rather low. However, Giovanni Branca's idea for a machine powered by vapor was remarkably close to that of a steam engine. Even if the system is simple, it is without any doubt the prototype of a machine moved by steam. This again was clearly a lost opportunity for an anticipated industrial revolution in Europe.

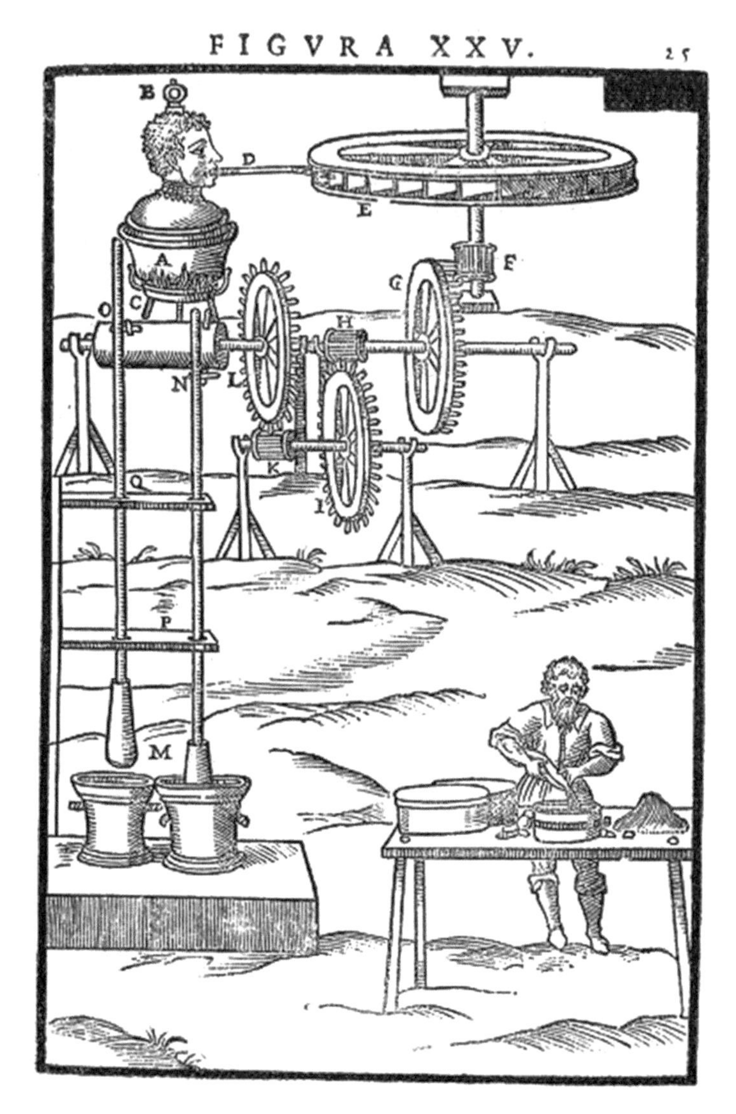

Fig. 15.9 Steam boiler (sufflator) which moves a double pestle. Giovanni Branca. Le machine, Fig. XXV. 1629 (Branca, G. *Le Machine*. Roma 1629. Available on line at: http://gallica.bnf.fr.). (This work is in the public domain in its country of origin and other countries and areas where the copyright term is the author's life plus 100 years or less)

Steam Power After the Renaissance

It is from the beginning of the eighteenth century that the use of steam power became widely diffused as source of motion for different types of machines such as pumps, factory machinery for textile production, and, from the nineteenth century on, steam-powered trains and ships. This was the industrial revolution ignited by the steam engines.

In the meantime, the idea of using steam as the source of propulsion for ejecting high-speed bullets was basically forgotten. With the arrival of the steam age, some attempts were realized to create a cannon based on this technology similar to the systems for steam locomotives. For example, one simple but quite effective model for a steam-powered weapon was the Holman Projector (model Mk II) developed as an antiaircraft gun for the Royal Navy during the Second World War (Fig. 15.10).

Fig. 15.10 A Holman Projector's prototype during a trial (1940). Illustration from "The Secret War," by Gerald Pawle, page 47. (This work is in the public domain in its country of origin and other countries and areas where the copyright term is the author's life plus 70 years or less. https://commons.wikimedia.org/wiki/File:Holman_Projector_in_action.jpg)

This cannon was not very precise but was able to hit at least 12 airplanes in its first year of operation. The steam cannons were used to equip cargo ships with a portable antiaircraft weapon. Their advantage lay in their low cost and low weight, as well as in the fact that it was not necessary to store and use gun powder.

Besides this successful use during the Second World War, the steam cannons were quickly forgotten, because, in comparison, firearms became so precise and devastating to relegate them into the museum as a part of the history of technology.

16

Beyond Leonardo

After a lifetime of wandering around Italy, Leonardo made his way to France in the autumn of 1516, accompanied by his servant Battista, and his beloved pupils Francesco Melzi and Salai. He was tired and sick and likely knew that this might be his final journey. Leonardo brought with him some of his most famous paintings, the *Gioconda* (Mona Lisa) and *San Giovanni Battista*, and moved into the castle of Cloux not far from Amboise, which was home to King Francis I of France. It is likely that Leonardo was partially paralyzed on his right side due to an ictus, but he never stopped studying the laws of nature and investigating the world of mechanics and machines. He remained until the very end a scientist and engineer.

On 23 April 1519, Leonardo dictated his will, leaving his written legacy to Francesco Melzi and his paintings to Salai. He died 9 days later, on 2 May 1519. Francesco sent a letter to Leonardo's brothers in Florence to inform them of the Maestro's death. It is a very touching letter, in which he wrote that the loss of such a unique man caused them all tremendous grief.

There is a famous 1818 painting by the French artist Jean Auguste Ingres (1780–1867), "Francis I receives the last breaths of Leonardo da Vinci" (Fig. 16.1). It depicts King Francis I conferring with the Maestro until the very last moment of his life. It is unlikely that the painting was based on any real encounter on the Maestro's deathbed. However, it is nice to believe that when he died, Leonardo was surrounded by the love and respect that his great life deserved—a passing fitting of the man who famously wrote, on folio 680r of the *Codex Atlanticus*, "Quando io crederò imparare a vivere, e io imparerò a morire" ("When I would believe that I have learned to live, I would learn to die").

P. Innocenzi, *The Innovators Behind Leonardo*, https://doi.org/10.1007/978-3-319-90449-8_16

Fig. 16.1 Francis I receives the last breaths of Leonardo da Vinci. Jean Auguste Dominique Ingres, 1818, oil on canvas, 40 × 50.5 cm. Petit Palais, Paris, France. (This work is in the public domain in its country of origin and other countries and areas where the copyright term is the author's life plus 70 years or less. https://commons.wikimedia.org/wiki/File:Francois_Ier_Leonard_de_Vinci-Jean_Auguste_Dominique_Ingres.jpg)

After Leonardo's death, his spiritual heir Francesco Melzi was left to handle the enormous but fragmentary and disorganized written material that had been left to him. As discussed previously, Leonardo always intended but never succeeded in preparing for publication several thematic treatises using his notes. Though a clever and loyal pupil, Francesco also found it impossible to achieve much in this direction, only completing the book dedicated to painting. As discussed in Chap. 3, Leonardo's drawings of anatomy, his engineering and architectural projects, and studies in physics, optics, and the flight of birds—all were relegated to the same destiny, remaining hidden and forgotten in the hands of various owners. These important contributions to scientific thought were lost for a long time.

After Leonardo's death, it was extremely difficult to organize his written material. He was likely the only person who could have succeeded in the Herculean task of systematizing the sheer volume of notes and sketches. As we have said, Leonardo employed an extremely complex method of taking notes. Not only did he employ mirror writing, but the language itself—a form of Italian which was not yet standardized—would also have been very difficult to understand. The text, which contains some orthographic errors and personal abbreviations, requires specialized knowledge to be clearly understood. The sustained effort of generations of dedicated scholars was necessary to create a comprehensive transcription of his notes.

The complexity of the written text is not mitigated by the accompanying drawings. Rather, the illustrations add their own level of complexity. Sketches related to the same project are sometime scattered across different pages, as are the notes regarding specific subjects. In some cases, Leonardo accompanied his sketches with explanations and observations, but frequently, difficult-to-interpret drawings were simply left without any annotation.

Many of the recent reconstructions of Leonardo's machines—such as the robot (see Chap. 7), the self-propelled cart, and the flying machines—rely on contemporary interpretations, with parts added from different drawings. Without these additions and creative interpretations to fill in the gaps, in many cases, Leonardo's machines simply do not work or cannot be built. This is also true of the work of other inventors and innovators of the Renaissance whom we have encountered in these pages. They were not able to bring their machines into the practical dimension, likely because most of these machines were simply too advanced for their time and, as such, far surpassed the inventors' capabilities as technical draftsmen.

It is often said that Leonardo was intentionally vague in the depiction of his ideas because he wanted to avoid disclosing key details, so that his projects could not be reproduced by potential competitors. However, because this is true of so many of his drawings, it is difficult to determine whether or not it was done intentionally. More likely, Leonardo, with his ever active imagination, was interested in an idea and wanted to quickly create a proof of concept. To his mind, this was enough, and it gave him sufficient time to plan the wide variety of machines that crowded his imagination and to explore the secrets of nature in a wider and more systematic manner.

We do not know how many—if any at all—of Leonardo's machines were actually tested in his lifetime. Constructing a working model would have been highly time consuming and expensive. Moreover, there is essentially no record of any physical machines, with the exceptions of the systems he created for theatrical performances. This approach to the act of invention does not seem strange if we put Leonardo in the proper context. Leonardo's behavior in this regard is exactly the same as that of other Renaissance engineers and inventors, such as Taccola, Francesco di Giorgio, and Antonio da Sangallo, and even the authors of the German school, such as Konrad Kyeser.

This had some important consequences, which become clearer when we consider the impact that the Renaissance engineers had on the history of technology. Two aspects in particular of their work styles had a direct influence on the future. Firstly, all of them preferred a theoretical approach to building real working machines. Secondly, none of them managed to have an edition of their notebooks printed. They remained in great part at the embryonic stage of a thematic treatise, as demonstrated by the scattered notes and drawings that form the largest part of the notebooks, with all of the limits on diffusion that plague works only available in manuscript copies. In some cases, the manuscripts contain only a collection of drawings without any explanatory text. Other times, the accompanying notes are simple labels that offer only minimal support for detailed interpretation.

These brilliant, innovative polymath engineers and artists, who were capable of dreaming up entirely new technological worlds, did not fully anticipate the arrival of the new age that was heralded by the invention of the printing press. Leonardo was born in Vinci on 15 April 1452. Around 1455 in Mainz, Johannes Gutenberg (c. 1390–1468) printed the first copies of his famous Bible, which ignited the printing revolution in Europe. Within a short time, the number of printing presses multiplied and spread throughout Europe. In 1489 there were 110 presses, 50 of them in Italy alone. Venice, in particular, quickly became the main center of printing in Europe, with 417 printers in 1500.

Leonardo was indeed quite attracted by the technical aspects of the invention of movable type. He designed a hand-press device to synchronize the forward movement of the print trolley with the descent of the printing press

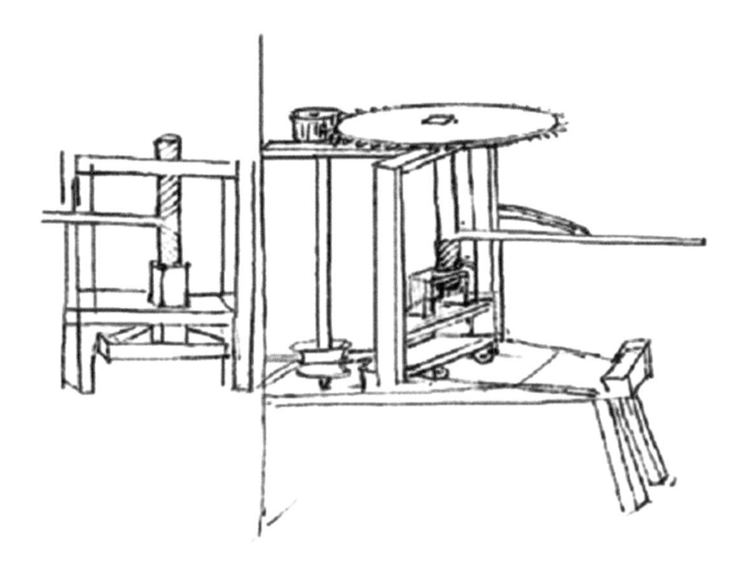

Fig. 16.2 Leonardo's printing system. Codex Atlanticus, folio 995r. (c. 1497). BAM

(Fig. 16.2). His interest, however, seems to have been limited to the mechanics of the system without any particular attention to the possibility of printing his own work.

Leonardo, therefore, even if he was born at the right moment and in the right place, did not take advantage of this unique opportunity. He was not alone. As we have seen, the same happened for most of the manuscripts of the Renaissance engineers, sharing the destiny of a quick oblivion until their more recent rediscovery by modern scholars. Indeed, many of the manuscripts related to machines and technologies remain hidden and unpublished to this day.

On the other hand, the very eclecticism of the Renaissance artist-engineers that captures our modern imagination also represents an inherent limitation of their approach. They were wearing several hats at the same time, simultaneously striving for excellence and expertise as artists, architects, military and civilian engineers, and so on. They probably considered their activity as inventors a secondary profession, or even a hobby. Their engineering projects also remained largely confined to the world of ideas, with the exception of hydraulics and military architecture, where most of their projects were directly related to the requests of their patrons or political leaders. In some cases, the notebooks represented a kind of catalog to interest a potential patron or to obtain a commission as an engineer.

Leonardo's case was a little different. At the very end, what he really liked was his world of machines and dreams. He was happy only when he could spend his time with mathematics, geometry, and mechanics, exploring the natural world using mathematics as a universal language, writing in his treatise on painting that "No human investigation can be defined real science if it does not pass through the mathematical proof…."[1]

Upon analyzing the work of the Renaissance engineers, putting aside the "theoretical" dimension of their projects, it is evident that (though they were unable to organize and print their work) their way of thinking heralded a new era. Their notebooks do not contain "magic" considerations, alchemical studies, drawings of monsters, or other fantastical sketches that were typical of technical manuscripts of the Middle Ages. These images were still present in the German school and in the works of some scholars who were working in relative isolation, such as Giovanni Fontana. Reading Leonardo and the Siena engineers' manuscripts leaves one with a strong impression of the fresh and creative environment of the Renaissance. There is a marked difference with

[1] Treatise on painting: "Nessuna umana investigazione si po dimandare vera scienza, s'essa non passa per le matematiche dimostrazioni…".

regard to past works and a sense of the scientific way of thinking that would dominate the future. They were, however, not completely aware of this. Only Leonardo was able to integrate scientific and technological studies, experiments, critical discussion of the results, and theoretical studies to give support to personal experience.

On the other hand, if we consider the amount of technical knowledge available during the Renaissance and the incredible concentration of brilliant minds, a simple question arises: Why didn't the industrial revolution begin in that moment, and why did it take another century before the scientific method was understood and applied? There is no simple explanation in answer to this question. There have also been several other moments of human history in which all the conditions for a scientific and technological revolution existed but never materialized. We have only to look to China during the flourishing of the Tang dynasty, to the Arab world in the Middle Ages, and to India during the Moghul Empire, which reached a very high level of both mathematical and technical knowledge.

The simpler answer is that such a "revolution" is in fact the result of a slow and continuous process that requires an extended period of time and a combination of numerous favorable conditions. Even today, the creation of seeding environments for creativity that boost innovation, such as Silicon Valley in California, is a common goal. But such an environment cannot be reproduced by simple imitation of a model. The easy diffusion of ideas, close contact of people, mobility of scholars, availability of economic resources, freedom of expression, and requests for innovation from society are just some of the conditions necessary to create a social environment ready to take advantage of the opportunities for innovation.

In Leonardo's time, Italy was at the center of the Western world, with its flourishing rival city states that competed to show off their wealth by raising splendid churches and palaces, drawing on the enormous talents of painters and sculptors of unrivalled ability. They were competing to realize masterpieces of unique beauty. The people were eager for novelty and ready to accept innovation. Money was no object, thanks to lucrative trade with the rest of the world. In fact, Italian bankers were financing most of the royal families of Europe.

Mobility of people, at least for the educated and the wealthy, was also favorable, despite the continuous wars that affected the Italian city states and the struggle for power within them. Leonardo moved freely during his life, visiting many cities and serving a number of patrons, such as the Duke of Milan and the Pope, in Italy and France. The same can be said for many of the other innovators of the Renaissance described in this book. Francesco di Giorgio,

Luca Pacioli, and Giuliano da Sangallo, to cite just a few, shared Leonardo's lifestyle, traveling around Italy and moving to wherever they could find the best opportunity to express their technical and artistic capabilities.

The Renaissance was a unique cultural and artistic moment in which the rediscovery of Classical knowledge—in mathematics and geometry especially—ignited new interests and opened new horizons for the fertile minds of the time. An industrial revolution, however, did not occur. When we consider the number of different Renaissance innovators whose studies show they recognized the potential of using steam for powering machines, it feels like a significant lost opportunity. However, this does not mean that the work performed by Leonardo and his peers was useless or had no impact. Their work represents the prologue to a long, slow process involving the development of a critical, experimentally based approach whose true beginning is marked by the work of Galileo Galilei. Among his peers, Leonardo was perhaps the closest to developing a scientific method before Galileo. He was, however, still bound by Greek philosophy—such as that of Plato and Aristotle—and also distracted by his wide-ranging interests and activities. He was unable to see beyond the ideas of his time regarding some natural phenomena—such as the use of the *impetus* for defining motion—which were still confused and rooted in a limited knowledge and comprehension of mechanics.

To what extent did those who came after Leonardo benefit from his legacy as an engineer and inventor? Unfortunately, such benefit was highly limited. As discussed, Leonardo's notebooks were difficult to read and understand; the material was unfinished and never systematically organized. This was an integral part of Leonardo's personality—we have to accept him as he was. He had a prolific imagination, enormous artistic talent, and was eminently capable, but he was unable to finish a large portion of his creative endeavors and never completed a technical or scientific treatise.

On the other hand, as we have just underlined, it seems he did not really grasp the importance of printing. His friend Luca Pacioli, an extremely clever individual, begged everywhere for money to print his books. While these books were largely the result of systematic collection from different sources (if not outright plagiarism), they nevertheless enjoyed enormous impact because they were formally printed. Pacioli copied and translated Piero della Francesca's work *De quinque corporibus regularibus* (for which, as we have seen, he was unjustly accused of plagiarism). Though he was not the original author, without his contribution, this important treatise would simply have lain buried somewhere in an Italian library for centuries. Instead, Pacioli's printed edition exerted a strong influence on the development of geometry.

As we have seen however, Luca Pacioli owes the greatest part of his fame not to his contributions in mathematics and geometry but rather as the father of financial accounting. His printed work on the subject likely relied on unpublished manuscripts from other unknown authors. But because he was the first to organize and print the material, it is he who is universally recognized for fundamental contributions as the founder of this new branch of mathematics.

The example of Luca Pacioli suggests an initial, simple answer to the question posed earlier as to why a scientific and technical revolution did not occur during the Renaissance despite favorable conditions. The circulation of ideas was still limited to a restricted community of scholars, and a large portion of scholarship remained hidden and unpublished. Indeed, the works of Francesco di Giorgio and Taccola enjoyed much greater influence than Leonardo's mechanical studies. Nonetheless, their engineering treatises experienced relatively limited circulation because only manuscript versions were available. Traces of their work can still be found, however, in most of the treatises on machines which were later compiled and printed. Even though Taccola and Francesco di Giorgio's volumes were not printed, they were better organized than Leonardo's and made to look like treatises. Several handwritten copies with catchy illustrations were produced. This was not sufficient to allow their work sustained impact beyond the Renaissance. But it was enough to avoid complete oblivion.

Taccola and Francesco di Giorgio's machines became part of the repertoire of inventions that comprised a popular form of technical literature known as the "theater of machines" during the late Renaissance. These "theaters" heavily relied on the material of the Siena engineers and were largely inspired by their work. We have seen that their mechanical devices even found their way into Chinese texts. The memory of Brunelleschi's construction machines survived thanks to their drawings and to those of Sangallo and Ghiberti. Conversely, Leonardo's scattered notes did not play a significant role in transmitting this knowledge.

We began this book with the work of Brunelleschi, but in some ways he was exactly the opposite of Leonardo and the other Renaissance engineers: He did not dedicate much time to writing theoretical treatises or to imagining new machines for hypothetical uses and applications. Brunelleschi was eminently practical as an architect, engineer, and artist. He constructed the Cupola of Florence, designing and building all the necessary construction machines along the way. Meanwhile, he also found time to devote to his work as a highly sophisticated clockmaker. His construction machines represented a concrete innovation, with their advanced ideas and the introduction of totally new mechanical devices, such as multispeed gears and brake controls. Brunelleschi reached such brilliant results because he was inspired by practical problems that he had to solve in order to construct his "impossible" structure. Brunelleschi's machines worked because they had to work. Leonardo and the other engineers faced no such need. Most of their work on mechanics was confined to the world of dreams and imagination. On the other hand, the machines during Renaissance were very much appreciated also as a source of fun and wonder.

Leonardo, for one, was a famous organizer of events. Thanks to his machines, his parties became fantastic attractions full of surprises, such as a mechanical lion. The playful character of the Italian Renaissance engineers should not be underestimated. Multifaceted personalities motivated by the pleasure of novelty, a sense of humor, and an innate spirit for innovation emerge from the pages of their notebooks. The machines, such as the automata, were used as instruments to create astonishment and to transpose the Renaissance preoccupation with beauty and aesthetics into a very different field.

It should not be forgotten, however, that they were all artists first and foremost. It is not by chance that the legacy of their mechanical work was largely reported in the several books dedicated to "theaters" of machines, to transmit the sense of a "show of technical skills and knowledge" which were very popular in the sixteenth century. In the late Renaissance, a new generation of engineers was still using machines to astonish and to display the wonders of mechanics. Jacques Besson (*Theatrum instrumentorum* (*Theater of instruments*)); Agostino Ramelli (*Le diverse et artificiose machine…* (*Different and artificial machines…*)), and Vittorio Zonca (*Novo teatro di machine et edificii…* (*New theater of machines and buildings*)) all created catalogs of fantastic machines which were likely never built or tested. This work, though it was still very much rooted in the past, played an important role in the development and diffusion of technical knowledge in Europe and beyond. The machines from the theaters of Besson, Ramelli, and Zonca were included in the Chinese encyclopedia of Western mechanics and machines, the *Diagrams and explanation of the wonderful machines of the Far West* (Yaunxi qiqi tushuo luzui), which was compiled by the Jesuit Johann Schreck (1576–1630) and translated into Chinese and published in 1627. It is interesting that the long titles of both Zonca and Ramelli's works appear as a kind of "manifesto," a sweeping ambition that well reflects the concept of wonder inherent in the Renaissance attitude.

At the end of the day, as modern readers, perhaps we do not really care so much whether the models of Leonardo and the other artist-engineers actually work. We are more fascinated by the general ideas behind them, by the fact that these visionaries were able to conceptualize new systems and machines so much in advance of their time. Ultimately, they had the right ideas without the proper materials or general conditions to realize them and were moreover not yet interested in taking advantage of their innovations in a practical manner.

Within the vast scope of work done by Leonardo and the other Renaissance engineers, this book only presents a limited number of the ideas to which they dedicated their time and attention. In particular, the military and hydraulic works have not been treated in detail. This is because, in the case of military machines and warfare technologies, Leonardo and the other Renaissance engineers only made incremental improvements. Meanwhile, the subject of hydraulic technologies is so vast that it would require a dedicated book of its own to explore all the connections between "classic" Greek and Roman knowledge, medieval scholars, and Renaissance innovators.

Regarding the ideas that were able to be presented in this book, special care has been taken to insert Leonardo and the Renaissance engineers' projects within their proper historical context. This shows that they were first and foremost the fruit of a continuous process of innovation that has characterized the history of science and technology in Europe. The Renaissance innovators were fully part of this important process, even if they were partially forgotten by history. The "rediscovery" of Leonardo in the last century has brought with it the creation of the myth of a singular genius—the cutting-edge inventor who stands alone—an image which we now know is a distortion of reality.

It is possible to attend Leonardo exhibitions or read books where a good portion of what is presented was not originally Leonardo's idea. Sometimes, one encounters outright fabrication, as in the case of Leonardo's alleged "bicycle." This image of Leonardo corresponds much more to a present need for the comforting and reassuring figure of a man who discovered and understood everything than to historical reality. As we have seen, Leonardo was just one of the many people who were able to dream and imagine a different future. If we are so fascinated by them, it is because their ideals and thirst for knowledge motivate us to conquer our own limits.

Appendix I: Notable Personages

Villard de Honnecourt (thirteenth century?)
- *Livre de portraiture* (ca. 1230)

Petrus Peregrinus of Maricourt (thirteenth century?)
- *Epistola Petri Peregrini de Maricourt ad Sygerum de Foucaucourt, militem, de magnete* (Letter of Peter Peregrinus of Maricourt to Sygerus of Foucaucourt, Soldier, on the Magnet) (1269)

Guido da Vigevano, (Pavia, ca. 1280–Paris, after 1349)
- *Texaurus regis Francie* (1335)
- *Liber notabilium illustrissimi principis Philippi septimi, Francorum regis, a libris Galieni per me Guidonem de Papia, medicum suprascripti regis atque consortis eius inclite Iohanne regine, extractus, anno Domini 1345* (1345)

Giovanni Dondi dell'Orologio (Chioggia, ca. 1330–Abbiategrasso, 1388)
- *Completion of the astronomical clock Astrarium* (1364)
- *Tractatus astrarii* (fourteenth century)

Konrad Kyeser (1366–after 1405)
- *Bellifortis* (ca. 1405)

Filippo Brunelleschi (Florence, 1377–1446)
- *Cupola (Dome) of Duomo of Florence* (1420–1436)

Anonymous of the Hussite Wars
- *Manuscript of the Anonymous of the Hussite Wars* (after 1472)

Mariano Daniello di Jacopo (Taccola) (Siena, 1381–ca. 1458)
- *De Ingeneis* (ca. 1419–1450)
- *De Machinis* (1449)

P. Innocenzi, *The Innovators Behind Leonardo*, https://doi.org/10.1007/978-3-319-90449-8

- *Copy of De Machinis from Paolo Santini* (colored manuscript version, second half of the fifteenth century) Ms. Lat 7239, Bibliothèque Nationale, Paris.

Giovanni Fontana (Venice, 1395?–after 1454)
- *Bellicorum instrumentorum liber* (ca. 1430)
- *Nova compositio horologi* 1418)
- *De horologio aqueo* (ca. 1417)

Leon Battista Alberti (Genoa, 1404–Rome, 1472)
- *Ludi mathematici* (1448)
- *De re aedificatoria* (1450)

Roberto Valturio (Rimini, 1405–1475)
- *De re militari* (1472) printed version in Latin
- *De re militari* (1483) printed version in Italian

Piero della Francesca (Borgo Sansepolcro, ca. 1416/1417–Borgo Sansepolcro, 1492)
- *De prospectiva pingendi* (ca. 1460–1480)
- *De quinque corporibus regularibus* (ca. 1472 and 1475)
- *Trattato d'abaco* (ca. 1450)

Giovanni Giocondo da Verona (Fra Giocondo) (Verona, ca. 1433–Romea, 1515)
- *Inlays of Santa Maria in Organo in Verona* (1520)

Francesco di Giorgio Martini (Siena, 1439–1501)
- *Codicetto* (drawings of machines) (Biblioteca Apostolica Vaticana, Rome) Codex Urb. Lat. 1757
- *Opusculum de architectura* (British Museum, London) Codex 187.b.21 (drawings without text, machines and fortresses, ca. 1476)
- *Trattato di architettura civile e militare (I version).* (1479–1484) Codex Ahsburnham 361, Biblioteca Laurenziana; Codex Saluzziano 148, Biblioteca Reale of Turin
- *Trattato di architettura civile e militare (II version).* (1485–1492) Codex Senese S.IV.4, Biblioteca Comunale di Siena; Codex Magliabechiano II.1.141, parte 1. Biblioteca Nazionale di Firenze
- *Traduzione Vitruviana* (Biblioteca Nazionale di Firenze) Codex Magliabechiano II.I.141, parte 2

Anonymous (after Mariano di Jacopo (Taccola) and Francesco di Giorgio)
- *Drawings of machines.* Late fifteenth century. Ms Palatino 767. Biblioteca Nazionale Centrale, Florence

- *Drawings of machines.* End of fifteenth century. Codex S.IV.5. Biblioteca Comunale, Siena
- *Book of machines* (Anonymous Sienese Engineer) Late fifteenth century. Ms Additional 34113. British Library London

Giuliano da Sangallo (Florence, ca. 14431516)
- *Sienese sketchbook* (*Taccuini Sienesi*) (1490s–1516)
- *Codex Barberini* (post 1464–ante 1516).

Luca Pacioli (Borgo Sansepolcro, ca. 1445–Rome, 1517)
- *Divina Proportione* (1509)
- *Translation into Latin of the Elements of Euclid* (1509)
- *De Ludo Scachorum* (manuscript, in Italian ca. 1500)
- *De Viribus quantitatis* (ca. 1500)
- *Summa de arithmetica, geometria, proportioni e proportionalità* (1494)

Philipp Mönch (?)
- *Kriegsbuch* (1496)

Lorenzo della Volpaia (Florence, 1446–1512)
- *The realization of the Orologio dei Pianeti* (Clock of the planets) 1510

Bonaccorso Ghiberti (Florence, 1451–1516)
- *Zibaldone* (ca. 1500)

Leonardo da Vinci (Vinci, 1452–Amboise, 1519)
- *Codex Atlanticus* (different topics: geometry, physics, architecture, 1480–1518) (Biblioteca Ambrosiana, Milan, Italy)
- *Codex Arundel* (different topics: geometry, warfare, architecture, mechanics of machines, botany, zoology, 1480–1518) (British Library, London, UK)
- *Codex Ashburnham* (assorted drawings, 1489–1492) (Institut de France, Paris)
- *Windsor folios* (drawings of different subjects: anatomy, maps, studies of human figures, 1478–1518) (Royal Library at Windsor Castle, UK)
- *Codex Forster I* (1490–1505), *Codex Forster II* (1495–1497), *Codex Forster III* (geometry, mechanics, hydraulic, 1490–1496) (Library of Victoria and Albert Museum, London, UK)
- *Codex Madrid I* (mechanics, 1490–1496) (Biblioteca Nacional, Madrid, Spain)
- *Codex Madrid II* (geometry, 1503–1505) (Biblioteca Nacional, Madrid, Spain)

- *Codex Trivulzianus* (architecture, religious themes, exercises, 1487–1490) (Biblioteca Trivulziana, Castello Sforzesco, Milan, Italy)
- *Codex on the Flight of Birds* (study of the flight of birds, 1505) (Biblioteca Reale, Turin, Italy)
- *Codex Leicester* (hydraulics and movement of water, 1505) (former Hammer Codex) (Bill Gates, Seattle, Washington, USA)
- *Manuscripts of France A-M* (various topics: optics, geometry, warfare, hydraulics, 1492–1516) (Institute de France)
- *Trattato della pittura* (Compiled by Francesco Melzi using 18 manuscripts of Leonardo; only six of them have been identified, A, E, F, G, L, Trivulziano and a Windsor folio)

Giovanni Battista della Valle (Venafro, ca. 1470–1550)
- *Il Vallo* (*The Book of Captains*) 1521.

Cesare Cesariano (Milan, 1475–1543)
- *Translation into Italian of Vitruvio's De Architectura* (1521)

Vannuccio Biringuccio (Siena, ca. 1480–ca. 1539)
- *Pirotechnia* (1534–1535), published posthumous in Venice, 1540

Damiano Zambèlli (fra Damiano da Bergamo) (Zogno (Bergamo) ca. 1490–Bologna 1549)
- *Inlays of the Basilica di San Domenico in Bologna* (1549)

Jamnitzer Wenzel (Wien, 1508–1585)
- *Perspectiva Corporum Regularium* (1568)

Giorgio Vasari (Arezzo, 1511–Florence, 1574)
- *Le vite de' più eccellenti pittori, scultori e architettori* (1550)

Daniele Barbaro (Venice, 1514–Venice, 1570)
- *Della Perspettiva di Monsignor Daniel Barbaro, Eletto Patriarca d'Aquileia. Opera molto utile a Pittori, a scultori & ad Architetti* (printed in Venice 1569).

Carlo Urbino (Crema, ca. 1510/20–Crema, after 1585)
- *Codex Huygens* (mid-sixteenth century)

Agostino Ramelli (Ponte Tresa, 1531–1608)
- *Le diverse et artificiose machine del Capitano Agostino Ramelli Dal Ponte Della Tresia.* (in French and Italian) 1588 Paris.

Jacques Besson (Colombière (Briançon) ca. 1540–England, 1573)
- *Theatrum instrumentorum* (1571–1572)

Fausto Veranzio (Sebenico, 1551–Venice, 1617)
- *Machinae novae* (printed in Venice 1606)

Vittorio Zonca (Padova, 1568–1602)
- *Novo teatro di machine et edificii. Per varie et sicure operationi con le loro figure tagliate in rame e la dichiaratione et dimostratione di ciascuna* (printed in Padua 1607, posthumous)

Giovanni Branca (Sant'Angelo in Lizzola, 1571–Loreto 1645)
- *Le machine: volume nuovo et di molto artificio da fare effetti maraviglio si tanto spirituali quanto di animale operatione arichito di bellissime figure con le dichiarationi a ciascuna di esse in lingua volgare et latina* (1629)

Lorenzo Sirigatti (after 1554–1625)
- *La pratica di prospettiva del cavaliere Lorenzo Sirigatti* (printed in Venice, 1596)

Giovanni Alfonso Borelli (1608 Naples–1679 Rome)
- *De Motu Animalium* (*On the Movements of Animals*) 1680, posthumous

Appendix II: A Short Biography of Leonardo da Vinci

15 April 1452

Leonardo was born in Anchiano, close to the Tuscan town of Vinci, which was at that time part of the Republic of Florence, ruled by the Medici family. He was the first son of Messer Piero Fruosino di Antonio da Vinci, a wealthy gentleman working as a legal notary in Florence, and a peasant woman named Caterina di Meo Lippi, a 15-years-old orphane. Because of their difference in social status, Piero da Vinci decided to take the child without marrying Caterina and found a suitable husband for Leonardo's mother.

Leonardo's grandfather carefully recorded his birth in the family diary: "*nacque un mio nipote, figliolo di ser Piero mio figliolo a dì 15 aprile in sabato a ore 3 di notte*" (A grandson of mine, son of my son Piero, was born on April 15, Saturday, 3 hours into the night). In modern terms with the introduction of the Gregorian calendar in October 1582, this would be 23 April around 9.30 pm.

Leonardo spent most of his childhood with his paternal grandfather, Antonio, also a legal notary, and his uncle Francesco. The two men provided for his education.

1464: To Florence

At the age of 12, Leonardo moved to Florence to live with his father.

1466–1476: Apprentice at Verrocchio's Workshop

At the age of 14, Leonardo joined the workshop of Andrea di Cione (known as Verrocchio), one of the most famous artists in Florence. During his apprenticeship, Leonardo produced his first known painting, the face of an angel appearing in Verrocchio's *Battesimo di Cristo* (Baptism of Christ).

Leonardo's first dated work (1473) is an ink drawing of the Arno valley.

P. Innocenzi, *The Innovators Behind Leonardo*, https://doi.org/10.1007/978-3-319-90449-8

1472: Member of Compagnia dei pittori (Painters' guild)
At the age of twenty, Leonardo was accepted as a member of Florence's guild of painters. However, he likely did not leave Verrocchio's workshop before 1477.

1482–1499: First Milanese Period (age 30–47)
This first Milanese period was one of the most fruitful in Leonardo's life. In Milan, he found a stimulating and exciting intellectual environment where he could dedicate himself to his myriad artistic and scientific activities. In 1496, Leonardo met the Mathematician Luca Pacioli, who would become one of his closest friends.

Ludovico il Moro commissioned a bronze equestrian monument from Leonardo to be dedicated to Francesco Sforza. Leonardo devoted enormous effort to solving the many problems related to its fabrication, but before he could cast the bronze to fashion the statue, Milan was invaded by King Charles VIII of France. Suddenly, bronze became a key resource that was instead needed to produce cannons for the defense of the city.

1500: A Year of Travels
When the city of Milan was conquered by the French army, Leonardo and his friend Luca Pacioli traveled to Mantua on the invitation of Isabella D'Este. After a short stay, Leonardo moved to the Republic of Venice in March of 1500, where he worked on a defense system to protect the city from the Turkish threat. After Venice, he visited Rome and Tivoli and in April 1501 found himself once again in Florence, many years after his initial departure.

1502: Working for Cesare Borgia
On 18 August 1502, Leonardo was appointed "Architecto et Ingegnero Generale" (architect and general engineer) by Cesare Borgia (Il Valentino), captain general of the Papal Army and son of Pope Alexander VI. Leonardo visited several cities in Central Italy, including Imola, Rimini, Pesaro, Cesena, Urbino, and Porto Cesenatico.

1503–1505: Florence Once More
Leonardo returned to Florence and joined the painters' confraternity, Compagnia di San Luca, (18 October 1503). He was paid for his hydraulic projects by the *Signoria* of Florence and opened a workshop in Santa Maria Novella to prepare Battaglia di Anghiari (Battle of Anghiari), a great fresco for the *Salone dei Cinquecento* in *Palazzo Vecchio*, which celebrated the victory of Florence against Milan in 1440. Leonardo likely started work on the portrait of Mona Lisa in 1503. During this second stay in Florence, he once more turned his attention to studies of flight, culminating in the *Codice sul volo degli uccelli* (Codex on the Flight of Birds), which he wrote in 1505.

1504: Father's Death

Leonardo's father died on 9 July 1504, an event he painstakingly recorded in his diary: "*Addi' 9 di luglio 1504 a ore 7 mori' Piero da Vinci notaio al palagio del potestà, mio padre a ore 7. Era d'età d'anni 80. Lasciò' 10 figlioli maschi e due femmine*" (On 9 July 1504 at 7 am, Piero da Vinci, legal notary at the palace of city administration and my father, passed away. He was 80 years old. He left 10 sons and 2 daughters). His father, however, did not name him as an heir, and Leonardo pursued legal action to obtain his part of the inheritance, though without success.

1506–1513: Second Milanese Period (age 54–61)

Charles d'Amboise, later the governor of Milan, wrote a letter to the *Signoria* to request the return of Leonardo to Milan. In early September, he moved to Milan with Salaì, Lorenzo, and Il Fanfoia. For a while, he traveled back and forth between Florence and Milan, finally settling in Milan in 1508. In September 1507, he met Francesco Melzi, the beloved pupil to whom Leonardo left the precious legacy of his manuscripts.

1511–1512: Retreat to Melzi's Villa

In December 1511, Swiss soldiers invaded the city of Milan. Leonardo fled to the Villa of his pupil Francesco Melzi in Vaprio d'Adda. He returned to Milan in March only to leave again in September of 1513, this time headed to Rome.

1513–1516: Rome

On 24 September 1513, Leonardo left Milan again with pupils Francesco Melzi and Salaì, moving to Rome to be part of the household of Giuliano de' Medici (brother of Pope Leo X). Leonardo set up a workshop in the Belvedere, a wing of the Vatican Palace, and was inscribed in the Roman confraternity of San Giovanni dei Fiorentini.

1516–1519: Final Years—France

In 1516, Leonardo was invited to France by King Francis I. He moved into the manor house Clos Lucé near the royal Castle of Amboise as *peintre du Roi* (royal painter). There, he spent the last 3 years of his life with his pupils Federico Melzi and Salaì and the servant Battista.

2 May 1519

Leonardo died at Clos Lucé on 2 May 1519, at the age of 67. He was buried in the chapel of Saint-Florentine in the Castle of Amboise, which was unfortunately destroyed in 1802. Some stone fragments found during an excavation of the site with the inscription "EO [...] DUS VINC" have led to the supposition that the partial skeleton and large skull found at the site are in fact Leonardo's remains. They were reinterred in another chapel of the Castle of Amboise dedicated to Saint-Hubert.

Printed in the United States
By Bookmasters